Bond Graph Model-based Fault Diagnosis of Hybrid Systems

Wolfgang Borutzky

Bond Graph Model-based Fault Diagnosis of Hybrid Systems

Springer

Wolfgang Borutzky
Bonn-Rhein-Sieg University of Applied Sciences
Sankt Augustin
Germany

ISBN 978-3-319-11859-8 ISBN 978-3-319-11860-4 (eBook)
DOI 10.1007/978-3-319-11860-4

Library of Congress Control Number: 2014951659

Springer Cham Heidelberg New York Dordrecht London

Printed on acid-free paper

Springer is part of Springer Science+Business Media (www.springer.com)

To my wife Heidrun

Preface

With an-ever increasing complexity today's engineering systems are built of interacting components that exchange not only physical quantities such as energy but also exchange and process information. A lot of sensors and controllers are in use in all kinds of complex mechatronic systems. Being equipped with sensors and embedded computers mechantronic systems become able to perform some tasks autonomously and are sometimes called intelligent mechatronic systems. However, faulty sensors may cause controllers to output not the signals needed, while faulty actuators and faulty system components downgrade the effect of a healthy control. The increase in complexity makes engineering systems vulnerable to all kinds of faults and make humans dependent on their reliability and safety. As a result, there is a need for implementing mechanisms that allow complex intelligent systems to autonomously detect, isolate and accommodate faults in real-time. Once a fault has occurred, a controlled system is no longer the one that was designed to serve a certain purpose and to meet certain requirements. Depending on the severity of a fault a system may continue operating at downgraded performance and with restricted functionality, or may require immediate reaction. Incipient faults may lead to the future failure of a component which in turn may require a shutdown if possible at all. Accommodation of a fault can mean that a fault tolerant control was designed that can deal with faults, or that hardware redundancy is available that can replace faulty components.

In any case, fault detection and isolation is a prerequisite for real-time system supervision. In order to ensure reliability and safety it is important to take into account detection and diagnosis of possible abnormal system behaviour and means for automatic correction already during an integrated, concurrent design of complex intelligent mechatronic system by deliberately injecting faults into a system model and to study their effects on the system's dynamic behaviour.

Fault diagnosis has been a research subject since more than three decades and has received quite some attention. Besides data-driven methods various model-based approaches to fault detection and isolation have been reported in the literature. One of these methods is based on bond graphs. They were introduced by Prof. H. Paynter at Massachusetts Institute of Technology, Cambridge, USA back in

1959. As this modelling methodology starts from physical first principles, in particular from considering the exchange of energy between system components and the conversion from one kind of energy into another, bond graphs are particularly well suited for the development of models with continuous variables for multidisciplinary systems in order to study their behaviour in time domain as well as in frequency domain.

In the course of the past 10 years, bond graphs have also been used for model-based approaches to fault detection and isolation of systems represented by a continuous time model. On the other side, modelling assumptions and the abstraction of state changes taking place virtually instantaneously lead to hybrid models that can appropriately and adequately accurate capture the dynamics of a complex system. Over the last decades, various approaches have been proposed on how to extend bond graph methodology so that systems represented by a hybrid model can also be covered. However, only recently, bond graphs have also been used for fault detection and isolation of such systems. The subject of this book is to demonstrate that bond graph methodology can well contribute to model-based fault detection and isolation for systems that are appropriately described by a hybrid model.

The book briefly recalls various bond graph representations of hybrid system models proposed in the literature. The development of hybrid models for the purpose of fault detection and isolation, in this book, makes use of conceptual non-ideal switches representing devices for which it is justified to abstract their fast state transitions into instantaneous discrete state switches and accounts for structural model changes by special sources that are switched on or off at the advent of a discrete event. As other possible approaches, this approach has its pros and cons. For illustration, the presented method is applied in a number of elaborated case studies that consider fault scenarios for switched power electronic systems that are commonly used in a variety of applications. Power electronic systems have been chosen because they may be appropriately described by a hybrid model and are well suited for application of the presented bond graph model-based approach to fault detection and isolation. The approach, however, is not limited to this kind of systems.

Beyond fault diagnosis, failure prognosis becomes more and more important with regard to maintenance. Taking actions only when a system component has run into a failure is not always an option. Periodical maintenance in certain time intervals is not cost efficient. A promising solution is to constantly monitor the health of a system in operation, to process and assess sensor measurements, and to estimate the so-called remaining useful life of component when an incipient fault has been detected. The book briefly shows that bond graphs can also support model-based failure prognosis.

Being active in the field of bond graph modelling for more than three decades, it has been a natural step for the author to use bond graphs also for model-based fault diagnosis of engineering systems and to show that bond graphs are not only useful for model development, model analysis, simulation and control of engineering systems but can also serve fault diagnosis of systems represented by hybrid models.

The book has arisen from the author's research and teaching experience and has been influenced by the work of other bond graph modellers and by the author's interactions with many leading personalities in this field. It addresses members of the world wide community of bond graph modellers, as well as students, researchers, and practicing engineers in industry concerned with fault diagnosis who might be interested to see how a graphical methodology such as bond graph modelling can support quantitative model-based fault diagnosis of engineering systems represented by a hybrid model. An appendix provides some fundamentals of bond graph methodology so that the use of bond graph modelling for FDI and prognosis in this book may be more easily followed. The book may be used in courses on fault diagnosis, as a supplementary text in modelling, simulation, and control and may also serve self-studies and as a reference.

The author wishes to thank Profs. A. Fakri, Université Paris-Est, France, Y. Merkuryev, Riga Technical University, Latvia, D. Murray-Smith, University of Glasgow, Scotland, UK and G. Romero, Universidad Politécnica de Madrid, Spain for their support of this project and for their time reading some chapters. Their valuable comments and suggestions are gratefully acknowledged.

Furthermore, I would like to thank the Editorial Team with Springer at New York, NY, USA, especially Brett Kurzman, Mary James, and Rebecca Hytowitz for their interest, support and help.

Last but not least, I wish to express my sincere thanks and appreciation to my wife for her support, and her patience with me.

Sankt Augustin, August 2014 Wolfgang Borutzky

Contents

Abbreviations

ARR	Analytical Redundancy Relation
BD	Block Diagram
BDF	Backward Differentiation Formula
BG	Bond Graph
BGD	Bond Graph in Preferred Derivative Causality
BGI	Bond Graph in Preferred Integral Causality
CBM	Condition-Based Maintenance
DAE	Differential-Algebraic Equation
DBG	Diagnostic Bond Graph
FDI	Fault Detection and Isolation
FSM	Finite Signature Matrix
FTC	Fault Tolerant Control
HDAE	Hybrid Differential-Algebraic Equation
HFSM	Hybrid Fault Signature Matrix
incBG	Incremental Bond Graph
IVP	Initial Value Problem
LFT	Linear Fractional Transformation form
LTI	Linear Time Invariant system
MBG	Multibond Graph
MIMO	Multiple Input Multiple Output system
NLSP	Nonlinear Least Square Problem
ODE	Ordinary Differential Equation
PID	Proportional, Integral and Derivative
PN	Petri Net
PWM	Pulse Width Modulation
RUL	Remaining Useful Life
SCAP	Sequential Causality Assignment Procedure
SPBG	Sensitivity Pseudo Bond Graph
SPWM	Sinusoidal Pulse Width Modulation
SPJs	Switched Power Junctions
VSI	Voltage Source Inverter

Chapter 1
Introduction

1.1 Motivation

Today's engineering systems, industrial systems, and technical processes become more and more complex. They encompass subsystems from various energy domains and more and more mechatronic systems are equipped with sensors, actuators, embedded digital circuitry and software, e.g. intelligent transport systems in harbours, that are meant to operate autonomously, or to reduce the burden of human operators. With regard to safety, reliability, availability, maintainability and in order to protect complex engineering systems as assets it becomes more and more important beyond complex automatic feedback control to provide autonomous health monitoring and supervisory systems that are able to quickly detect abnormal system behaviour after its occurrence, isolate causes for malfunctions, faults, or even failures, generate alarms and initiate appropriate timely measures, e.g. to continue the operation of a process if it turns out that the identified abnormal dynamic behaviour is still within acceptable bounds, or to use fault tolerant control (FTC), to use redundant hardware if available, or to shut-down a system in a controlled manner if possible. Controllers aim to keep outputs of a multiple input, multiple output (MIMO) system within prescribed bounds even in the presence of one or more simultaneous faults, so the assessment of abnormal behaviour must include controller outputs. Controllers themselves obtain information from sensors and deliver their control signals of low energy level to actuators while sensors as well as actuators can be become faulty. Hence, controllers can get faulty inputs and provide faulty control signals to the system. As a result, system behaviour may deteriorate or may even become instable. Clearly, in complex safety critical systems such as aircrafts, nuclear power plants, or chemical engineering processes automatic real-time supervisory systems assisting human operators are indispensable.

For the above reasons, and as a faulty design of system components may lead to a reduced availability of a system, entail unwanted repair costs, or may lead to failures with consequences that can be detrimental or even catastrophic, it is important to take into account detection, diagnosis of possible abnormal system behaviour and

W. Borutzky, *Bond Graph Model-based Fault Diagnosis of Hybrid Systems*,
DOI 10.1007/978-3-319-11860-4_1

means for automatic correction already during an integrated, concurrent design of an intelligent mechatronic system including its supervisory system before it is built and used in industry. That is, modelling and off-line simulation of various possible fault scenarios become an essential part during the design of autonomous health monitoring systems as part of complex intelligent mechatronic systems.

Due to their importance, fault detection, fault diagnosis, and fault accommodation have been subject of intensive research and various approaches have been proposed. As to fault detection and isolation (FDI) being a prerequisite for fault diagnosis and any automatic process supervisory system, approaches can be categorised into data driven ones and model-based approaches. In this book, quantitative physical models are developed for FDI using bond graph methodology. That is, a theoretical approach to the development of models based on physical first principles is used. As engineering systems subject to FDI are, in general, multidisciplinary and are built of components from various energy domains, bond graph modelling is considered most suitable because it starts from considering the exchange of energy between power ports of system component, enables to analyse a model with regard to required control properties such as structural observability and structural controllability and facilitates the generation of fault indicators.

As energy exchange and conversion from one form into another happen continuously with respect to time, at first, bond graphs were used to develop continuous time models for the purpose of model analysis and simulation of the dynamic behaviour. However, fast switching devices such as electronic diodes and transistors, hydraulic check valves, mechanical clutches, or phenomena such as stick-slip effects or hard stops in mechanical motion suggest to make use of the model abstraction that for some devices transitions from one state to another take place instantaneously, that some state variables change discontinuously at some time instances. Neglecting the dynamics of such fast transients can be justified as they are irrelevant with the regard to the time scale of the system behaviour. It is the change of state in such a device that matters. As a result of this abstraction, a model captures discrete events that are either controlled by local automata or take place autonomously and cause system mode changes and the continuous time behaviour of a system in various modes. Such models are usually called *hybrid* system models. An example of a system that may be represented by a hybrid model is a gearbox. Depending on the level of detail required in a simulation study, the behaviour in different gears may be captured by continuous time models and the shifts from one gear into another may be considered discrete events.

During the past decades various approaches have been proposed to extend bond graph methodology also to hybrid system models. The modelling of fast switching devices by ideal on-off switches, on the one hand, avoids small time constants and the generation of stiff equations and thus can avoid long simulation runs. On the other hand, the opening and closing of switches changes the model structure. Hence, structural model properties, i.e. structural observability and structural controllability, the form of the mathematical model and its characteristics such as the index of a set of differential algebraic equations (DAEs) as well as the equations for the fault indicators are affected and become system mode dependent.

Beyond model development for the purpose of analysis and simulation of the dynamic or the frequency domain behaviour, bond graphs have also proven useful as a tool for model-based FDI in engineering systems represented by continuous time models. Only recently, bond graph model-based FDI has also been used for systems described by a hybrid model. Following common terminology a system will be called a hybrid system for short if its dynamic behaviour is appropriately described by a hybrid model. In Chap. 2.1 of [1] Kowalewski rightly remarks:

> A hybrid system is usually defined as a system which combines continuous and discrete dynamics. This definition is superficial. To be more precise, the term hybrid systems refers to models, not systems as such. A system is not hybrid by nature, but it becomes hybrid by modelling it this way. Whether it makes sense to build a hybrid model depends not only on the system, but also on the application and the purpose of the model.

Considerations on what a hybrid system is may be found in [2–4]. Provan gives a formal definition in [5] that has been provided in Appendix A.2.

The bond graph model-based approach to FDI in hybrid systems in this book represents fast switching devices by non-ideal switches and accounts for structural changes by special sources that are switched on or off at discrete events. The bond graphs established are invariant for all system modes with regard to their structure and with regard to computational causalities. This way, the handling of variable computational causalities due to mode changes can be avoided. Modification of the standard widely used Sequential Causality Assignment Procedure (SCAP) that aims at keeping causality changes caused by switch state changes local is not needed. As a result, system mode dependent model equations and mode dependent equations for fault indicators can be derived and can be solved numerically simultaneously. The equations determining fault indicators are termed *Analytical Redundancy Relations* (ARRs). The results of their evaluation are called *residuals* and serve as fault indicators. In an off-line simulation, a bond graph of the faulty system is coupled to a bond graph of the non-faulty system. By this way, the effect of various deliberately introduced faults and their detection and isolation can be studied without any risk. In real-time simulation, the faulty system model is to be replaced by properly filtered measurements from the real system that are fed into ARRs derived from a bond graph of the healthy system.

In order to avoid that false alarms are reported to a supervisor system or that faults remain undetected, fault indicators should be significantly sensitive to actual faults but insensitive to parameter variations given uncertain system parameter values. Fault indicators robust with regard to parameter uncertainties can be obtained by introducing appropriate thresholds. As long as values of a fault indicator are within small prescribed bounds, no fault is reported. The system is considered to be healthy. As the dynamic behaviour of hybrid systems can be quite different in different system modes, thresholds adapted to system modes are needed. One way to obtain such bounds is to use *incremental* bond graphs. They can be systematically constructed from an original bond graph with nominal parameters and have proven useful for various tasks such as the determination of frequency domain parameter sensitivities of state and output variables of linearised systems, parameter sensitivities of transfer functions, or of fault indicators. The motivation to use incremental bond graphs also for the determination of adaptive fault thresholds in hybrid system models is that in an incremental bond graph, bonds carry variations of power variables and in switched linear time-invariant (LTI) systems, variations of fault indicators are a weighted sum of variations of power variables. The weights are parameter sensitivities. An upper bound can be easily obtained by application of the triangle inequality.

In online model-based FDI, ARRs are evaluated using measurements from the real system being subject to disturbances. The time evolution of ARR residuals serve as fault indicators. For hybrid systems, ARRs are system mode dependent. Hence, an unobserved mode change invalidates the actual set of ARRs. As a result, computed values of fault indicators may exceed current thresholds indicating faults in some system components that have not happened. ARR residuals derived from a bond graph can not only serve as fault indicators but may also be used for model-based system mode identification.

Finally, with regard to condition based maintenance (CBM) of engineering systems it is important to detect the initiation of incipient faults, to assess the current health of a system and to predict the remaining useful life (RUL) of a component that has become faulty. As bond graph modelling enables the systematic generation of ARRs for FDI, it can also support failure prognosis.

1.2 Organisation of the Book

The next section briefly consideres different types of faults and possible causes for them before a brief survey of fault detection methods is given. The purpose of the overview is to classify quantitative bond graph model-based FDI in the context of fault detection methods.

Section 1.2 gives a literature review of various approaches to a bond graph representation of hybrid system models. In particular, the section discusses the representation of fast switching devices by means of non-ideal switches with fixed mode-independent causality and addresses the equation formulation as a DAE system.

Clearly, for FDI it is necessary that a system is structurally observable. As switches temporarily disconnect and reconnect model parts they change the structure of a hybrid system model. Consequently, control properties, i.e. structural observability and structural controllability as well as characteristics of the mathematical model derived from the bond graph, i.e. the number of state variables, or the index of a DAE system become system mode dependent. Chapter 3 briefly addresses these issues by confining to switched LTI systems and provides some small illustrating examples.

Chapter 4 presents bond graph model-based quantitative FDI for hybrid systems. As the approach uses ARRs and their residuals as fault indicators, the generation of

ARRs from a bond graph representing a hybrid system model and the condensation of structural information in a system mode dependent structural fault signature matrix are presented. The chapter concludes by addressing off-line numerical computation of ARR residuals. To that end, a model of a system subject to faults is coupled to a bond graph of the healthy system by means of so-called residual sinks. This way, faults may be deliberately introduced without any risk and their effect, detection and isolation may be studied.

As fault indicators should be sensitive to real faults but insensitive to parameter uncertainties, adaptive system mode dependent thresholds are needed for FDI in hybrid systems robust with regard to parameter uncertainties. Chapter 5 demonstrates that *incremental* bond graph can serve this purpose for switched LTI systems. To that end some basics of incremental bond graphs are recalled. It is shown how parameter sensitivities of ARRs and ARR thresholds can be obtained. A small example illustrates the approach.

Chapter 6 addresses the case in which several faults may occur simultaneously. As long as some of them do not cancel each other, they may be detected. Isolation, however, is a problem. One way to isolate multiple simultaneous faults is to formulate an optimisation problem and to apply parameter estimation on a set of ARRs. This is illustrated by a small example and the use of Scilab function `optim()` [6].

Chapter 7 demonstrates that ARR residuals cannot only serve as fault indicators but may also be used for system mode identification in online FDI. First, the general case of switched LTI systems is considered and it is assumed that ARRs can be expressed in explicit form relating known system inputs and inputs either obtained by measurements or by simulation of the real system behaviour. A small example illustrates bond graph based system mode identification using ARRs.

In Chap. 8, fault scenarios in small switched power electronic systems are studied for illustration. The bond graph model-based approach to FDI in hybrid systems is, however, not limited to this kind of systems.

Chapter 9 briefly consideres the use of ARRs derived from a bond graph for the purpose of failure prognosis for hybrid systems.

The book concludes with a discussion of the presented bond graph model-based quantitative approach to FDI in hybrid systems and recommends as a possible topic for further research the use of bond graph modelling in fault tolerant control of hybrid systems.

Appendix A compiles definitions of key terms used throughout this book that are in accordance with the outcome of a standardisation effort of the IFAC SAFEPROCESS Technical Committee. Appendix B provides some fundamentals of bond graph modelling so that the use of bond graph modelling for FDI and prognosis in this book can be more easily followed. Appendix C briefly recalls some mathematical background. Finally, a glossary lists explanations of some key notions.

Throughout the text, a square, □, at the end of a line denotes the end of a remark or the end of a definition.

1.3 Fault Detection Techniques

Fault detection means to distinguish between normal and faulty system operation. If a system is healthy, the time evolution of measurements is within given bounds for the desired system behaviour. Values of system parameters are within given thresholds around their nominal value. If values of observed variables or of parameters exceed an admissible range then there is a *fault*. The cause may be a system component or device that is temporarily malfunctioning. In that case it may be possible to continue using the identified malfunctioning component and by changing control laws which is known as fault tolerant control (FTC). The faulty system is no longer the original one and requires a different control. A *failure* has occurred if it is not possible to recover from a fault without system reconfiguration, i.e. by replacing an identified faulty system component or a device by redundant healthy hardware that can provide similar functionality if available. Faults, if not accommodated, can cause undesired or uncontrollable system behaviour.

Fault diagnosis comprises

- fault detection,
- fault isolation, and
- fault identification and estimation.

Fault detection means to decide whether a fault has happened or not and to determine the time instant at which a fault has occurred. Fault isolation aims at finding the component in which a detected fault has occurred. Once a fault has been detected and isolation, fault identification and estimation will determine its kind and its severity in order to be able to decide how to accommodate the fault appropriately.

1.3.1 Causes for Faults and Types of Faults

Supervision of a complex system has to account for faults in the system itself, in its sensors, actuators and controllers. Faults are usually classified into

- parametric faults,
- structural changes,
- sensor and actuator faults.

For instance, contamination of the fluid in a hydraulic system may result in a partial blockage of a valve. In the constitutive relation of the valve, this expresses as a change in the value of the discharge coefficient. Examples of structural changes are broken electronic switching devices, or broken mechanical transmission systems, a leakage in a hydraulic tank or a stuck valve. As a result, the structure of a system model used by a model-based diagnosis system has to be adapted. Sensors may malfunction due to external environmental disturbances such as changes in temperature, or internal disturbances caused by a change in power supply. As a result, readings provided by a sensor may exhibit a constant offset, change in gain, or in hysteresis. Due to saturation actuators may not be able to provide a required output signal. Friction or backlash in the mechanical part may be further causes for a malfunction of an actuator.

Beyond these causes, wear and tear during normal system operation, human operator errors, postponed maintenance, or a faulty design of a system component, or errors in the assembly of system components from different engineering domains may result in gradually or suddenly occurring faults with possibly disastrous consequences.

With regard to their occurrence, faults are usually classified into

- abrupt,
- progressive,
- incipient, and
- intermittent

faults.

There is an abrupt fault if at some time instant a system component suddenly turns from normal into abnormal operation. A fault is of progressive nature if instead the behaviour of a component gradually deviates from that of normal operation. If the behaviour of a system component does not match the desired behaviour but is still within given bounds then this is an indication for an incipient fault. As incipient faults develop slowly, are small in comparison to abrupt faults and may be obscured by disturbances, they are rather difficult to detect. They are usually due to wear and tear, and to ageing in normal operation and, although the system behaviour is still acceptable, they should be detected so that an identified component causing an incipient fault can be replaced before the component's behaviour turns into a non-acceptable malfunction that can cause a failure.

In the case of an intermittent fault, the dynamic behaviour of some system component is abnormal for short time intervals. The occurrence and the length of these time intervals are unpredictable which makes it difficult to react appropriately. After detection of faulty values, fault diagnosis takes some time, but then the intermittent fault may have gone for an unpredictable period of time.

The case studies in Chap. 8 consider abrupt faults.

1.3.2 Modelling Faults

Assume that the dynamic behaviour of a process is within a neighbourhood of an operating point and can be described sufficiently accurate by a linear time-invariant state space model. Then sensor and actuator faults, e.g. leakage from a tank, are additional external input signals to the process. They are commonly taken into account as additive terms in the state space equations and are classified as *additive faults* [7, 8].

Let $\mathbf{A}, \mathbf{B}, \mathbf{C}, \mathbf{D}, \mathbf{E}, \mathbf{F}, \mathbf{G}, \mathbf{K}$ be constant coefficient matrices of appropriate dimensions and let $\boldsymbol{x}$ denote the state vector, $\boldsymbol{u}$ the vector of known inputs, $\boldsymbol{y}$ the vector of measured outputs, $\boldsymbol{f}(t)$ additive faults and $\boldsymbol{d}(t)$ disturbances. The dynamic behaviour of a process subject to additive faults can then be described by the linear state space model

$$\dot{\boldsymbol{x}}(t) = \mathbf{A}\boldsymbol{x}(t) + \mathbf{B}\boldsymbol{u}(t) + \mathbf{E}\boldsymbol{d}(t) + \mathbf{K}\boldsymbol{f}(t)\,, \quad \boldsymbol{x}(0) = \boldsymbol{x}_0 \tag{1.1a}$$

$$\boldsymbol{y}(t) = \mathbf{C}\boldsymbol{x}(t) + \mathbf{D}\boldsymbol{u}(t) + \mathbf{F}\boldsymbol{d}(t) + \mathbf{G}\boldsymbol{f}(t) \tag{1.1b}$$

The healthy system is described by the matrices $\mathbf{A}$, $\mathbf{B}$, $\mathbf{C}$, $\mathbf{D}$. Their coefficients are nonlinear functions of the nominal system parameters $\boldsymbol{\Theta}_n$. Note that the entries of the system matrices remain fixed in the case of additive faults. In bond graph models, additive faults may be captured by modulated sources (cf. Fig. 5.15).

Non-additive faults that express themselves as changes in the system parameters such as contamination of a fluid through an orifice are usually termed *multiplicative* faults. Let $\boldsymbol{\Theta}$ denote the actual system parameters that may have been changed by faults and let $\Delta\boldsymbol{\Theta} := \boldsymbol{\Theta} - \boldsymbol{\Theta}_n$. In this case, the state space model of a faulty system is of the form [8]

$$\dot{\boldsymbol{x}}(t) = \mathbf{A}(\boldsymbol{\Theta})\boldsymbol{x}(t) + \mathbf{B}(\boldsymbol{\Theta})\boldsymbol{u}(t)\,, \quad \boldsymbol{x}(0) = \boldsymbol{x}_0 \tag{1.2a}$$

$$\boldsymbol{y}(t) = \mathbf{C}(\boldsymbol{\Theta})\boldsymbol{x}(t) + \mathbf{D}(\boldsymbol{\Theta})\boldsymbol{u}(t) \tag{1.2b}$$

In fault diagnosis and in robust control of systems with parameter uncertainties (1.2a, 1.2b) is commonly used in a form in which the matrices are decomposed into the sum of two matrices [9–11]. One of them only depends on the nominal system parameters $\boldsymbol{\Theta}_n$, the other one accounts for parametric faults or for model parameter uncertainties $\Delta\boldsymbol{\Theta}$ and may be constant or time varying.

$$\dot{\boldsymbol{x}}(t) = [\mathbf{A}(\boldsymbol{\Theta}_n) + \Delta\mathbf{A}(\boldsymbol{\Theta}_n, \Delta\boldsymbol{\Theta}]\boldsymbol{x}(t) + [\mathbf{B}(\boldsymbol{\Theta}_n) + \Delta\mathbf{B}(\boldsymbol{\Theta}_n, \Delta\boldsymbol{\Theta}]\boldsymbol{u}(t) \tag{1.3a}$$

$$\boldsymbol{y}(t) = [\mathbf{C}(\boldsymbol{\Theta}_n) + \Delta\mathbf{C}(\boldsymbol{\Theta}_n, \Delta\boldsymbol{\Theta}]\boldsymbol{x}(t) + [\mathbf{D}(\boldsymbol{\Theta}_n) + \Delta\mathbf{D}(\boldsymbol{\Theta}_n, \Delta\boldsymbol{\Theta}]\boldsymbol{u}(t) \tag{1.3b}$$

Equations (1.3a, 1.3b) are called the canonical form of uncertain system state equations.

Parametric faults are called multiplicative because they contribute terms to the state space equations that are the product of one of the usually time constant matrices $\Delta\mathbf{A}$, $\Delta\mathbf{B}$, $\Delta\mathbf{C}$, $\Delta\mathbf{D}$ with the state vector $\boldsymbol{x}(t)$ or the vector of known inputs $\boldsymbol{u}(t)$, i.e. with a time-varying vector [10].

This decomposition of the matrices may be cumbersome and costly as the matrix entries, in general, are nonlinear functions of the system parameters. Two approaches have been reported in the literature that can be used to set up the matrices $\Delta\mathbf{A}$, $\Delta\mathbf{B}$, $\Delta\mathbf{C}$, $\Delta\mathbf{D}$ in symbolic form from a bond graph [12, 13].

1.3.3 Classification of Fault Detection Methods

Fault detection and isolation are a prerequisite of fault diagnosis necessitating system observability and proper signal measurement and processing in the presence of noise. The thus obtained information then is to be processed in real time by a decision support software system that takes into account that sensors themselves can be faulty. According to Venkatasubramanian et al. [14–16], fault detection and diagnosis methods may be classified into

- process history based methods,
- qualitative model and search methods, and
- quantitative model-based methods.

Process History Methods

Process history based methods extract fault characteristics, i.e. occurrence time, fault size, from huge amounts of measured data in order to determine the state of a process and whether a fault can be accommodated so that the process can be continued. Fault characterisation is part of the a priori knowledge needed by a diagnosis system. The extraction from process history data is known as *feature* extraction which can be either qualitative or quantitative. The latter may use statistical methods. In their book [17], Basseville and Nikiforov present algorithms and statistical methods for the detection of abrupt changes in the parameter vector of a dynamic statistic systems (see also [18]). Gertler and Cao use the Principle Component Analysis (PCA) method as a tool for fault detection [19]. Starting from the covariance matrix of measured data obtained from a non-faulty system, linear relations among variables are identified and are used as a reference in order to detect if measured data deviate from that of the non-faulty process behaviour. The PCA technique has recently also been used by Ding et al. for fault detection and isolation [20].

Model-Based Fault Detection Methods

Model-based fault detection methods make use of a model to obtain the a priori knowledge needed for fault diagnosis. The development of a model can be a challenging task. Modelling assumptions, decisions on which effects shall be taken into account and how to model them clearly affect the quality of diagnosis results. Model-based FDI depends on the model accuracy. However, once an appropriate model has been developed it allows to obtain information about the expected system behaviour and to compare it with measurements of the real system behaviour. The diagnosis techniques and the diagnosis precision depend on the kind of model used.

Qualitative models consider cause and effect relationships, express component malfunctions in a qualitative manner and link them with deviations in measurement data. Qualitative fault diagnosis is based on fault tree analysis [15, 21], or uses qualitative simulation [22, 23], or combines qualitative reasoning with bond graph-based physical system modelling [24, 25].

Quantitative models describing the dynamic behaviour of a system are often developed from first principles and establish mathematical relations between system inputs and outputs. The computed outputs of a behavioural model, however, will never perfectly match measured outputs of the real system even if no fault has occurred for the following reasons.

- There is always some modelling uncertainty due to the simplifying assumptions taken and due to the fact that some observed effects can only be approximately modelled and some dynamics are not captured at all.
- The actual values of system parameters are not exactly known. A parameter estimation process may be erroneous. That is, modelling and parameter uncertainties reduce the accuracy of the system model.
- On the other side, the real world system may be subject to unknown disturbances and measurements carry noise.

Residual Generation

Once a quantitative system model is available, different methods can be used for the generation of a fault indicator or a residual as a primary step in FDI. These methods are either based on observers or a bank of observers [26, 27], on parity relations [28, 29], or on analytical redundancy relations [30, 31], or on parameter estimation [32, 33]. In case a fault has occurred in the system, the time evolution of some residuals must deviate distinctly from that during normal healthy system operation.

Observer-Based Residual Generation

Real-time observer-based residual generation uses an analytical model that is subjected to the same input as the real system. An observer-based residual is obtained by comparison of the measured real system output subject to faults with that of the observer as depicted in Fig. 1.1.

Consideration of the equations of a faulty LTI system and of a Luenberger state variable observer reveals that any faults affecting the system have an affect on the observer output error which, after transients have settled, can be used as a fault indicator ([34], Sect. 5.2.2). Assume that the dynamics of a system may be represented by the linear time-invariant state space model

$$\dot{\tilde{\boldsymbol{x}}}(t) = \mathbf{A}\tilde{\boldsymbol{x}}(t) + \mathbf{B}\boldsymbol{u}(t) + \mathbf{E}\boldsymbol{d}(t) + \mathbf{K}\boldsymbol{f}(t)\,, \quad \tilde{\boldsymbol{x}}(0) = \tilde{\boldsymbol{x}}_0 \tag{1.4a}$$

$$\tilde{\boldsymbol{y}}(t) = \mathbf{C}\tilde{\boldsymbol{x}}(t) + \mathbf{F}\boldsymbol{d}(t) + \mathbf{G}\boldsymbol{f}(t) \tag{1.4b}$$

where $\tilde{\boldsymbol{x}}$ denotes the state vector, $\tilde{\boldsymbol{y}}$ the output vector, $\boldsymbol{f}(t)$ additive faults and $\boldsymbol{d}(t)$ disturbances. The constant coefficient matrices **A**, **B**, **C**, **E**, **F**, **G**, **K** are of appropriate dimensions. The term $\boldsymbol{f}_a(t) := \mathbf{K}\boldsymbol{f}(t)$ accounts for actuator and components faults

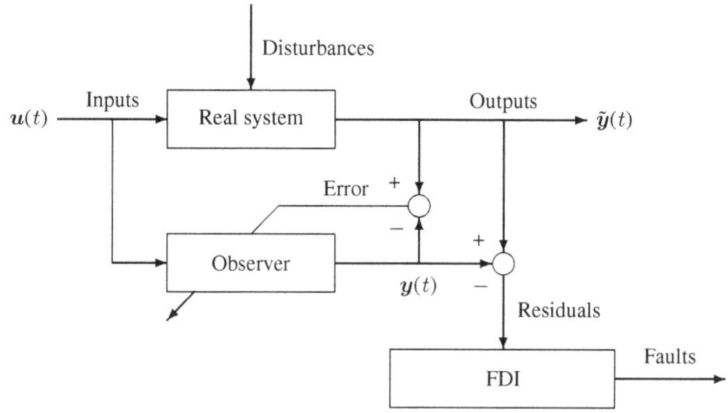

Fig. 1.1 Observer-based residual generation

while $\boldsymbol{f}_s(t) := \mathbf{G}\boldsymbol{f}(t)$ models sensor faults. An observer is then given by the equations

$$\begin{aligned}\dot{\boldsymbol{x}}(t) &= \mathbf{A}\boldsymbol{x}(t) + \mathbf{B}\boldsymbol{u}(t) + \mathbf{L}(\tilde{\boldsymbol{y}}(t) - \boldsymbol{y}(t)) \\ &= (\mathbf{A} - \mathbf{LC})\boldsymbol{x}(t) + \mathbf{B}\boldsymbol{u}(t) + \mathbf{L}\tilde{\boldsymbol{y}}(t)\,, \quad \boldsymbol{x}(0) = \boldsymbol{x}_0 \qquad (1.5a)\\ \boldsymbol{y}(t) &= \mathbf{C}\boldsymbol{x}(t) \qquad (1.5b)\end{aligned}$$

where $\mathbf{L}$ is a feedback matrix which is chosen so that in the case of no disturbances and no faults the state estimation error $\boldsymbol{e}_x(t) := (\tilde{\boldsymbol{x}}(t) - \boldsymbol{x}(t)) \rightarrow \mathbf{0}$ for $t \rightarrow \infty$, i.e. all eigenvalues of the matrix $(\mathbf{A} - \mathbf{LC})$ have a negative real part. Let $\boldsymbol{e}_y(t) := \tilde{\boldsymbol{y}}(t) - \boldsymbol{y}(t)$ be the output estimation error. Subtracting the equations of the observer from the equations of the system gives for the errors

$$\begin{aligned}\dot{\boldsymbol{e}}_x(t) &= (\mathbf{A} - \mathbf{LC})\boldsymbol{e}_x(t) + (\mathbf{E} - \mathbf{LF})\boldsymbol{d}(t) + (\mathbf{K} - \mathbf{LG})\boldsymbol{f}(t) \qquad (1.6a)\\ \boldsymbol{e}_y(t) &= \mathbf{C}\boldsymbol{e}_x(t) + \mathbf{F}\boldsymbol{d}(t) + \mathbf{G}\boldsymbol{f}(t) \qquad (1.6b)\end{aligned}$$

The output estimation error $\boldsymbol{e}_y(t)$ is a good candidate for a residual as it depends on the faults $\boldsymbol{f}(t)$ and the disturbances $\boldsymbol{d}(t)$ but not on the input values $\boldsymbol{u}(t)$. In the case of no faults and no disturbances this fault indicator apparently vanishes for $t \rightarrow \infty$.

A disadvantage of real-time observer-based residual generation is that for large behavioural models, or for stiff model equations, the numerical computation of the estimated response to system inputs may take a significant part of the sampling interval [34].

Residual Generation Based on Parameter Estimation

Instead of using state estimation for fault detection also parameter identification may be used [33]. Assume that a relation between inputs $\boldsymbol{u}(t)$ and outputs $\tilde{\boldsymbol{y}}(t)$ of the form

$$\sum_{\nu=0}^{n} a_\nu \tilde{\boldsymbol{y}}^{(\nu)}(t) = \sum_{\mu=0}^{m} b_\mu \boldsymbol{u}^{(\mu)}(t) \tag{1.7}$$

can be established by theoretical modelling. Furthermore, assume that an invertible function $\boldsymbol{f}$ exists that relates the nominal physical parameters $\boldsymbol{p}$ to the nominal model parameters $\boldsymbol{\Theta}$.

$$\boldsymbol{\Theta} = f(\boldsymbol{p}) \tag{1.8}$$

Then the known inputs $\boldsymbol{u}(t)$ and measured outputs $\tilde{\boldsymbol{y}}(t)$ can be used to estimate the uncertain model parameters $\tilde{\boldsymbol{\Theta}}$. The estimate $\hat{\tilde{\boldsymbol{\Theta}}}$ of the vector of model parameters can be translated into the vector of physical parameters $\tilde{\boldsymbol{p}}$.

$$\tilde{\boldsymbol{p}} = f^{-1}(\hat{\tilde{\boldsymbol{\Theta}}}) \tag{1.9}$$

Finally, changes $\Delta \boldsymbol{p}$ in the physical system parameter can serve as fault indicators. The procedure is displayed in Fig. 1.2.

The method is particulary suited for the detection of incipient faults as it can pinpoint small parameter changes. In Chap. 6, least squares parameter estimation is considered with to regard to the isolation of multiple parametric faults.

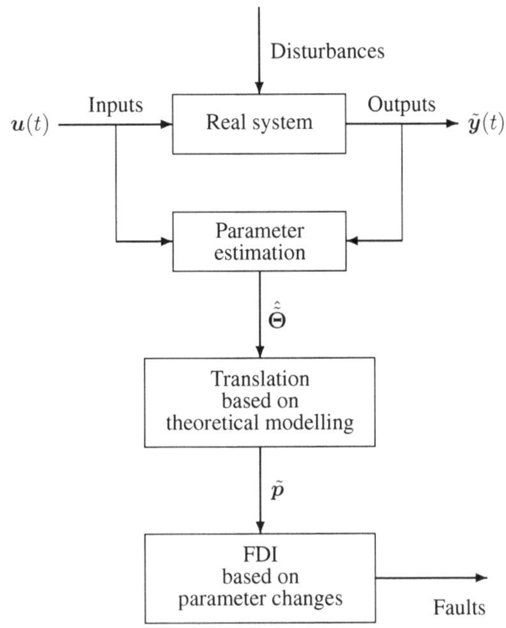

Fig. 1.2 Residual generation based on parameter estimation

Parity Space Residuals

The parity space approach to FDI is based on a comparison of the behaviour of a real process with that of a model that describes the non-faulty process. Any discrepancies between the two are described by residuals. The development of parity relations for residuals using a state space model goes back to Chow and Willsky [35] and has been presented in various publications, e.g. [8, 34, 36]. In the following, only the basic idea is outlined.

Suppose that the dynamic behaviour of a system can be described by a linear LTI model (1.4a, 1.4b). Let n be the order of the system, n_i the number of inputs, n_o the dimension of the output vector $\boldsymbol{y}$ and $q \leq n$. The idea of the parity space approach to FDI is then to choose a $n_r \times (q+1)n_o$ matrix $\mathbf{W}$ such that the n_r dimensional parity vector

$$\boldsymbol{r}(t) = \mathbf{W}\boldsymbol{y}_q(t) - \mathbf{W}\mathbf{Q}_u\boldsymbol{u}_q(t) \neq \mathbf{0} \tag{1.10}$$

for a faulty system and vanishes for a healthy system. In (1.10), $\boldsymbol{y}_q(t)$ denotes the vector composed of the output $\boldsymbol{y}(t)$ and the derivatives of $\boldsymbol{y}(t)$ up to the q-th order, i.e.

$$\boldsymbol{y}_q(t) := \begin{bmatrix} \boldsymbol{y}(t) \\ \dot{\boldsymbol{y}}(t) \\ \vdots \\ \boldsymbol{y}^{(q)}(t) \end{bmatrix} \tag{1.11}$$

The vector $\boldsymbol{u}_q(t)$ is likewise defined and $\mathbf{Q}_u$ is a $(q+1)n_o \times (q+1)n_i$ matrix.

The direction and the magnitude of the parity vector depend on the faults that have occurred. All parity vectors build a n_r dimensional so-called *parity space*. Any linear combination of rows in (1.10) is called a *parity relation* [8].

Expression (1.10) for the parity vector $\boldsymbol{r}(t)$ can be obtained in the following way. Successive differentiation of (1.4b) up to the order q and substitution of (1.4a) gives

$$\boldsymbol{y}_q(t) = \mathbf{T}\boldsymbol{x}(t) + \mathbf{Q}_u\boldsymbol{u}_q(t) + \mathbf{Q}_d\boldsymbol{d}_q(t) + \mathbf{Q}_f\boldsymbol{f}_q(t) \tag{1.12}$$

where

$$\boldsymbol{d}_q(t) = \begin{bmatrix} \boldsymbol{d}(t) \\ \dot{\boldsymbol{d}}(t) \\ \vdots \\ \boldsymbol{d}^{(q)}(t) \end{bmatrix}, \quad \boldsymbol{f}_q(t) = \begin{bmatrix} \boldsymbol{f}(t) \\ \dot{\boldsymbol{f}}(t) \\ \vdots \\ \boldsymbol{f}^{(q)}(t) \end{bmatrix} \tag{1.13}$$

$$\mathbf{T} = \begin{bmatrix} \mathbf{C} \\ \mathbf{CA} \\ \mathbf{CA}^2 \\ \vdots \\ \mathbf{CA}^q \end{bmatrix}, \quad \mathbf{Q}_u = \begin{bmatrix} \mathbf{0} & \mathbf{0} & \mathbf{0} & \dots & \mathbf{0} \\ \mathbf{CB} & \mathbf{0} & \mathbf{0} & \dots & \mathbf{0} \\ \mathbf{CAB} & \mathbf{CB} & \mathbf{0} & \dots & \mathbf{0} \\ \vdots & \vdots & & \ddots & \vdots \\ \mathbf{CA}^{q-1}\mathbf{B} & \mathbf{CA}^{q-2}\mathbf{B} & \dots & \mathbf{CB} & \mathbf{0} \end{bmatrix} \tag{1.14}$$

$$\mathbf{Q}_d = \begin{bmatrix} \mathbf{F} & \mathbf{0} & \mathbf{0} & \dots & \mathbf{0} \\ \mathbf{CE} & \mathbf{F} & \mathbf{0} & \dots & \mathbf{0} \\ \mathbf{CAE} & \mathbf{CE} & \mathbf{F} & \dots & \mathbf{0} \\ \vdots & \vdots & & \ddots & \vdots \\ \mathbf{CA}^{q-1}\mathbf{E} & \mathbf{CA}^{q-2}\mathbf{E} & \dots & \mathbf{CE} & \mathbf{F} \end{bmatrix} \tag{1.15}$$

$$\mathbf{Q}_f = \begin{bmatrix} \mathbf{G} & \mathbf{0} & \mathbf{0} & \dots & \mathbf{0} \\ \mathbf{CK} & \mathbf{G} & \mathbf{0} & \dots & \mathbf{0} \\ \mathbf{CAK} & \mathbf{CK} & \mathbf{G} & \dots & \mathbf{0} \\ \vdots & \vdots & & \ddots & \vdots \\ \mathbf{CA}^{q-1}\mathbf{K} & \mathbf{CA}^{q-2}\mathbf{K} & \dots & \mathbf{CK} & \mathbf{G} \end{bmatrix} \tag{1.16}$$

Remark 1.1 For $q = n$ matrix $\mathbf{T}$ is the observability matrix (cf. (3.2)). □

As the state $\boldsymbol{x}(t)$ and the disturbances $\boldsymbol{d}(t)$ are unknown in (1.12) a $1 \times (q+1)n_o$ vector $\boldsymbol{w}_1^T$ is chosen such that

$$\boldsymbol{w}_1^T \mathbf{T} = \mathbf{0} \quad \text{and} \quad \boldsymbol{w}_1^T \mathbf{Q}_d = \mathbf{0} \tag{1.17}$$

Then

$$\begin{aligned} \boldsymbol{r}_1(t) &= \boldsymbol{w}_1^T \boldsymbol{y}_q(t) - \boldsymbol{w}_1^T \mathbf{Q}_u \boldsymbol{u}_q(t) \\ &= \boldsymbol{w}_1^T \mathbf{Q}_f \boldsymbol{f}_q(t) \end{aligned} \tag{1.18}$$

As (1.18) indicates, residual $\boldsymbol{r}_1(t)$ is independent of the states $\boldsymbol{x}(t)$ and the unknown disturbances $\boldsymbol{d}(t)$. It depends only on the faults $\boldsymbol{f}(t)$ and if there are none, the residual vanishes. Besides $\boldsymbol{w}_1^T$, further vectors $\boldsymbol{w}^T$ can be chosen that satisfy (1.17); n_r such vectors constitute the $n_r \times (q+1)n_o$ matrix $\mathbf{W}$ in (1.10).[1] Residuals generated in this way are called parity space residuals [28, 37].

Equation (1.18) further shows that the computation of parity space residuals requires the derivatives of measured inputs and outputs up to the order q. As these measured signals carry noise, they need to be filtered which can be done by so-called

[1] n_r is the dimension of the left nullspace of the matrix $[\mathbf{T} \mid \mathbf{Q}_d]$, i.e. $n_r = (q+1)n_o - \text{rank}[\mathbf{T} \mid \mathbf{Q}_d]$. That is, this subspace is not empty [8].

state variable filters. Such filters are low-pass filters that filter the disturbance signals and provide the derivatives [38].

ARR-Based Residual Generation

If algebraic or dynamic contraints between known system inputs and measured outputs can be derived in closed symbolic form from a system model, that is, if nonlinear constitutive element relations permit to eliminate unknown variables in constraints, then even large sets of such contraints called ARRs can be evaluated in real-time in parallel on a multicore processor or on a multiprocessor system. Residuals as a result of that evaluation indicate a fault if their time evolution deviates distinguishably from that obtained during non-faulty system operation. Figure 1.3 illustrates the scheme of an ARR-based residual evaluation.

In Fig. 1.3, the real system includes actuators and sensors. The ARRs have been obtained from a model of the system that may include models of sensors and actuators depending on whether possible faults in these components are considered or not. In the FDI block, ARR residuals are used to detect and to isolate faults. As known inputs and measurements from the real system are inputs into the ARRs, their evaluation must take place in real-time. The ARRs, however, can be generated off-line by deducing them from a so-called diagnostic bond graph model as indicated in Fig. 1.4.

In this book, a quantitative ARR based approach to FDI for systems described by a hybrid model will be pursued and hybrid models are represented by bond graphs.

1.3.4 Bond Graph Model-Based Fault Detection Methods

A methodology especially suited for the development of models for multidisciplinary engineering systems that can be used for FDI is based on bond graphs [12, 39–41]. Beyond the purpose of simulating the dynamic behaviour of systems represented by

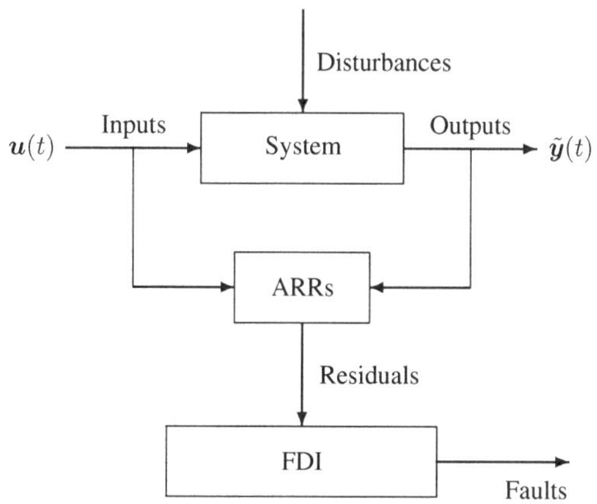

Fig. 1.3 ARR-based residual generation

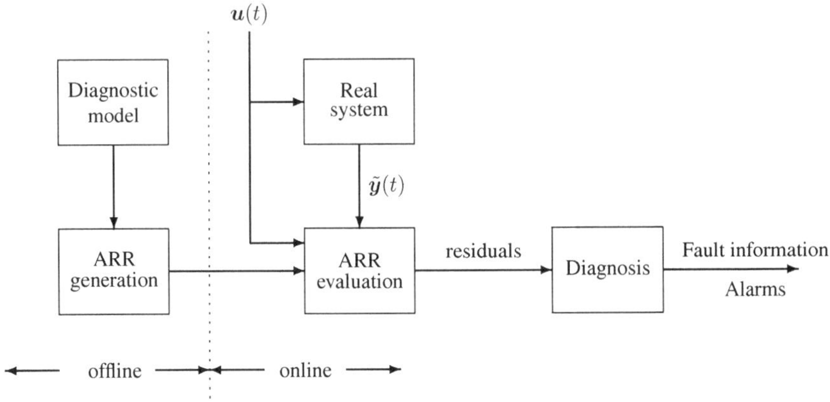

Fig. 1.4 Offline ARR generation and online residual generation

continuous time or hybrid models respectively and beyond the purpose of model-based control, a lot of research has been devoted to the application of bond graphs for qualitative as well as quantitative model-based FDI.

As to bond graph-based qualitative FDI, a number of approaches have been proposed by exploiting causal information in a bond graph and by using tree graphs [42], or temporal causal graphs [43]. In [44], Daigle presents a qualitative event-based approach to fault diagnosis of hybrid systems.

Bond graph-based quantitative FDI methods use ARRs [45–48], parameter estimation in the case of multiple simultaneous faults [46, 49], or observers [34]. Recently, bond graph-based FDI has been applied to various systems such as an industrial steam generator [50], an industrial chemical reactor [51], a mobile robot test bed [52], or an intelligent autonomous vehicle [53].

A survey of fault diagnosis using analytical model-based and knowledge-based redundancy has been given by Frank [31]. More recent presentations of fault diagnosis techniques including model-based fault diagnosis may be found in the textbooks by Ding and Isermann respectively [36, 54, 55]. In [34], Samantaray and Ould Bouamama give a detailed coverage of bond graph model-based quantitative as well as of qualitative fault diagnosis of systems presented by continuous time models. A recent book by Wang et al. uses bond graphs for model-based health monitoring of hybrid systems [56].

1.4 Summary

With the ever increasing complexity of modern engineering systems comes the demand for the ability of a system to autonomously monitor its performance, to detect, isolate and identify faults in critical components. This does not only mean to develop fault diagnosis and supervisory systems for existing systems and processes but to include modelling and off-line simulation studies of possible faults and failures already into the concurrent design of intelligent systems as part of a design for autonomy. As many engineering systems are mechatronic systems incorporating components involving various energy domains, bond graphs are considered most suitable for the development of models. Furthermore, as fast switching such as transistors, or hydraulic check valves are part of a system, hybrid models are usually used that abstract fast dynamic state transitions of switching devices to instantaneous state switches. This book presents a quantitative model-based approach to FDI of mechatronic systems described by a hybrid model that is represented by a bond graph. The accuracy of a behavioural model is not only essential for the simulation of the transient system behaviour or for its behaviour in the frequency domain but also for FDI and depends on modelling uncertainties, modelling assumptions, or inaccurate parameter values. An essential part of FDI is the generation of residuals. To that end, in this book, system mode dependent ARRs are derived from bond graphs that are invariant for all system modes with regard to their structure and computational causalities.

References

1. Kowalewski, S. (2002). Introduction to the Analysis and Verification of Hybrid Systems. In S. Engell, G. Frehse & E. Schnieder (Eds.). *Modelling, Analysis, and Design of Hybrid Systems* (pp. 153–171), vol. *279* of Lecture Notes in Control and Information Science. Verlag: Springer.
2. Lunze, J. (2002). What Is a Hybrid System? In S. Engell, G. Frehse & E. Schnieder (Eds.), *Modelling, Analysis, and Design of Hybrid Systems* (pp. 3–14). vol. *279* of Lecture Notes in Control and Information Sciences. Berlin Heidelberg: Springer.
3. Heemels, W. P. M. H., Lehmann, D., Lunze, J., & De Schutter, B. (2009). Introduction to hybrid systems. In J. Lunze & F. Lamnabhi-Lagarrigue (Eds.), *Handbook of Hybrid Systems Control: Theory, Tools, Applications* (pp. 4–30). Cambridge: Cambridge University Press.
4. van der Schaft, A. J., & Schuhmacher, H. (2000). *An introduction to hybrid dynamical systems*. vol. 251. in Lecture Notes in Control and Information Sciences. London: Springer.
5. Provan, G. (2009). Model abstractions for diagnosing hybrid systems. In: *Proceedings of the 20th International Workshop on Principles of Diagnosis, DX-09* (pp. 321–328). Stockholm, Sweden. Retrieved from http://www.cs.ucc.ie/ccsl/GP-papers/2009/Provan_DX_2009_2.pdf.
6. Scilab Enterprises. Scilab;. 78000 Versailles, France. Retrieved from http://www.scilab.org/.
7. Isermann, R. (2005). Model-based fault-detection and diagnosis - status and applications. *Annual Reviews in Control*, *29*, 71–85.
8. Blanke, M., Kinnaert, M., Lunze, J., & Staroswiecki, M. (2006). *Diagnosis and Fault-Tolerant Control*. Berlin: Springer.
9. Andrianiaina, P. J., Seuret, A., & Simon, D. (2011). Robust control under weakened real-time constraints.In *2011 50th IEEE Conference on Decision and Control and European Control Conference (CDC-ECC)* (pp. 2016–2021). Orlando: IEEE.
10. Frank, P. M. (2004). Control systems, robotics and automation. In H. Unbehauen (Ed.), *Fault diagnosis for linear systems*, vol. XVI. Encyclopedia of Life Support Systems (EOLLS) ed. Oxford: EOLLS Publishers. Retrieved from http://www.eolss.net.
11. Xie, L., & Soh, Y. C. (1997). Robust Control of Systems with Generalized Postive Real Uncertainty. *Automatica*, *33*(5), 963–967.
12. Borutzky, W. (2010). *Bond Graph Methodology - Development and Analysis of Multidisciplinary Dynamic System Models*. London, UK: Springer-Verlag. ISBN : 978-1-84882-881-0.

13. Kam, C. S., & Dauphin-Tanguy, G. (2005). Bond graph models of structured parameter uncertainties. *Journal of the Franklin Institute*, *342*(4), 379–399.
14. Venkatasubramanian, V., Rengaswamy, R., Yin, K., & Kavuri, S. N. (2003). A review of process fault detection and Diagnosis, Part I: Quantitative model-based methods. *Computers and Chemical Engineering*, *27*, 293–311.
15. Venkatasubramanian, V., Rengaswamy, R., Yin, K., & Kavuri, S. N. (2003). A review of process fault detection and diagnosis, Part II: Qualitative models and search strategies. *Computers and Chemical Engineering*, *27*, 313–326.
16. Venkatasubramanian, V., Rengaswamy, R., Yin, K., & Kavuri, S. N. (2003). A review of process fault detection and diagnosis, Part III: Process history based methods. *Computers and Chemical Engineering*, *27*, 327–346.
17. Basseville, M., & Nikiforov, I. V. (1993). *Detection of Abrupt Changes - Theory and Application*. Prentice-Hall, Inc. Retrieved from http://www.irisa.fr/sisthem/michele/.
18. Basseville, M. (2009). Statistical methods for change detection. In H. Unbehauen (Ed.), *Control systems, robotics and automation* (*XVI*, pp. 130–145). Encyclopedia of Life Support Systems (EOLLS) ed. Oxford: EOLSS Publishers. Retrieved from http://www.eols.net.
19. Gertler, J., & Cao, J. (2004). PCA-Based Fault Diagnosis in the Presence of Control and Dynamics. *AIChE Journal*, *50*(2), 388–402.
20. Ding, S. X., Zhang, P., Ding, E., Naik, A., Deng, P., & Gui, W. (2010). On the application of PCA technique to fault diagnosis. *Tsinghua Science and Technology*, *15*(2), 138–144.
21. Hurdle, E., Bartlett, L., & Andrews, J. (2007). System fault diagnosis using fault tree analysis. *Proceedings of the Institution of Mechanical Engineers, Part O, Journal of Risk and Reliability*, *221*(1), 43–55.
22. Cellier, F. E. (1991). *Continuous System Modeling*. New York, Berlin, Heidelberg: Springer-Verlag.
23. Kuipers, B. (1987). Qualitative simulation as causal explanation. *IEEE Transactions on Systems Man and Cybernetics*, *17*(3), 432–445.
24. Linkens, D., Xia, S., & Bennett, S. (1993). A computer-aided qualitative modelling and analysis environment using unified principles (QREMS). In J. Granda & F. Cellier (Eds.), *Proceedings of the International Conference on Bond Graph Modeling (ICBGM'93)* (vol. *25*(2), pp. 53–58). of Simulation Series. SCS. SCS Publishing.
25. Xia, S., Linkens, D. A., & Bennett, S. (1992). Integration of qualitative reasoning and bond graphs: an engineering approach. In P. C. Breedveld & G. Dauphin-Tanguy (Eds.), *Bond graphs for engineers* (pp. 323–332). North-Holland.
26. Patton, R., & Chen, J. (1997). Observer-based fault detection and isolation: Robustness and applications. *Control Engineering Practice*, *5*, 671–682.
27. Frank, P. M., & Ding, X. (1997). Survey of robust residual generation and evaluation methods in observer-based fault detection systems. *Journal of Process Control*, *7*(6), 403–424.
28. Patton, R., & Chen, J. (1991). A review of parity space approaches to fault diagnosis.In *Proceedings of the SAFEPROCESS 91* (*vol. 1* pp. 239–255).
29. Gertler, J., & Singer, J. C. (1990). A new structural framework for parity equation-based failure detection and isolation. *Automatica*, *26*(2), 381–388.
30. Staroswiecki, M., & Comtet-Varga, G. (2001). Analytical redundancy relations for fault detection and isolation in algebraic dynamic systems. *Automatica*, *37*, 687–699.
31. Frank, P. (1990). Fault Diagnosis in Dynamic Systems Using Analytical and Knowledge-based Redundancy - A Survey and Some New Results. *Automatica*, *26*, 459–474.
32. Bloch, G., Ouladine, M., & Thomas, P. (1995). On-line fault Diagnosis of dynmic systems via robust parameter estimation. *Control Engineering Practice*, *12*(3), 1709–1711.
33. Isermann, R. (1984). Process fault detection based on modelling and estimation methods: A survey. *Automatica*, *20*, 387–404.
34. Samantaray, A. K., & Ould Bouamama, B. (2008). *Model-based Process Supervision - A Bond Graph Approach. Advances in Industrial Control*. London: Springer.
35. Chow, E. Y., & Willsky, A. S. (1984). Analytical redundancy and the design of robust failure detection systems. *IEEE Transactions on Automatic Control*, *29*, 603–614.

36. Isermann, R. (2006). *Fault-Diagnosis Systems An Introduction from Fault Detection to Fault Tolerance*. Verlag: Springer.
37. Gertler, J. (1997). Fault detection and isolation using parity relations. *Control Engineering Practice*, *5*, 653–661.
38. Young, P. (1970). An instrumental variable method for real-time identification of a noise process. *Automatica*, *31*, 134–144.
39. Merzouki, R., Samantaray, A., & Pathak, P. (2013). *Ould Bouamama B*. Intelligent Mechatronic Systems: Springer.
40. Mukherjee, A., Karmakar, R., & Samantaray, A. K. (2006). *Bond Graph in Modeling, Simulation and Fault Identification*. New Delhi, India: I.K. International Publishing House.
41. Karnopp, D. C., Margolis, D. L., & Rosenberg, R. C. (2005). *System dynamics - modeling and simulation of mechatronic systems*. (4th ed.). John Wiley & Sons Inc. ISBN: 0-471-70965-4.
42. Kohda, T., Inoue, K., & Asama, H. (2001) Computer-aided failure effect analysis using system bond graphs. In j. Granda & G. Dauphin-Tanguy (Eds.), *Proceedings 2001 International Conference on Bond Graph Modeling and Simulation (ICBGM '01)* (vol. 33(1), pp. 71–76). of Simulation Series. Phoenix, Arizona: SCS.
43. Feenstra, P., Mosterman, P., Biswas, G., & Breedveld, P. (2001). Bond graph modeling procedures for fault detection and isolation of complex flow processes. In J. Granda & G. Dauphin-Tanguy (Eds.), *Proceedings 2001 International Conference on Bond Graph Modeling and Simulation* (vol. 33(1), pp. 77–82). of Simulation Series. Phoenix, Arizona: SCS.
44. Daigle, M. A. (2008). *Qualitative event-based approach to fault diagnosis of hybrid systems [PhD thesis]*. Vanderbilt University, Nashville, Tennessee.
45. Borutzky, W. (2011). Analytical redundancy relations from bond graphs of hybrid system models. In A. Bruzzone, G. Dauphin-Tanguy, S. Junco & M. A. Piera (Eds.), *Proceedings of the 5th International Conference on Integrated Modeling and Analysis in Applied Control and Automation (IMAACA 2011)* (pp. 43–49). Rome.
46. Ghoshal, S. K. (2006). *Model-based fault diagnosis and accommodation using analytical redundancy: A bond graph approach [PhD thesis]*. Department of Mechanical Engineering, Indian Institute of Technology. Kharagpur, India.
47. Medjaher, K., Samantaray, A.K., & Ould Bouamama, B. (2005). Diagnostic bond graphs for direct residual evaluation. In J. J. Granda & F. E. Cellier (Eds.). *Proceedings of the International Conference on Bond Graph Modeling, ICBGM'05* (vol. 37(1), pp. 307–312). Simulation Series New Orleans, Louisiana: SCS Publishing. ISBN: 1-56555-287-3.
48. Ould Bouamama, B., Samantaray, A.K., Staroswiecki, M., & Dauphin-Tanguy, G. (2003). Derivation of constraint relations from bond graph models for fault detection and isolation. In J. J. Granda & F. E. Cellier (Eds.). *Proceedings of the International Conference on Bond Graph Modeling, ICBGM'03* (vol. 35(2), pp. 104–109). Simulation Series Orlando, Florida: SCS Publishing. ISBN: 1-56555-257-1.
49. Samantaray, A. K., & Ghoshal, S. K. (2007). Sensitivity bond graph approach to multiple fault isolation through parameter estimation. *Proceedings of the Institution of Mechanical Engineers Part I: Journal of Systems and Control Engineering*, *221*(4), 577–587.
50. Djeziri, M. A., Ould Bouamama, B., & Merzouki, R. (2009). Modelling and robust FDI of steam generator using uncertain bondgraph model. *Journal of Process Control*, *19*, 149–162.
51. El Habibi, R., Ould Bouamama, B., Ben Gayed, M. K., & Abdelkrim, M. N. (2010). Pseudo bond graph for fault detection and isolation of an industrial chemical reactor. In F. E. Cellier & J. J. Granda (Eds.), *Proceedings of Simulation Series* (vol. 42(2), pp. 180–189). Orlando: SCS, San Diego.
52. Ming, Yu. (2012). *Fault diagnosis and prognosis of hybrid systems using bond graph models and computational intelligence [phd thesis]*. School of Electrical & Electronic Engineering. Nanyang Technological University, Singapore. Available from https://dr.ntu.edu.sg/handle/10356/50480.
53. Bera, T. K., Merzouki, R., Ould Bouamama, B., & Samantaray, A. K. (2012). Design and validation of a reconfiguration strategy for a redundantly actuated intelligent autonomous vehicle. *Proceedings of the Institution of Mechanical Engineers, Part I: Journal of Systems and Control Engineering*, *226*(8), 1060–1076

54. Ding, S. X. (2008). *Model-Based Fault Diagnosis Techniques - Design Schemes*. Algorithms and Tools: Springer-Verlag.
55. Isermann, R. (2011). *Fault diagnosis applications - model-based condition monitoring: actuators, drives, machinery, plants, sensors, and fault-tolerant systems*. Springer.
56. Wang, D., Yu, M., Low, C., & Arogeti, S. (2013). *Model-based health monitoring of hybrid systems*. Springer.

Chapter 2
Bond Graph Representations of Hybrid System Models

This chapter addresses the modelling abstraction of fast dynamic state transitions into instantaneous discrete state changes and its consequences and surveys various bond graph representations of hybrid system models proposed in the literature. The bond graph model-based quantitative approach to FDI in hybrid systems presented in Chap. 4 uses a bond graph representation with system mode independent computational causalities. As a result, a single set of equations in the form of a DAE system can be derived that holds for all system modes. The chapter concludes by addressing the index of the DAE system as it is an information that is relevant with regard to a symbolic and numerical processing of the DAE system.

2.1 Hybrid System Models

The way an engineering system is modelled depends on the purpose of a modelling and simulation study. Depending on the kind of problem or on the design stage modellers take different views of a system that either exists or is yet to be developed. For instance, it is well known that integrated digital circuits are modelled and simulated at different levels of abstractions. A hierarchical approach may start at the circuit level by accounting for the electrical dynamical behaviour of transistors, resistors and capacitors. One level up the hierarchy, the electrical behaviour is still of interest but transistors are abstracted into switches. State transitions are accounted for by rise and fall times. At the gate level the focus is on the logic behaviour of the circuits. The fact that a logic gate is built my means of electrical elements is taken into account by the time a signal needs to pass the logic gate. A top-down hierarchical design of complex digital circuitry starts at even higher levels of abstraction such as the register transfer level.

Besides the fast switching of transistors, electrical diodes or thyristors, fast changing phenomena such as the opening or closing of hydraulic check valves, the engagement-disengagement of clutches, stick-slip effects or stops in mechanical motion give rise to the modelling abstraction that such mode changes take place instantaneously and may be considered discrete events that either happen autonomously caused by internal continuous time variables that have crossed some

W. Borutzky, *Bond Graph Model-based Fault Diagnosis of Hybrid Systems*,
DOI 10.1007/978-3-319-11860-4_2

threshold values or are due to external control signals. Such a modelling abstraction is justified because the continuous time evolution of fast mode transitions during short time intervals is not relevant for the overall dynamic system behaviour. That is, there are discrete events at which switch state variables instantaneously switch between discrete values. The set of all discrete switch states $m_j(t),\ j = 1, \ldots, n_f$ characterises a system mode at time instant $t > 0$. For the time span of a system mode, the discrete switch state variables take constant values while the system behaviour is described by a continuous time model that is mode specific. Hence, if the abstraction of instantaneous mode switches is adopted then a model encompasses discrete events at which mode changes instantaneously take place and continuous time models for the dynamic system behaviour in each mode. Such models are usually called *hybrid* system models.

Adopting a hybrid system model entails a number of consequences.

- If fast switching devices are modelled as switches then parts of an overall system model are instantaneously disconnected or reconnected for a while. That is, an overall model is of variable structure. The result of such a structural change is a continuous time model that holds for the resulting system mode.
- The number of state variables of an overall model is not time-invariant but mode dependent. Due to a connection of model parts storage elements may become dependent, e.g. if a clutch is engaged or if rigid bodies stick together for a while due to Coulomb friction.
  In [1], Cellier and Kofman call a model a *variable structure model* if structural properties such as the number of differential equations depend on the discrete state of some switches.
- The formulation of a mathematical model becomes mode dependent. In one mode it may be an explicit state space model. If storage elements become dependent the model turns into a set of differential-algebraic equations (DAEs).
- The numerical integration of the model equations often requires the determination of the time instant of the discrete events and a reinitialisation. Hence, the numerical computation of a hybrid system model may be viewed as the solution of a sequence of initial value problems (IVPs). Modern numerical solvers for DAE systems such as IDA [2] from the SUNDIALS suite or DASRT [3] provide a root finding feature such that the time instances of mode switches can be located.
  If an *event condition* rather than the time instant of an event is known, the event is commonly denoted as a *state event*. The event condition is specified by a function of continuous variables and a state event occurs when the function crosses zero. If the time instant of a discrete event is known in advance, the event is called a *time event*. In that case, the event can be scheduled to happen by entering its time and its type into a calendar of forthcoming events.

The computation of models with variable structure is still a subject of ongoing research as latest publications show [4, 5]. With regard to a modelling by means of bond graphs the question arises how to represent a hybrid system model in a bond graph framework.

2.2 Bond Graph Representations of Hybrid System Models

Bond graph modelling starts from considering the exchange of energy between system components and energy conversion. As the exchange and the conversion of energy happen continuously with respect to time, bond graph modelling was, at first, limited to the representation of continuous time models. Bond graphs were mainly developed to obtain a mathematical model in the form of a system of ODEs or DAEs for the simulation of the dynamic behaviour of a system or to obtain transfer functions for the analysis of the frequency domain behaviour, for the analysis of structural observability and controllability, or for the design of a controller. Besides, various proposals to extend the bond graph modelling framework to capture discrete discontinuous events have evolved during the past two decades. So far, none of the reported approaches has attained common usage.

Combination of Petri Nets and Bond Graphs

One approach is to combine the advantages of established graphical representations by capturing physically feasible system modes and discrete transitions between them in a Petri net (PN) and by modelling the dynamic behaviour in each system mode by a bond graph or set of disjoint bond graphs composed of standard elements [6, 7]. Accordingly,

- the dynamic system behaviour is modelled by a set of bond graphs each of which has time-invariant computational causalities,
- each bond graph holds for the time period between two discrete events,
- for each system mode, the structure of the bond graph can be different,
- the mathematical model is a set of initial value problems.

Other approaches prefer a uniform model representation and suggest to let bond graphs capture also instantaneous discontinuous state changes although power conservation means that it takes a finite time to change the content of an energy storage element. Bond graph representations of hybrid models can be categorised into those that aim to keep computational causalities system mode independent and those accepting variable causalities.

Switching Off and On Degrees of Freedom by Means of Switched Residual Sinks

One way to keep computational causalities time-invariant as proposed in [8] is to use sinks of invariant causality in conjunction with a modulated transformer MTF : $b(t)$ that switches off and on degrees of freedom.

Example: Bouncing Ball

Figure 2.1 depicts the well known bouncing ball problem and a bond graph representation of a hybrid model.

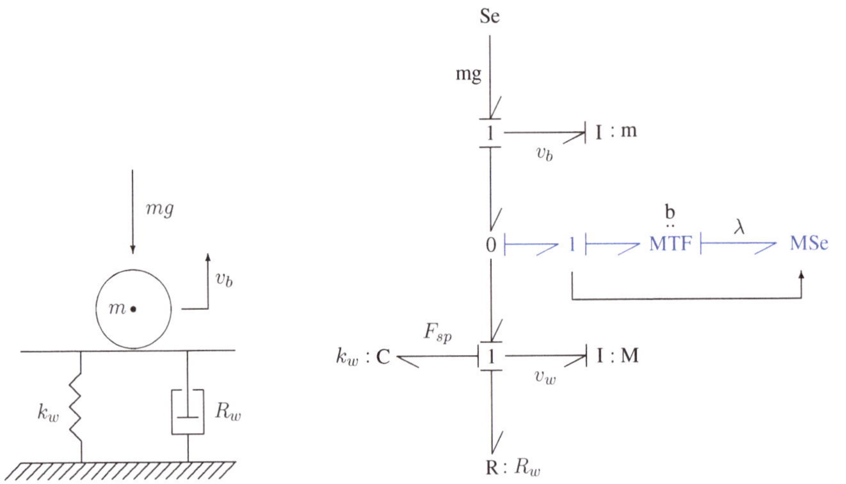

Fig. 2.1 Switching off and on degrees of freedom by sinks of invariant causality [6]

In the bond graph of Fig. 2.1, $b(t) \in \{0, 1\}\ \forall\, t \geq 0$. For $b = 0$, there is no force between the floor and the ball. The ball is freely moving in the air. If, however, the ball is in contact with the ground ($b = 1$) then the modulated effort source MSe imposes a constraint force λ which enforces that the ground and the ball move with the same velocity. Both inertia elements are in fixed integral causality. When the ball drops to the floor numerical integration restarts with new initial conditions. Modulated sinks imposing either an effort or a flow so that their input variable vanishes are termed *residual sinks*. As to the representation of the residual sink in Fig. 2.1, one might ask whether the relative velocity $v_b - v_w$ is fed into the modulated effort sink as a signal when this information is already delivered into the sink by the power bond attached to it. A standard effort sink Se does not serve the intended purpose as its output is independent of the input. In this case, a sink is needed that delivers an output so that the input vanishes. To that end, the input is sensed and fed as a signal into a modulated sink. The presentation has been adopted from [9] where it is introduced to display controlled constraint forces in multibond graphs of rigid multibody systems. Actually, these modulated sinks represent Lagrange multipliers that are not limited to mechanical systems modelling and can be used for various other purposes as well, e.g. to add tearing information into a bond graph [10]. In the context of FDI, residual sinks can be used to couple a behavioural bond graph model of a system subject to faults to a bond graph of the faultless system with nominal parameters [11]. In Chap. 4, they are used in this way to numerically determine ARR residuals. For the sake of a concise bond graph representation residual sinks are denoted by the standard symbol of a source or sink prefixed by the letter ‘r’, e.g. rSe denotes a residuals effort sink.

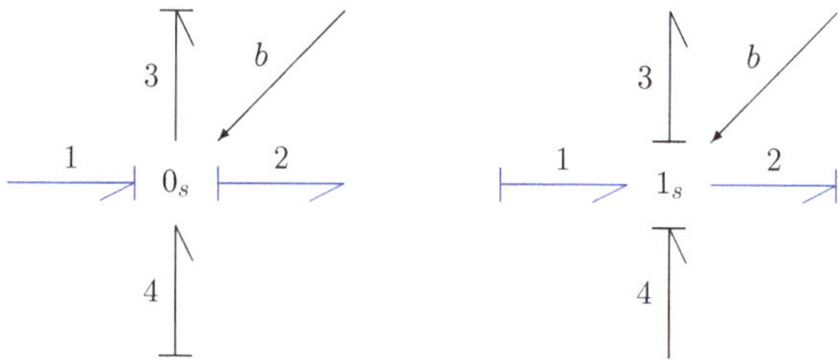

Fig. 2.2 Switched power junctions [6]

Switched Power Junctions (SPJs)

In order to account for the abstraction of ideal, no power consuming switching in a bond graph with invariant causalities that holds for all system modes, Umarikar extended 0- and 1-junctions by allowing for more than one bond to impose an effort on a 0-junction and more than one bond imposing a flow on a 1-junction with the constraint that only one of the causality imposing bonds is active at a time instant [12, 13]. These extensions are called *switched power junctions* and are not to be confused with *controlled junctions* to be referred to subsequently. Figure 2.2 illustrates the idea.

For a switched power junction holds

$$\forall\, t \geq 0 \quad \exists\, j \in \mathbb{N} \quad (m_j(t) = 1 \wedge m_i(t) = 0 \quad \forall\; i \neq j)$$

The equations for a switched power 0-junction read

$$e_3(t) = m_1(t)e_1(t) + m_2(t)e_2(t) = e_4(t) \tag{2.1a}$$
$$f_1(t) = m_1(t)(f_3(t) - f_4(t)) \tag{2.1b}$$
$$f_2(t) = m_2(t)(f_4(t) - f_3(t)) \tag{2.1c}$$

Accordingly, equations can be formulated for a switched power 1-junction.

Example: Boost Converter

The bond graph of a boost converter in Fig. 2.3 illustrates the use of these extended junctions.

Summation of efforts at the encircled left 1-junction and summation of flows at the rightmost encircled 0-junction yields mode dependent state equations (2.2a)–(2.2b)

$$1: \quad L\frac{d}{dt}i_L = E - m_2 V_C - R_L i_L \tag{2.2a}$$
$$0: \quad C\frac{d}{dt}V_C = m_2 i_L - \frac{V_C}{R} \tag{2.2b}$$

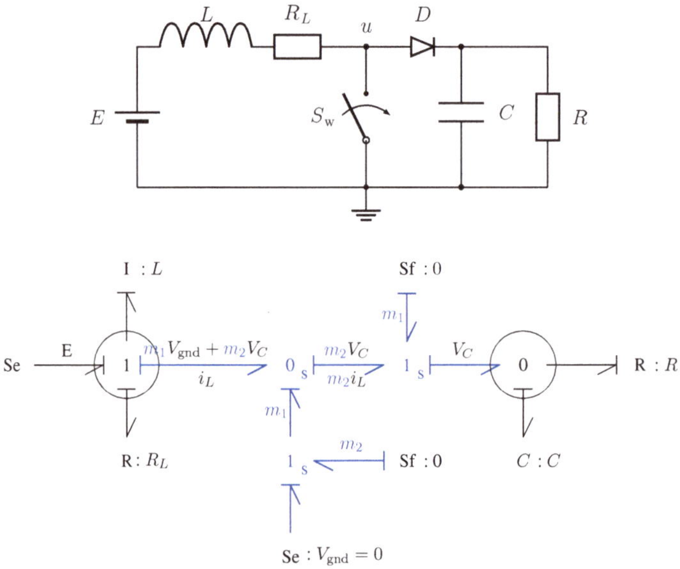

Fig. 2.3 Bond graph with SPJs of a boost converter [12]

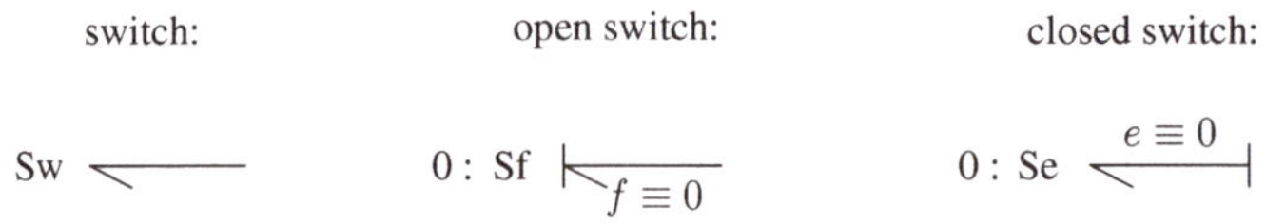

Fig. 2.4 The ideal switch as an additional bond graph element

The Ideal Switch as a Bond Graph Element

If one is willing to accept that computational causalities in bond graphs are mode dependent and if one neglects the energy loss associated with an instantaneous change of a switch state then fast switching devices may be modelled as an ideal switch [14–18] and represented by a bond graph switch element Sw or by controlled junctions [19–21]. In the case of an open switch, the bond graph switch Sw becomes a zero flow source Se : 0. In the case of a closed switch the bond graph switch element turns into a zero flow source Sf : 0. Hence, the causality at a switch port is mode dependent as displayed in Fig. 2.4.

Remark 2.1 Van der Schaft and Schuhmacher call variables such as the power variables of the bond graph switch element Sw *complementary* variables in the sense that for the two of them an inequality holds and for all times at least one of them is strictly an equality. Systems in which mode switching is determined by *complementarity conditions* they call *complementary systems* [22]. In the case that both variables can be assumed to be nonnegative, the complementarity condition is often expressed as

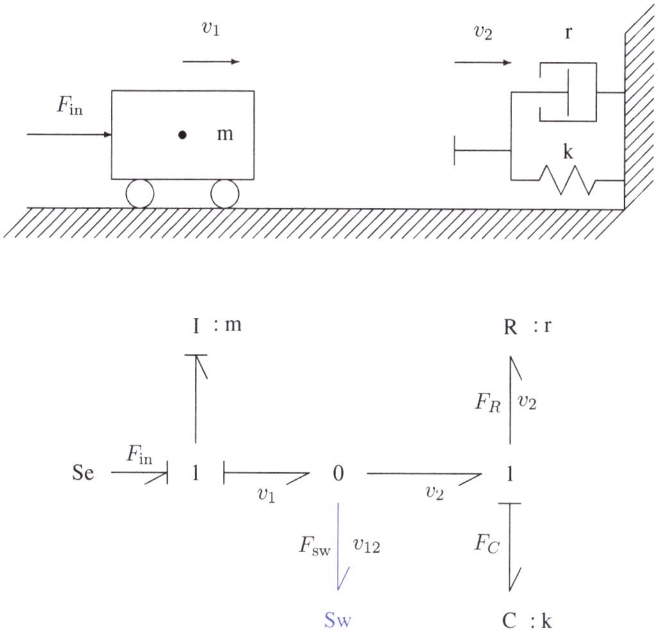

Fig. 2.5 Modelling a mechanical stop by means of an ideal switch [6, 18]

$$0 \leq f(t) \perp e(t) \geq 0 \quad \text{or} \quad 0 \leq e(t) - f(t) \geq 0 \tag{2.3}$$

In general, the complementary variables can be vectors if inequalities hold componentwise.

Let $\mathbf{M}$ be a $k \times k$ matrix and $\boldsymbol{q}$ a k-dimensional real vector. The so-called *linear complimentary problem* then consists in finding k-dimensional vectors $\boldsymbol{\lambda}$ and $\boldsymbol{y}$ such that

$$\boldsymbol{y} = \mathbf{M}\boldsymbol{\lambda} + \boldsymbol{q} \tag{2.4a}$$

$$\mathbf{0} \leq \boldsymbol{y} - \boldsymbol{\lambda} \geq \mathbf{0} \tag{2.4b}$$

This problem has a unique solution if the matrix fulfills a certain condition [22]. □

Example: Mechanical Stop

In the bond graph of Fig. 2.5, an ideal switch Sw is used to model an elastic mechanical stop.

Clearly, if the cart is not in contact with the stop then $F_{sw} = 0 \wedge v_{12} := v_1 - v_2 \geq 0$. If the cart is in contact with the stop then $F_{sw} \geq 0 \wedge v_{12} = 0$. That is, $0 \leq F_{sw} \perp v_{12} \geq 0$. The force F_{sw} may be thought of as a Lagrange multiplier λ which is either zero or takes a value to be computed.

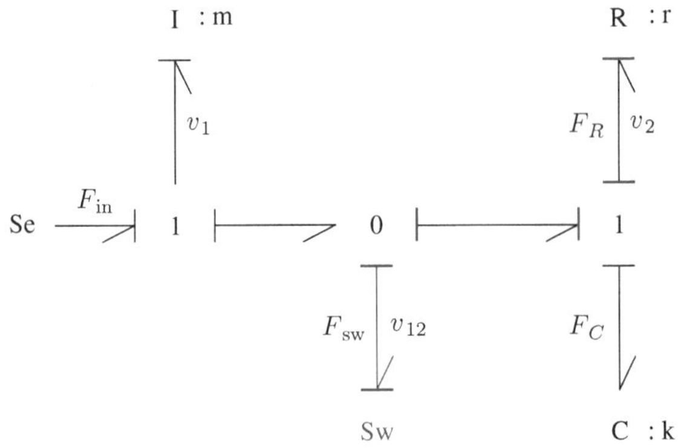

Fig. 2.6 Propagation of causality changes at the switch port

A causality change at a switch port propagates at least locally into the bond graph and affects the causality at the ports of other elements as indicated in Fig. 2.6 for the example of the mechanical stop.

Causality changes due to a mode change can be captured by using a 'Boolean' variable b in the constitutive equation of the switch.

$$0 = b \cdot F_{sw} + \bar{b} \cdot (v_1 - v_2) \tag{2.5}$$

where $b(t) \in \{0, 1\}\ \forall\, t \geq 0$ and $\bar{b} := 1 - b$. Then the following DAE system can be deduced from the bond graph in Fig. 2.6.

$$\left[\begin{array}{cc|cc} m & 0 & 0 & 0 \\ 0 & 1 & 0 & 0 \\ \hline 0 & 0 & 0 & 0 \\ 0 & 0 & 0 & 0 \end{array}\right] \begin{bmatrix} \dot{v}_1 \\ \dot{F}_c \\ \hline \dot{v}_2 \\ \dot{F}_{sw} \end{bmatrix} + \left[\begin{array}{cc|cc} 0 & 0 & 0 & 1 \\ 0 & 0 & -k & 0 \\ \hline 0 & 1 & r & -1 \\ \bar{b} & 0 & -\bar{b} & b \end{array}\right] \begin{bmatrix} v_1 \\ F_c \\ \hline v_2 \\ F_{sw} \end{bmatrix} = \begin{bmatrix} F_{in} \\ 0 \\ 0 \\ 0 \end{bmatrix} \tag{2.6}$$

In each of the two modes, DAE system (2.6) reduces to a simple ODE system.

Mosterman argues that ideal switches are not energy elements, that they represent control aspects rather than physical concepts ([21], p. 53).

Causality Resistors

Back in 1993, Asher proposed to assist an ideal switch by a resistor he called 'causality resistor' that adapts its causality to causality changes at the switch port so that the rest of the bond graph remains causally unaffected [23]. As long as the simulated dynamic behaviour is not significantly affected, the parameter value of a causality resistor can be chosen within reasonable limits but may lead to stiff model equations and thus may give rise to an increase of computational costs.

The physical meaning of a causality resistor depends on the application area and how it is used in conjunction with an ideal switch. For instance, the bond graph in

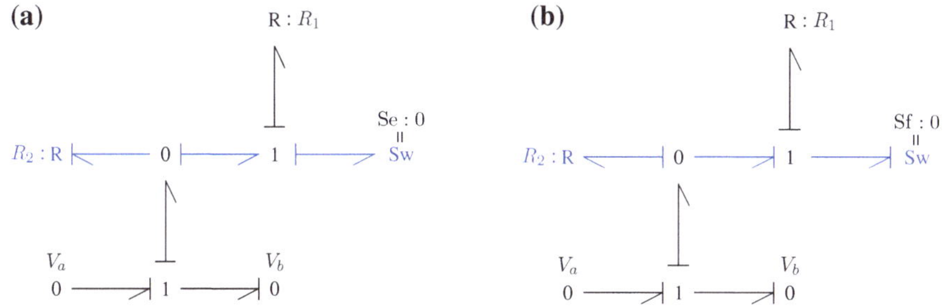

Fig. 2.7 Bond graph model of an electrical diode using an ideal switch Sw and a 'causality resistor' R : R_2. **a** Ideal switch closed. **b** Ideal switch open

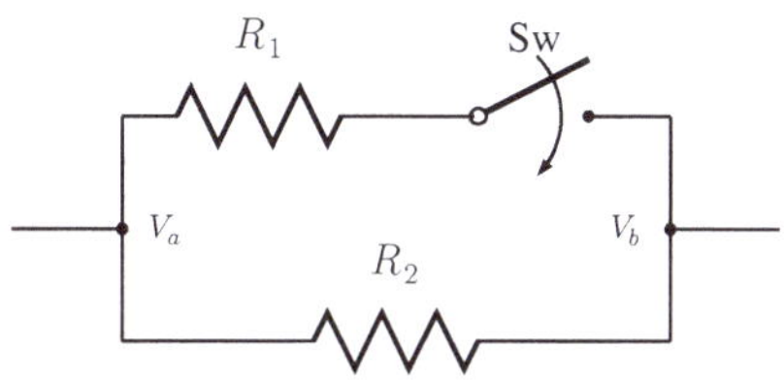

Fig. 2.8 Piecewise linear model of an electrical diode

Fig. 2.7 may be a piecewise linear model of an electrical diode. The corresponding circuit schematic is displayed in Fig. 2.8.

The causality resistor R : R_2 clearly avoids the propagation of causality changes at the port of the ideal switch into the rest of the bond graph and captures the diode's high resistance R_{off} in reverse mode. The resistor R : R_1 represents the diode's small ON-resistance R_{on}.

Let i_D denote the current through the diode and $\Delta V := V_a - V_b$. Then from the bond graph in Fig. 2.7a the equation

$$i_D = \left(\frac{1}{R_1} + \frac{1}{R_2}\right) \Delta V = \left(\frac{1}{R_{\text{on}}} + \frac{1}{R_{\text{off}}}\right) \Delta V \approx \frac{1}{R_{\text{on}}} \Delta V \tag{2.7}$$

is immediately deduced, where $R_{\text{off}} \gg R_{\text{on}}$. Likewise, the causal bond graph in Fig. 2.7b gives

$$i_D = i_{R_2} = \frac{1}{R_2} \Delta V = \frac{1}{R_{\text{off}}} \Delta V \tag{2.8}$$

As in the case of a mechanical stop (cf. Fig. 2.6), the bond graphs for both modes in Fig. 2.7 can be merged into one single bond graph from which equations valid for both modes can be deduced.

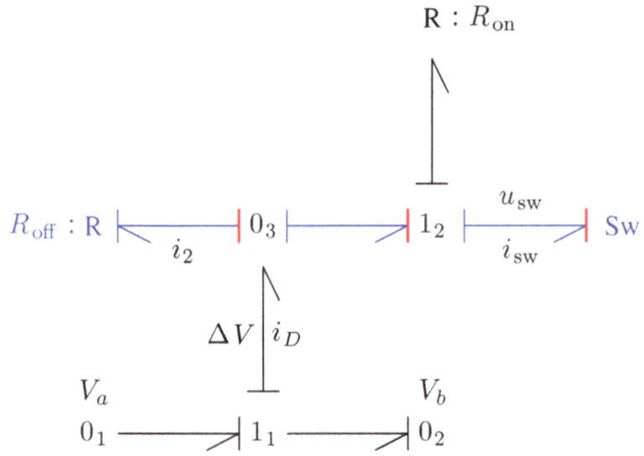

Fig. 2.9 Bond graph of the diode model capturing both modes

From the bond graph in Fig. 2.9 the following equations can be derived.

$$i_2 = \frac{\Delta V}{R_{\text{off}}} \tag{2.9a}$$

$$i_D = i_2 + i_{\text{sw}} \tag{2.9b}$$

$$\Delta V = R_{\text{on}} i_{\text{sw}} + u_{\text{sw}} \tag{2.9c}$$

The constitutive equation of the switch may be expressed by means of a variable $b(t) \in \{0, 1\}\ \forall\, t \geq 0$. Let $\bar{b} := 1 - b$. Then

$$0 = b\, i_{\text{sw}} + \bar{b}\, u_{\text{sw}}. \tag{2.10}$$

That is, $b(t) = 0$ means that the switch is closed, $b(t) = 1$ indicates that the switch is open.

From this set of equations an equation for the current i_D can be obtained that holds for both modes.

$$i_D = \left(\frac{1}{R_{\text{off}}} + \frac{\bar{b}}{\bar{b}\, R_{\text{on}} - b} \right) \Delta V \tag{2.11}$$

Controlled Junctions

As early as 1974, Thoma introduced the concept of *time dependent junctions* [24] in order to switch off and on connections between power ports. Mosterman picked up this idea and introduced *controlled junctions* [21]. In contrast to switched power junctions, a local control algorithm associated with a controlled junction switches off all adjacent bonds of a controlled junction when a switching device considered as an ideal switch turns off and re-activates all bonds when the switch is closed. That is, an ideal switch in ON-mode is represented by a standard 0- or 1-junction. In OFF-mode the junction is replaced by a source of value zero as shown in Fig. 2.10.

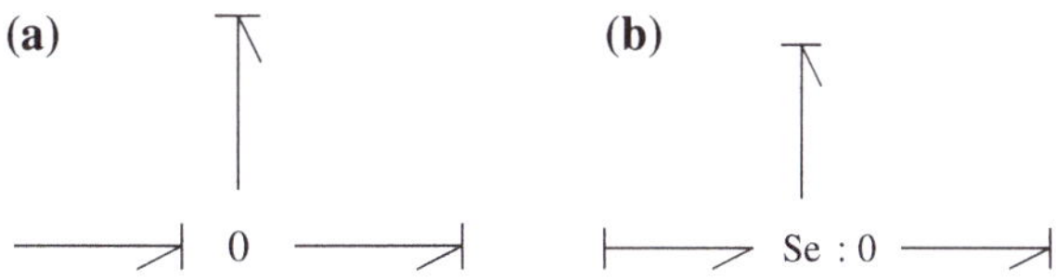

Fig. 2.10 Controlled 0-junction. **a** ON mode. **b** OFF mode

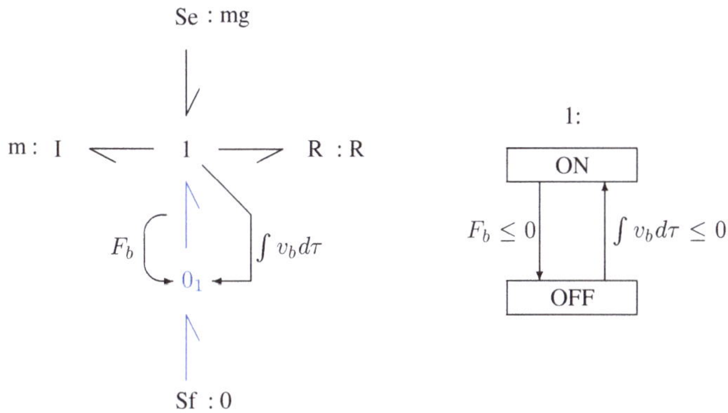

Fig. 2.11 Hybrid bond graph of the bouncing ball problem and a finite state automaton associated with the controlled 0-junction 0_1 [6, 21]

As Fig. 2.10 shows, a discrete switch state change entails a change in causality that is to be propagated into the bond graph. Bond graphs using controlled junctions are usually called *Hybrid Bond Graphs* (HBGs).

Figure 2.11 displays a hybrid bond graph for the well known bouncing ball problem given the assumption of a perfect non-elastic collision between the ball and the floor. As the focus of the bond graph methodology is to model the continuous time exchange of energy between system components and possible energy conversion and since none of the proposed extensions to include discrete events has become a widely accepted standard, there is not much bond graph software that support hybrid bond graphs. Mosterman has developed a software HYBrSIM especially suited for continuous-discrete event bond graph modelling using controlled junctions [25]. In the tool suite MoTHS, hybrid bond graphs are transformed into block diagrams by using block diagram models that can be efficiently reconfigured for those parts of a hybrid bond graph that need a re-assignment of causality after a system mode change has happened [20]. Wang and his co-workers [26] also map hybrid bond graphs onto block diagrams for simulation with Simulink®[1] [27].

---

[1] Simulink is a registered trademark of The MathWorks, 3 Apple Hill Drive, Natick, MA, USA, http://www.MathWorks.com.

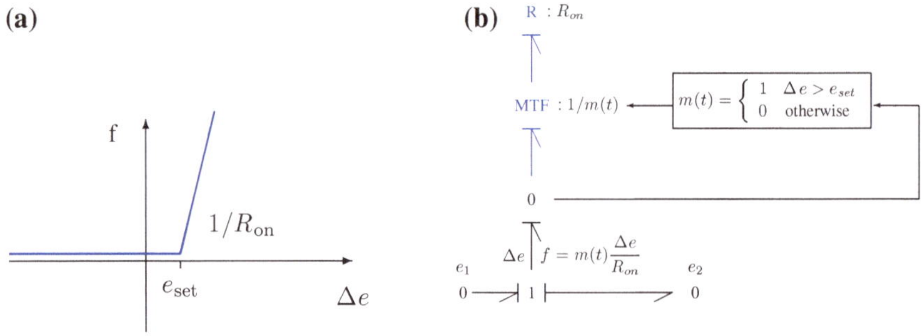

Fig. 2.12 Bond graph representation of a non-ideal switch. **a** Piecewise linear approximation of the characteristic of a diode. **b** Bond graph model of a switching device with R : R_{on} in fixed conductance causality

Non-ideal Switches

Finally, another way to keep computational causalities in a bond graph system mode independent is to model fast switching devices as non-ideal switches and to account for a resistance in ON-mode. The functionality of electronic diodes or hydraulic check valves suggests to represent a non-ideal switch in a bond graph by means of a resistor with conductance causality. In [28], the nonlinear characteristic of a diode is approximated by a piecewise linear one. The switching is explicitly represented by decomposing the resistor into an MTF with a modulus $m(t) \in \{0, 1\} \; \forall \, t \geq 0$ and a resistor R : R_{on} accounting for the ON-resistance as displayed in Fig. 2.12. Depending on the value of the transformer modulus $m(t)$, model parts linked by such MTFs are either connected or disconnected. For s switches in a hybrid model, there are $s_f \leq 2^s$ physically feasible switch state combinations. Each one of them constitutes a system mode.

In this book, the bond graph representation of a non-ideal switch proposed by Ducreux et al. as early as 1993 is used in a bond graph model-based approach to FDI in hybrid systems. This approach offers the following advantages.

- Switching devices, e.g. semiconductor switches in power electronic systems that are explicitly represented by a component model can be more easily identified in a bond graph in comparison to a representation that accounts for switches by controlled junctions. An initial not simplified bond graph is more close to a system schematic (cf. Chap. 4, Figs. 4.2 and 4.3).
- One single bond graph can be set up that holds for all system modes.
- The standard Sequential Causality Assignment Procedure (SCAP) can be applied without any modification resulting in time-invariant computational causalities.
- A single set of model equations as well as equations for fault indicators valid for all system modes can be (automatically) derived from the bond graph. They contain the discrete moduli $m_i(t) \in \{0, 1\}, i = 1, \ldots, s$, of the switching MTFs. Inserting values for the moduli belonging to a feasible switch state combination gives the equations that hold for the associated system mode.

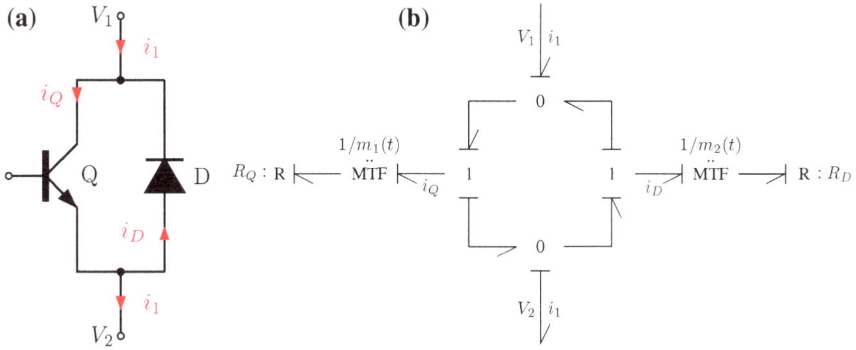

Fig. 2.13 Bipolar transistor with a reversed biased diode in parallel often used in power electronic converters and its bond graph representation. **a** Transistor-diode pair. **b** Bond graph representation

- Existing software such as SYMBOLS™[2] [29] could be used to generate a set of ARRs.

Semiconductor switches in power electronic inverters are commonly made up of a transistor together with a diode in anti-parallel connection as depicted in Fig. 2.13a for a bipolar transistor to provide a path for an inductive load current when conducting switches are turned off and thus to avoid damage of the transistors in an inverter. MOSFET transistors have a built-in diode. Figure 2.13b shows a bond graph model of such a transistor-diode pair.

The time invariant conductance causality of the ON resistor of non-ideal switches may lead to conflicts at junctions that can be resolved by adding a storage element. Equations derived from a bond graph may be symbolically formulated in such a way that the parameter of these additional storage elements can be set to zero as well as the small ON resistance of some switches turning them into ideal switches and avoiding small time constants. That is, ideal switches may be considered the limit case of non-ideal switches. Avoiding stiff equations is of interest in the case when an explicit formulation of ARRs is not possible so that the model equations need to be solved numerically in order to determine the residuals of ARRs. In [30], Buisson et al. compare the use of ideal and non-ideal switches in a bond graph framework in the light of perturbation theory.

For the sake of a concise representation in the bond graphs of the examples in this book, a switched MTF : $m(t)$, $m \in \{0, 1\}$, in conjunction with the ON resistance R : $\mathrm{R_{on}}$ of a switch is represented as a component model denoted by Sw : m as shown in Fig. 2.14.

Instantaneous Structural Changes

If phenomena such as the engagement of a clutch or a change from slip to stick friction is considered as a discrete event then the instantaneous mode change results

[2] Symbols is a trademark of HighTech Consultants, STEP, I.I.T.Kharagpur - 721 302, India, http://www.htcinfo.com/.

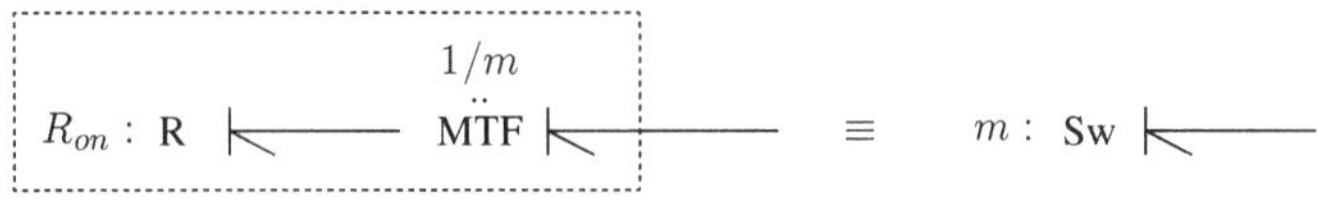

Fig. 2.14 Bond graph model of a non-ideal switch denoted as a component model Sw : m

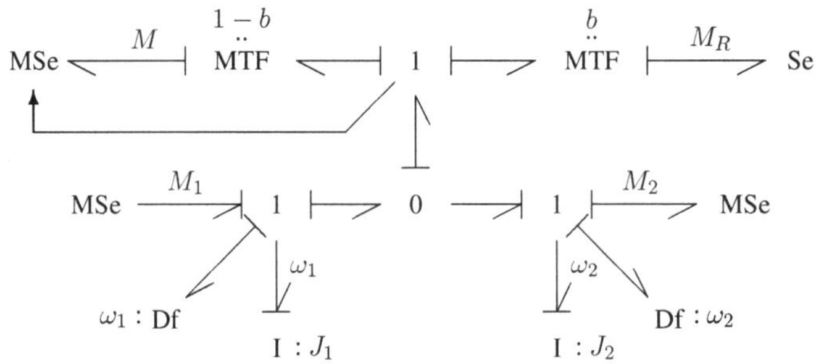

Fig. 2.15 Bond graph of a clutch with fixed mode independent causalities

in a change of the number of state variables. As long as two moving rigid bodies stick together due to friction they build a single body for some time span with one single state variable. In power electronic systems, the OFF state of a switch may cause two inductor currents to become equal. This means that in a bond graph representation of a hybrid model causality at some storage ports may change from integral to derivative causality. Preferred integral causality may, however, be preserved by using a modulated sink that either impose an effort or a flow on two dependent storage elements so that their output variables become equal. They are to be activated whenever a system mode change happens that causes two storage elements to become dependent. Their activation can be achieved by a 'Boolean' controlled transformer as depicted for the example of a mechanical clutch.

Example: Mechanical Clutch

Figure 2.15 depicts a bond graph representation of a hybrid model of a clutch.

The left upper modulated effort source provides a moment M that enforces that the plates of the clutch stick together when the clutch is in mode 'engaged', i.e. their angular velocities are equal. This moment is imposed on the two I-elements in integral causalities. At the time instant the clutch instantaneously switches from disengagement to engagement, numerical integration must be re-initialised. At this discrete event, the modulated sink is activated and the output of both I-elements jump to a common value and integration restarts from this value.

2.3 Equations Formulation for Switched LTI Systems

The use of such modulated sinks that are activated and deactivated at discrete time instances means that the underlying mathematical model is of the form of a DAE

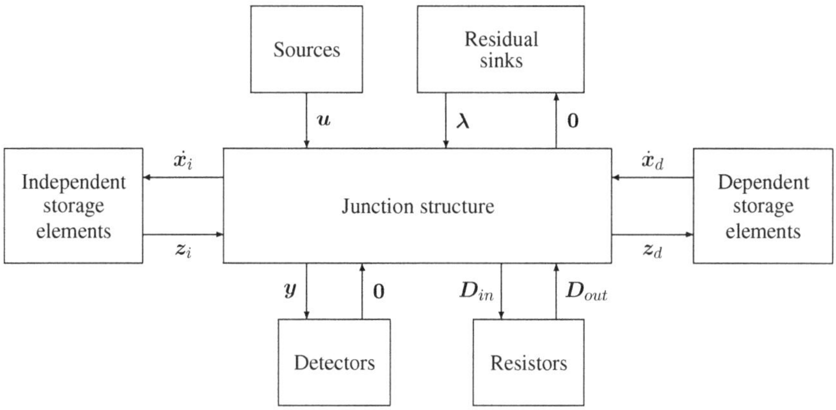

Fig. 2.16 Partitioning of a bond graph model into fields

system. These sinks enforce that algebraic constraints for the output variables of some storage elements are met but there is no differential equation for their output. When the modulated sinks are active, the DAE system is of index 2.

Let the moduli $m_j(t) \in \{0, 1\}\ \forall\, t \geq 0$, $j = 1, \ldots, n_s$, of the n_s switching MTFs be grouped into a vector $\boldsymbol{m}(t)$, the outputs z_i of all storage elements in integral causality, the inputs z_d of all storage elements in derivative causality and the outputs of resistors and switches be grouped into a descriptor vector $\boldsymbol{x}$. The outputs of all modulated sinks activated at switching events are combined into a vector $\boldsymbol{\lambda}$. Let $\boldsymbol{u}$ denote the vector of all system inputs.

If it is assumed that the system under consideration can be described by a linear time-invariant model for the time spans between two discrete switching events then, starting from the partitioning of a bond graph model into fields as displayed in Fig. 2.16, a DAE system in the form

$$\left[\begin{array}{c|c} \boldsymbol{E} & \boldsymbol{0} \\ \hline \boldsymbol{0} & \boldsymbol{0} \end{array}\right] \frac{d}{dt} \left[\begin{array}{c} \boldsymbol{x}(t) \\ \hline \boldsymbol{\lambda}(t) \end{array}\right] = \left[\begin{array}{c|c} \boldsymbol{A}_{11}(\boldsymbol{m}(t)) & \boldsymbol{A}_{12}(\boldsymbol{m}(t)) \\ \hline \boldsymbol{A}_{21}(\boldsymbol{m}(t)) & \boldsymbol{0} \end{array}\right] \left[\begin{array}{c} \boldsymbol{x}(t) \\ \hline \boldsymbol{\lambda}(t) \end{array}\right] + \left[\begin{array}{c} \boldsymbol{B}_1(\boldsymbol{m}(t)) \\ \hline \boldsymbol{B}_2(\boldsymbol{m}(t)) \end{array}\right] \boldsymbol{u}(t) \tag{2.12}$$

can be established. In (2.12), the matrices are of proper dimensions and $\boldsymbol{E} = \begin{bmatrix} \boldsymbol{E}_1 \\ \boldsymbol{0} \end{bmatrix}$.

If, for some time intervals, no residual sinks are switched on via an MTF, then the mathematical model takes the form

$$\boldsymbol{E}\dot{\boldsymbol{x}}(t) = \boldsymbol{A}_{11}(\boldsymbol{m}(t))\boldsymbol{x}(t) + \boldsymbol{B}_1(\boldsymbol{m}(t))\boldsymbol{u}(t) \tag{2.13}$$

for these time intervals. If the set of algebraic constraints can be solved then the DAE system can be reduced to an explicit linear state equation.

For controlled switching elements, the time instances of the discrete state transitions are known a priori. For other events such as the transition from slip to stick friction, the value of the MTF moduli involved is the result of a local automaton evaluating constraints on model variables. In that case, the time instant of the event is to be determined and numerical integration is to be reset in general.

For the clutch example, the DAE system derived from the BG in Fig. 2.15 reads

$$\left[\begin{array}{cc|c} J_1 & 0 & 0 \\ 0 & J_2 & 0 \\ \hline 0 & 0 & 0 \end{array}\right] \frac{d}{dt} \left[\begin{array}{c} \omega_1 \\ \omega_2 \\ \hline M \end{array}\right] = \left[\begin{array}{cc|c} 0 & 0 & -(1-b) \\ 0 & 0 & (1-b) \\ \hline (1-b) & -(1-b) & 0 \end{array}\right] \left[\begin{array}{c} \omega_1 \\ \omega_2 \\ \hline M \end{array}\right] + \begin{bmatrix} 1 & 0 & -b \\ 0 & -1 & b \\ 0 & 0 & 0 \end{bmatrix} \begin{bmatrix} M_1 \\ M_2 \\ M_R \end{bmatrix} \tag{2.14}$$

If the clutch is disengaged ($b = 1$) then the DAE system reduces to

$$\begin{bmatrix} J_1 & 0 \\ 0 & J_2 \end{bmatrix} \frac{d}{dt} \begin{bmatrix} \omega_1 \\ \omega_2 \end{bmatrix} = \begin{bmatrix} 0 & 0 & 0 \\ 0 & 0 & 0 \end{bmatrix} \begin{bmatrix} \omega_1 \\ \omega_2 \end{bmatrix} + \begin{bmatrix} 1 & 0 & -1 \\ 0 & -1 & 1 \end{bmatrix} \begin{bmatrix} M_1 \\ M_2 \\ M_R \end{bmatrix} \tag{2.15}$$

and could be formulated as an explicit state space equation.

A different bond graph approach to hybrid system modelling recently proposed by Margetts [19] inspired by the work of Buisson [31] and of Mosterman [21] also leads to an implicit formulation of system equations that holds for all system modes and can be systematically constructed from a hybrid bond graph. This approach makes use of controlled junctions and partitions the field of storage elements and the field of resistors into a field with elements having static causalities and another field with elements having causalities that dynamically change. Margetts terms the latter dynamic causalities and expresses them on a hybrid bond graph by a dashed perpendicular causal stroke in addition to the solid line causal stroke assigned by application of the SCAP.

2.4 Index of a DAE System Derived from the Bond Graph of a Hybrid Model

In this book, switching devices such as electrical diodes and transistors, or hydraulic valves are modelled as non-ideal switches represented by a bond graph component model Sw that is composed of a switched MTF and a resistor in fixed conductance causality. The choice of fixed conductance causality is motivated by the fact that it is the flow through the element that is determined by the discrete switch state.

Structural changes such as the engagement and disengagement of a clutch or the change between slip and stick mode for two rigid bodies considered as instantaneous discrete events cause the number of states in one system mode to change instantaneously at the advent of a discrete event. If two rigid bodies stick together for the duration of that system mode, the two of them may be considered as one body, or one inertia is dependent of the other so that one degree of freedom is temporarily lost. As presented in Sect. 2.3, such structural changes are modelled in a bond graph by means of a switched MTF and a residual sink that supplies an output variable so that its input vanishes. As a result, storage elements that have been assigned integral causality remain in integral causality for all system modes. If linear constitutive element equations are assumed, the equations derived from such a bond graph with static, system mode independent causalities can be formulated as a linear implicit DAE (2.12).

A characteristic of DAEs besides their form is their differentiation index [32]. For a definition and an example see Appendix C. It is an indicator for the problems to be encountered with the numerical solution of a set of DAEs. Systems of index >1 are usually called higher index DAEs and the higher the index the more severe numerical difficulties can be. As the mathematical description of problems in various disciplines often leads to DAE system, they have been a research subject for more than two decades. A large body of publications and a number software programs for their numerical solution have emerged. DAE systems of index 1 can be safely numerically computed by means of the backward differentiation formula (BDF) [33, 34] implemented in solvers such as the well known and widely used DASSL code [35].

For the solution of higher index problems, index reduction techniques [34] such as the graph-based Pantelides algorithm [36] or the dummy derivative method [37] have been reported in the literature and are used in some software programs. Both methods are based on the symbolic differentiation of some of the constraint equations of a DAE system and thus introduce equations with higher order derivatives of unknown variables with respect to time. This differentiation entails a problem in its own to find a set of consistent initial values. The Pantelides algorithm aiming at the determination of a set of consistent initial conditions can and is used to determine which equations are to be differentiated how many times in order to reduce the index to one or zero. The algorithm assumes that the index does not change. A direct initialization technique for DAE systems has been presented in [38]. Reference [39] consideres the problem of consistent initial conditions for switched linear passive network.

Index reduction entails another well known problem, namely that of numerical drift. Original constraint equations get lost by differentiation and cannot be taken into account in the numerical solution of the reduced index system. Hence, the numerical solution of the reduced index problem can only approximate the original constraints. This suggests to keep the original constraints and differentiated equations in the numerical solution of the reduced index problem resulting in more equations than unknowns in the reduced index problem. The dummy derivative method addresses this problem by considering the derivatives of some variables as new independent algebraic variables called *dummy derivatives* so that the number of unknowns matches the number of equations. This approach, however, requires to decide which variables are selected as dummy derivatives and which ones as states.

Hybrid models may be described by a single set of DAEs including discrete state variables, e.g. switch state variables that change their discrete values at discrete events. That is, for each set of values of the discrete state variables representing a physically feasible system mode a set of DAEs is obtained that describes the dynamic behaviour in that system mode. Such a set of n_f DAE systems

$$\boldsymbol{F}_i(\boldsymbol{x}(t), \dot{\boldsymbol{x}}(t), \boldsymbol{u}(t), \boldsymbol{y}(t)) = \boldsymbol{0}\,, \quad i = 1, \ldots, n_f \tag{2.16}$$

together with a set of conditions $i = C(t, \boldsymbol{x})$ is termed a system of *hybrid* DAEs (HDAEs) where $\boldsymbol{F}$ denotes a vector-valued function, t the time, $\boldsymbol{x}$ the vector of unknown variables composed of state variables and algebraic variables, $\boldsymbol{u}$ the vector of known inputs and $\boldsymbol{y}$ the vector of outputs. Evaluation of the function C at the advent of a discrete event determines which DAE system is to be used to describe the dynamic behaviour for the duration of the next system mode. At discrete events parts of the overall DAE system are deactivated or re-activated so that the index of a DAE system for subsequent system mode may be different from the one of the DAE system for the current system mode. For the duration of a system mode, however, the DAE index remains constant. A mixed symbolic and numerical solution of HDAEs involves

- index reduction and a selection of states for the reduced index problem augmented by initial constraints and differentiated equations,
- discrete event detection during evaluation of a continuous time model for the current system mode,
- initialisation at start time and re-initialisation at discrete events.

Algorithms for these tasks have been implemented, for instance, in the open-source OpenModelica software environment [40, 41].

As problems are to be encountered with the numerical solution of higher index DAEs and as there is no general solver for them, given a causal bond graph, it is an important question of what index a DAE system derived from the bond graph is. In [42], van Dijk classifies causal paths and determines the index of bond graphs of continuous time models with a given type of causal paths. For hybrid models, a bond graph representation with time-invariant causalities is used in this book. Causal paths through switches, however, depend on the switch state.

First, it is assumed that all storage elements are in integral causality for all system modes. That is, no residual sinks are switched on at discrete events to keep some storage elements in integral causality that otherwise would be become dependent and would get derivative causality accordingly. Causal paths between resistive ports are allowed. As the switch model contains a resistor in conductance causality, there may also be causal paths between a resistor and a switch or between switches. If a switch in one of these causal paths is in OFF mode, the switch and the causal path can be disregarded. Causal paths between resistors mean that their outputs are determined by a set of algebraic relations. Let $\boldsymbol{a}$ denote the vector of the outputs of resistors and of switches, then the DAE of a switched LTI system is of the form

$$\begin{bmatrix} \mathbf{I} & \mathbf{0} \\ \mathbf{0} & \mathbf{0} \end{bmatrix} \begin{bmatrix} \dot{x}(t) \\ \dot{a}(t) \end{bmatrix} = \begin{bmatrix} \mathbf{A}_{11}(\boldsymbol{m}(t)) & \mathbf{A}_{12}(\boldsymbol{m}(t)) \\ -\mathbf{A}_{21}(\boldsymbol{m}(t)) & \mathbf{I} - \mathbf{A}_{22}(\boldsymbol{m}(t)) \end{bmatrix} \begin{bmatrix} x(t) \\ a(t) \end{bmatrix} + \begin{bmatrix} \mathbf{B}_1(\boldsymbol{m}(t)) \\ -\mathbf{B}_2(\boldsymbol{m}(t)) \end{bmatrix} [\boldsymbol{u}] \tag{2.17}$$

where $\boldsymbol{m}(t)$ denotes the vector of switch states $m_j(t) \in \{0, 1\}$, $j = 1, \ldots, n_s$, at time instant $t > 0$. For the time interval between two discrete events, $\boldsymbol{m}(t)$ is constant. Differentiation of the algebraic equation

$$\mathbf{0} = -\mathbf{A}_{21}\boldsymbol{x} + (\mathbf{I} - \mathbf{A}_{22})\boldsymbol{a} - \mathbf{B}_2\boldsymbol{u} \tag{2.18}$$

with respect to time yields the new DAE system

$$\underbrace{\begin{bmatrix} \mathbf{I} & \mathbf{0} \\ -\mathbf{A}_{21} & \mathbf{I} - \mathbf{A}_{22} \end{bmatrix}}_{\mathbf{E}^1} \begin{bmatrix} \dot{x}(t) \\ \dot{a}(t) \end{bmatrix} = \begin{bmatrix} \mathbf{A}_{11}(\boldsymbol{m}(t)) & \mathbf{A}_{12}(\boldsymbol{m}(t)) \\ -\mathbf{0} & \mathbf{0} \end{bmatrix} \begin{bmatrix} x(t) \\ a(t) \end{bmatrix} + \begin{bmatrix} \mathbf{B}_1(\boldsymbol{m}(t))\boldsymbol{u} \\ -\mathbf{B}_2(\boldsymbol{m}(t))\dot{\boldsymbol{u}} \end{bmatrix} \tag{2.19}$$

In [42], van Dijk has shown that the determinant of $\mathbf{E}^1$, $\det(\mathbf{E}^1) = \det(\mathbf{I} - \mathbf{A}_{22})$, is non-zero for bond graphs with causal paths between resistive ports. That means that the inverse of $\mathbf{E}^1$ exists and that differentiation of the algebraic equation (2.18) is sufficient to transform the DAE system (2.17) into a set ODEs. Accordingly, (2.17) is a DAE system of index 1.

Example: Electrical Network with Two Independent Switches

This is checked for the example of the switched network in Fig. 2.17.

The two switches may be operated independently. That is, there are four physical feasible switch state combinations constituting four system modes. The circuit

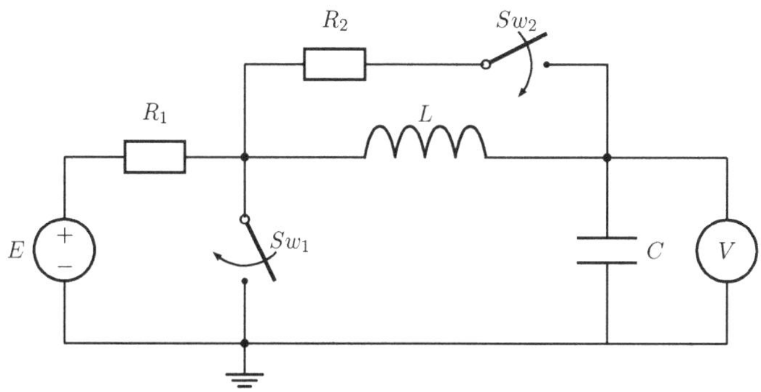

Fig. 2.17 Electrical circuit with two independent switches

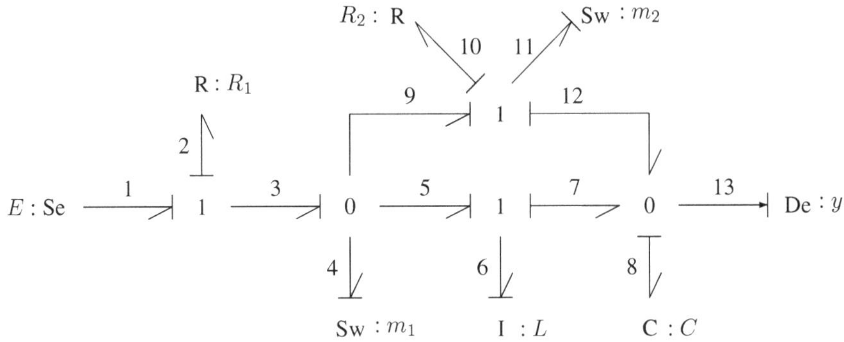

Fig. 2.18 Bond graph with system mode independent causalities of the electrical circuit in Fig. 2.17

diagram is easily transformed into the causal bond graph in Fig. 2.18. The switches are considered non-ideal and are represented by a model Sw containing a switched MTF: $m(t)$ in conjunction with an ON resistor in fixed conduction causality. The two storage elements are in integral causality in all four system modes.

The bond graph in Fig. 2.18 contains three causal paths between resistors and switches.

$$\begin{aligned}
&\mathrm{R} : R_1 \to e_2 \to e_3 \to e_4 \to \mathrm{Sw} : m_1\\
&\mathrm{R} : R_1 \to e_2 \to e_3 \to e_9 \to e_{11} \to \mathrm{Sw} : m_2\\
&\mathrm{R} : R_2 \to e_{10} \to e_{11} \to \mathrm{Sw} : m_2
\end{aligned}$$

The DAE system for that example reads

$$\frac{d}{dt}\begin{bmatrix} i_L \\ u_C \end{bmatrix} = \begin{bmatrix} 0 & -\frac{1}{L} \\ \frac{1}{C} & 0 \end{bmatrix}\begin{bmatrix} i_L \\ u_C \end{bmatrix} + \begin{bmatrix} -\frac{1}{L} & 0 & 0 & 0 \\ 0 & \frac{1}{C} & 0 & 0 \end{bmatrix}\begin{bmatrix} u_{R_1} \\ u_{R_2} \\ i_{Sw_1} \\ i_{Sw_2} \end{bmatrix} + \begin{bmatrix} \frac{1}{L} \\ 0 \end{bmatrix}[E] \quad (2.20)$$

and

$$\underbrace{\begin{bmatrix} u_{R_1} \\ u_{R_2} \\ i_{Sw_1} \\ i_{Sw_2} \end{bmatrix}}_{\boldsymbol{a}} = \underbrace{\begin{bmatrix} R_1 & 0 \\ 0 & 0 \\ 0 & 0 \\ 0 & -\frac{m_2}{R_o} \end{bmatrix}}_{\mathbf{A}_{21}}\underbrace{\begin{bmatrix} i_L \\ u_C \end{bmatrix}}_{\boldsymbol{x}} + \underbrace{\begin{bmatrix} 0 & 0 & R_1 & R_1 \\ 0 & 0 & 0 & R_2 \\ -\frac{m_1}{R_o} & 0 & 0 & 0 \\ -\frac{m_2}{R_o} & -\frac{m_2}{R_o} & 0 & 0 \end{bmatrix}}_{\mathbf{A}_{22}}\underbrace{\begin{bmatrix} u_{R_1} \\ u_{R_2} \\ i_{Sw_1} \\ i_{Sw_2} \end{bmatrix}}_{\boldsymbol{a}} + \underbrace{\begin{bmatrix} 0 \\ 0 \\ \frac{m_1}{R_o} \\ \frac{m_2}{R_o} \end{bmatrix}}_{\mathbf{B}_2}\underbrace{[E]}_{\boldsymbol{u}} \quad (2.21)$$

where R_o denotes the ON resistance of the switches. Transformation to triangular form shows that the matrix $(\mathbf{I} - \mathbf{A}_{22})$ is non-singular independent of the switch

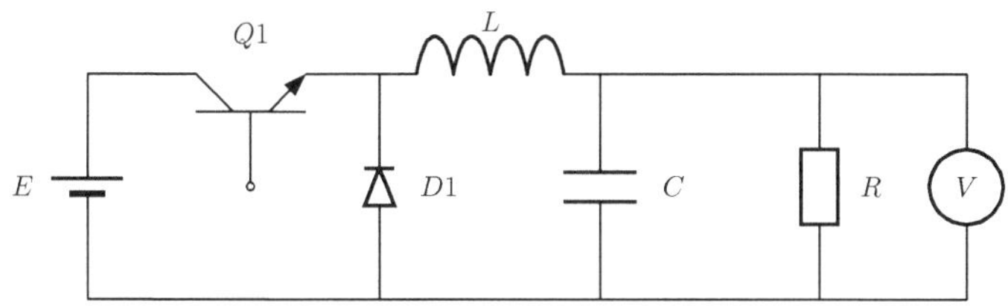

Fig. 2.19 Circuit schematic of a DC-DC buck converter

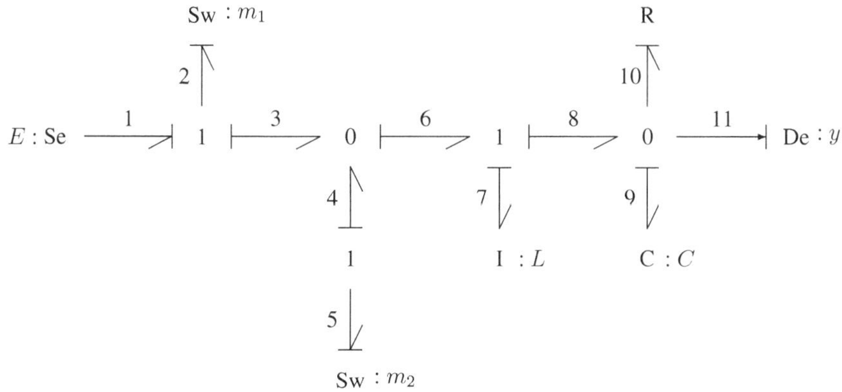

Fig. 2.20 Bond graph in preferred integral causality of the buck converter in Fig. 2.19

states $m_1(t), m_2(t)$. That is, the DAE system describing the dynamic behaviour of the switched network is of index 1 for all four system modes.

Example: DC-DC Buck Converter

Figure 2.19 depicts the circuit schematic of a buck converter.

If the transistor Q_1 and the diode D_1 are modelled as non-ideal switches the bond graph in Fig. 2.20 is obtained.

The two switches commutate conversely. There are two system modes, in which one switch is on while the other one is off. A third feasible mode, in which both switches are off is not considered. If one of the switches is on, conductance causality of its ON resistance can be changed. The result is a causal path between the two non-ideal switches. In Fig. 2.21, it has been assumed that Sw : m_2 is on.

However, as one of two switches is off, i.e. the current through that switch vanishes, the causal path can be disregarded. That is, in system modes $m_1 = 1 \wedge m_2 = 0$ and $m_1 = 0 \wedge m_2 = 1$, there are no causal paths between resistive ports and no dependent storage elements giving rise to an algebraic constraint between state variables. Hence, the dynamic behaviour in these modes is described by a set of two ODEs. In the context of DAEs, the special case of ODEs without any additional algebraic constraints is termed a DAE system of index 0.

If there are storage elements with derivative causality in all system modes then causal paths may exist to other storage elements of the same type in integral causality.

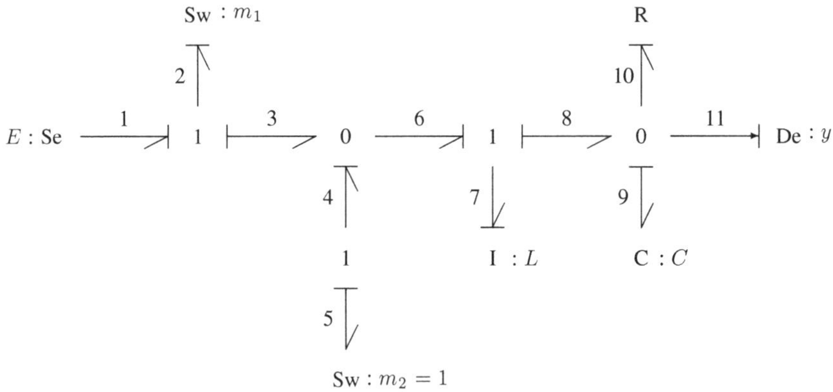

Fig. 2.21 Bond graph in preferred integral causality of the buck converter in case Sw : m_2 is on

This means that the state variable of a storage element in derivative causality is algebraically related to the state variables of a storage elements in integral causality. Also in this case, the DAE system for the time interval between two discrete events is of index 1 [42].

Structural Changes

For some systems with structural changes such as a clutch, storage elements may temporarily become dependent for the duration of a system mode. In such a case, a residual sink may be switched on that delivers a power variable so that the conjugate power variable vanishes and storage elements can keep integral causality. As their state variables jump to a new joint value, numerical integration has to be re-initialised at such a discrete event.

For a system mode, in which such switched residual sinks provide a non-zero output $\boldsymbol{\lambda}$, the DAEs for a switched LTI system are of the form

$$\begin{bmatrix} \mathbf{I} & \mathbf{0} \\ \mathbf{0} & \mathbf{0} \end{bmatrix} \frac{d}{dt} \begin{bmatrix} \boldsymbol{x} \\ \boldsymbol{\lambda} \end{bmatrix} = \begin{bmatrix} \mathbf{A}_{11} & \mathbf{A}_{12} \\ \mathbf{A}_{22} & \mathbf{0} \end{bmatrix} \begin{bmatrix} \boldsymbol{x} \\ \boldsymbol{\lambda} \end{bmatrix} + \begin{bmatrix} \mathbf{B}_1 \\ \mathbf{0} \end{bmatrix} [\boldsymbol{u}] \tag{2.22}$$

The entries of the matrices depend on the values of the current switch states. As the variable $\boldsymbol{\lambda}$ is absent in the algebraic constraint

$$\mathbf{0} = \mathbf{A}_{22}\boldsymbol{x} \tag{2.23}$$

differentiation of (2.23) does not give an equation for $\dot{\boldsymbol{\lambda}}$ but an algebraic equation that is called a *hidden constraint*.

$$\mathbf{0} = \mathbf{A}_{22}[\mathbf{A}_{11}\boldsymbol{x} + \mathbf{A}_{11}\boldsymbol{\lambda} + \mathbf{B}_1\boldsymbol{u}] \tag{2.24}$$

If the matrix $\mathbf{A}_{22}\mathbf{A}_{11}$ is non-singular, (2.24) differentiated with respect to time can be solved for $d\boldsymbol{\lambda}/dt$. According to the definition of the differentiation index, the DAE in this system mode is of index 2.

Remark 2.2

1. Initial conditions must also satisfy hidden constraints. It is these hidden constraints that pose problems for DAE-solvers.
2. The DAE form (2.22) is known as Hessenberg index-2 form. This form can be directly numerically solved by the DASKP 3.0 solver [43, 44] or the MEBDFI [45] solver available in Maple [46]. Both solvers use the BDF method. □

As a result, if the dynamic behaviour of a system can be described by a switched LTI system, the DAE system derived from the bond graph is of index ≤ 1 as long as no structural changes occur. If a structural change modelled by a switched residual sink happens, the DAE index jumps to two.

Example: Clutch

As an example, consider the bond graph of a clutch in Fig. 2.15. There are no causal paths between resistors and no dependent storage elements. Clearly, as long as the clutch is disengaged, the DAE system is of index 0. In the case when the clutch is engaged, the unknown constraint force M between the two plates keeps their inertia elements in integral causality and at the same time ensures that the algebraic constraint

$$0 = \omega_1 - \omega_2 \tag{2.25}$$

holds. Clearly, differentiating this constraint with respect to time does not give the derivative of the unknown constraint force M. Substituting $d\omega_1/dt$, $d\omega_2/dt$ in the derivative of the constraint allows to obtain dM/dt after differentiation of the resulting hidden constraint

$$0 = -\frac{1}{J_1}M + \frac{1}{J_1}M_1 - \frac{1}{J_2}M + \frac{1}{J_2}M_2 \tag{2.26}$$

Hence, once the clutch is engaged, the DAE is of index 2.

2.5 Discrete Event Simulation of Hybrid Systems

Instead of discretising the time and using a BDF-based method for the numerical computation of a continuous-time model one may think of quantising the state variables. That is, instead of using a multistep method to compute an approximation of the value $x(t_{k+1})$ of a state variable x at time t_{k+1}, the question then is at what time the state x will deviate from its current value $x(t_k)$ by more than a given quantum ΔQ. In other words, the task is to find the smallest time step h so that

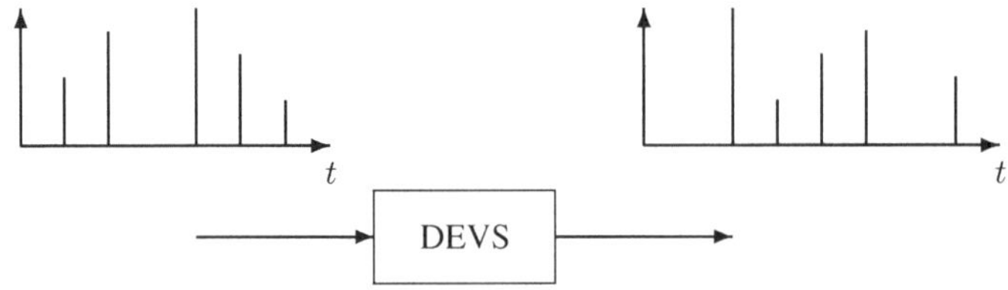

Fig. 2.22 Input/output behaviour of a DEVS model [1, 53]

$$x(t_k + h) = x(t_k) \pm \Delta Q \tag{2.27}$$

This approach replaces the traditional discrete time based computation of continuous time models by a discrete event simulation that advances the time from the time point of an event to the time of the next event which is attractive for the computation of hybrid models because discrete events, i.e. discontinuous mode changes, and the continuous time behaviour during system modes can be uniformly processed in the framework of the well-known Discrete EVent System (DEVS) specification introduced by Ziegler [47, 48]. Moreover, the DEVS formalism is supported by software libraries such as adevs [49, 50] or simulation environments such as PowerDEVS [51, 52].

A DEVS model of a system takes a sequence of discrete events, i.e. of instantaneous system changes, as inputs and produces an output sequence of events according its initial conditions. Figure 2.22 displays this input/output behaviour of a DEVS model. Events can be characterised by a value and the time point of their occurrence. Accordingly, they are indicated by perpendicular strokes in Fig. 2.22. A sequence of events is called an event trajectory. It is assumed that the number of state changes in any finite time interval is finite.

An *atomic* DEVS model is defined as a tuple of sets and functions (cf. Appendix A.3). If the output events of an atomic DEVS model are converted into input events of another atomic DEVS model, i.e. if atomic DEVS models are coupled, then the result defines a new DEVS model. That is, complex systems can be modelled in the DEVS framework in a hierarchical manner. The DEVS formalism is widely used in computer science. Its application to the numerical solution of continuous-time models, however, is much less common.

The quantised state system (QSS) method introduced by Kofman [53] allows for a discrete event simulation of hybrid systems. The method starts from the observation that a piecewise constant trajectory can be represented by sequences of events. The reader is referred to the literature, e.g. references [1, 53, 54] for details. In the following, only the basic idea is outlined in a simplified manner.

Consider the state space equation

$$\dot{x}(t) = f(x(t), u(t)) \tag{2.28}$$

where $x(t)$ is the state and $u(t)$ the input given by a piecewise constant function.

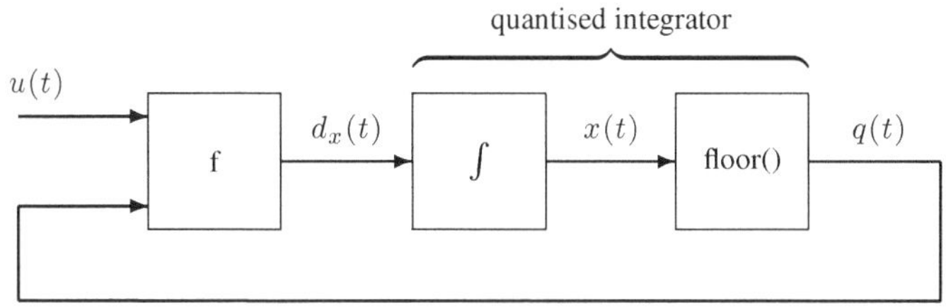

Fig. 2.23 Block diagram of a quantised state system

Let floor : $\mathbb{R} \to \mathbb{Z}^+$ denote a function that returns the integer part of its positive real-valued argument. The QSS-method then solves the approximate system

$$\dot{x}(t) = d_x(t) \tag{2.29a}$$

$$q(t) = \text{floor}(x(t)) \tag{2.29b}$$

$$d_x(t) = f(q(t), u(t)) \tag{2.29c}$$

The variable q is called a *quantised* variable and Eq. (2.29) a *quantised state system* (QSS). equations 2.29a–2.29b constitute a dynamic subsystem with a piecewise constant input trajectory $d_x(t)$ and a piecewise constant output trajectory $q(t)$. The third Eq. 2.29c is a static relation. The quantised state system (2.29) can be displayed by the block diagram in Fig. 2.23.

The static block and the quantised integrator can be represented by a DEVS model. Their coupling again builds a DEVS model. The DEVS models can be translated into PowerDEVS models. The software environment PowerDEVS provides a library with a number of predefined atomic and coupled DEVS models and a graphical user interface that permits to build coupled DEVS models.

Given a time instant t_k and a constant quantum ΔQ, then the next time step h that satisfies (2.27) can be determined if the derivative of the state variable in (2.29a) is approximated by a difference quotient.

$$\frac{x(t_k + h) - x(t_k)}{h} = \frac{\pm \Delta Q}{h} = d_x(t_k) = f(q(t_k), u(t_k)) \tag{2.30}$$

The QSS method outlined so far bears a problem because it cannot be guarantied that a DEVS model resulting from the coupling of atomic DEVS models will perform a finite number of transitions in any finite time interval. DEVS models with this ability are said to be *legitimate*. For illegitimate DEVS models the simulation will get stuck. Kofman has shown that the problem can be overcome by introducing hysteresis in the relationship between a state variable x and its quantised variable q [53]. However, the method is only first-order accurate. A quantisation for which the hysteresis value equals the quantum value produces a piecewise constant output trajectory that only changes when the difference between input and output exceeds the quantum. As a result, the state variables have a piecewise linear trajectory. To improve accuracy,

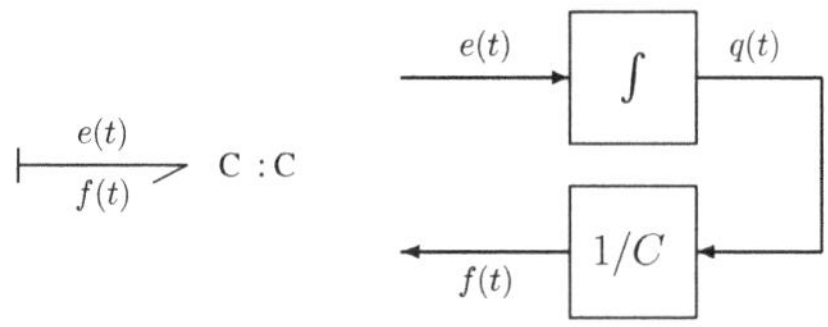

Fig. 2.24 Block diagram of a bond graph C storage element in integral causality

Kofman proposed a second-order accurate QSS2 method [55]. For this method, the quantisation that does not need hysteresis produces piecewise linear output trajectories having discontinuities whenever the absolute value of the difference between a state variable and its quantised variable reaches the quantum. If the QSS2 method is applied to LTI systems then state variables have piecewise parabolic trajectories provided input trajectories are piecewise constant.

The outlined quantised state integration can be applied to systems of coupled ODEs with piecewise constant input functions. Moreover, it can be extended so that hybrid DAE systems can be solved by discrete event simulation [54].

Bond Graph Models and Discrete Event Simulation

As can be seen from Fig. 2.23, a coupled DEVS model is obtained from a continuous time model by feeding the output of an integrator into a quantising block turning the integrator into a so-called *quantised integrator* provided input trajectories are piecewise constant. This observation suggests that discrete event simulation can be applied to systems represented by a bond graph because a causal bond graph is equivalent to a block diagram in which the block diagram replacement of each energy storage element includes an integrator (Fig. 2.24).

It is sufficient to quantise the static relation between the energy variable of a storage element and its output power variable. Consider, e.g. the linear 1-port C element in integral causality in Fig. 2.24 and let $q_q(t) = \text{floor}(q(t))$ the quantisation of its displacement q. The continuous constitutive equations of a C element with the capacitance C

$$\dot{q}(t) = f(t) \tag{2.31a}$$

$$e(t) = \frac{1}{C} q(t) \tag{2.31b}$$

are then replaced by

$$\dot{q}(t) = f(t) \tag{2.32a}$$

$$e(t) = \frac{1}{C} q_q(t) = \Phi_q(q(t)) \tag{2.32b}$$

That is, the characteristic of a storage element is approximated by a piecewise constant function $\Phi_q()$ (with hysteresis). The other bond graph elements are static elements. A 2-port transformer, for instance, takes two piecewise constant inputs and delivers two piecewise constant outputs. Bond graphs in which the characteristic of the storage elements has been quantised are called quantised bond graphs (QBGs) in [53, 56]. The elements of a quantised bond graph can be represented by DEVS models. The result of their coupling is a DEVS model of the QBG. Hence, this approximation of an initial continuous-time bond graph model can be computed by means of discrete event simulation.

Another approach that also allows to approximate each continuous time element of a bond graph by a DEVS model so that a DEVS simulation can be performed has been reported in [57, 58]. The task is to transform piecewise continuous input and output trajectories of a bond graph element into discrete event trajectories. To translate, e.g. the continuous time model of a C element in integral causality into a discret event model, the input trajectory of the flow $f(t)$ between two time instances t_i and t_j is approximated by a linear function $f(t) = a_1 t + a_0$. The output trajectory of $e(t)$ is a second order polynomial $e(t) = b_2 t^2 + b_1 t = (a_1/2C)t^2 + (a_0/C)t$. This polynomial is approximated by a linear function $\tilde{e}(t) = \tilde{b}_1 t + \tilde{b}_0$. Its coefficients are determined by the value of $e(t)$ at the beginning and at the end of the time interval $[t_i, t_j]$.

$$\tilde{e}(t) = \frac{e(t_j) - e(t_i)}{t_j - t_i} t + e(t_i) \tag{2.33}$$

A discrete event is considered to take place whenever the linear trajectory for the time interval $[t_i, t_j]$ crosses a threshold or when there is a significant change of its slope. This threshold crossing or a gradient change determines the time instant of the next event and thus the length of the time interval of the current linear trajectory.

An attractive feature of a discrete event simulation of hybrid systems is that the simulation time advances from discrete event to discrete event. For the QSS method, discontinuities in the inputs and the quantised variables dictate the time advance. No iteration is necessary to locate the time point of a discontinuity allowing for an efficient processing of models with discontinuities. Discrete event simulation using the quantised-based integrations needs much less simulation steps than a numerical integration method of comparable accuracy based on time-discretisation. Accordingly, computational costs are saved. Nevertheless, there are still some problems to be tackled with the QSS approach as detailed in [1, Chap. 12.11].

2.6 Summary

The survey in this chapter shows that there are various bond graph approaches to hybrid system modelling having their pros and cons. A comparison of approaches using ideal switches versus those using non-ideal switches has been given in [30].

In this book, a bond graph representation of hybrid system models is chosen as basis for bond graph model-based FDI in Chap. 4. The representation is system mode independent with regard to computational causalities and allows for deriving a set of model equations in the form of a DAE system that holds for all system modes. This is achieved by modelling fast switching devices by an MTF associated with a resistor in fixed conductance causality and by means of MTFs switching on and off residual sinks in the case of structural changes. The residual sinks allow for keeping integral causality at storage elements that become dependent by a system mode change and would take derivative causality.

If the dynamic behaviour of a system can be described by a switched LTI system, a linear implicit DAE system can be derived from the bond graph. The entries of its matrices depend on the discrete switch states. As long as no structural changes occur, i.e. no residual sinks are switched on, the DAE system is of index ≤ 1. For system modes in which residuals sinks are switched on, the DAE system is of index 2. There are solvers available for its direct numerical computation that are based on the BDF-method. An alternative may be to perform a DEVS simulation that uses quantised based integration.

References

1. Cellier, F. E., & Kofman, E. (2006). *Continuous system simulation*. New York: Springer.
2. Hindmarsh, A., & Taylor, A. (1999). *User documentation for IDA, a differential-algebraic equations solver for sequential and parallel computers*. Lawrence Livermore National Laboratory, Center for Applied Scientific Computing. UCRL-MA-136910.
3. Brenan, K. E., Campbell, S. L., & Petzold, L. R. (1989). *Numerical solution of initial-value problems in differential-algebraic equations*. New York: North-Holland.
4. Zimmer, D. (2013). A new famework for the simulation of equation-based models with variable structure. *Simulation*, *89*(8), 935–963.
5. Zimmer, D. (2010). Equation-based modeling of variable-structure systems. *Ph.D. Dissertation No. 18924*, ETH Zürich.
6. Borutzky, W. (2010). *Bond graph methodology—Development and analysis of multidisciplinary dynamic system models*. London: Springer. ISBN : 978-1-84882-881-0.
7. Borutzky, W., Broenink, J. F., & Wijbrans, K. C. J. (1993). Graphical description of physical system models containing discontinuities. In A. Pavé (Ed.), *Modelling and Simulation 1993, Proceedings of the 1993 European Simulation Multiconference* (pp. 208–214). Lyon: SCS Publishing.
8. Borutzky, W. (1995). Representing discontinuities by means of sinks of fixed causality. In F. E. Cellier & J. J. Granda (Eds.), *1995 International Conference on Bond Graph Modeling, ICBGM'95, Proceedings of the 1995 Western Simulation Multiconference* (Vol. 27, No. 1, pp. 65–72). Simulation Series. Lyon: SCS Publishing. ISBN: 1-56555-037-4.
9. Bos, A. M. (1986) Modelling multibody systems in terms of multibond graphs with application to a motorcycle. *Ph.D. Thesis*. The Netherlands: University of Twente Enschede.
10. Borutzky, W., & Cellier, F. E. (1996). Tearing algebraic loops in bond graphs. *Transactions of the SCS*, *13*(2), 102–115.
11. Borutzky, W. (2009). Bond graph model-based fault detection using residual sinks. *Proceedings of the Institution of Mechanical Engineers Part I Journal of Systems and Control Engineering*, *223*(3), 337–352.
12. Umarikar, A. C. (2006). Modelling of Switched mode power converters: A bond graph approach. *Ph.D. Thesis*. Bangalore: Centre for Electronics Design and Technology, Indian Institute of Science.
13. Umarikar, A. C., & Umanand, L. (2005). Modelling of switched systems in bond graphs using the concept of switched power junctions. *Journal of the Franklin Institute*, *342*, 131–147.

14. Buisson, J., Cormerais, H., & Richard, P. Y. (2002). Analysis of the bond graph model of hybrid physical systems with ideal switches. *Proceedings of the Institution of Mechanical Engineers Part I: Systems and Control Engineering*, *216*(1), 47–63.
15. Buisson, J. (2002). *Modélisation, Analyse et Commande des Systèmes en Commutation*. Rennes: Université de Rennes 1. l'habilitation à diriger des recherches (HDR).
16. Edström, K. (1999). Switched bond graphs: Simulation and analysis. *Ph.D. Thesis*. Linköping: Linköping University.
17. Strömberg, J. E. (1994). A mode switching modelling philosophy. *Ph.D. Thesis*. Linköping: Linköping University.
18. Strömberg, J. E., Top, J., & Södermann, U. (1993). Variable causality in bond graphs caused by discrete effects. In J. J. Granda & F. E. Cellier (Eds.), *International Conference on Bond Graph Modeling, ICBGM'93, Proceedings of the 1993 Western Simulation Multiconference* (Vol. 25, No. 2, pp. 115–119). Simulation Series. Lyon: SCS Publishing. ISBN: 1-56555-019-6.
19. Margetts, R. (2013). Modelling & analysis of hybrid dynamic systems using a bond graph approach. *Ph.D. Thesis*. University of Bath.
20. Roychoudhury, I., Daigle, M., Biswas, G., & Koutsoukos, X. (2010). Efficient simulation of hybrid systems: A hybrid bond graph approach. *Simulation: Transactions of the Society for Modeling and Simulation International*, *87*(6), 467–498.
21. Mosterman, P. J. (1997). Hybrid dynamic systems: a hybrid bond graph modeling paradigm and its application in diagnosis. *Ph.D. Thesis*. Nashville: Vanderbilt University.
22. van der Schaft, A. J., & Schuhmacher, H. (2000). An introduction to hybrid dynamical systems. No. 251 in Lecture Notes in Control and Information Sciences. London: Springer.
23. Asher, G. M. (1993). The robust modelling of variable topology circuits using bond graphs. In J. J. Granda & F. E. Cellier (Eds.), *International Conference on Bond Graph Modeling, ICBGM'93, Proceedings of the 1993 Western Simulation Multiconference* (Vol. 25, No. 2, pp. 126–131). Simulation Series. San Diego: SCS Publishing. ISBN: 1-56555-019-6.
24. Thoma, J. U. (1975). *Introduction to bond graphs and their appliations*. Oxford: Pergamon Press.
25. Mosterman, P. J. (2002). HYBrSIM—A modelling and simulation environment for hybrid bond graphs. *Proceedings of the Institution of Mechanical Engineers, Part I: Journal of Systems and Control Engineering Systems*, *216*(1), 35–46.
26. Wang, D., Yu, M., Low, C., & Arogeti, S. (2013). *Model-based health monitoring of hybrid systems*. New York: Springer.
27. The Mathworks. *Simulink—Simulation and model-based design*, from http://www.mathworks.com/products/simulink/.
28. Ducreux, J. P., Dauphin-Tanguy, G., & Rombaut, C. (1993). Bond graph modelling of commutation phenomena in power electronic circuits. In J. J. Granda & F. E. Cellier (Eds.), *International Conference on Bond Graph Modeling, ICBGM'93, Proceedings of the 1993 Western Simulation Multiconference* (Vol. 25, No. 2, pp. 132–136). Simulation Series. San Diego: SCS Publishing. ISBN: 1-56555-019-6.
29. HighTec Consultants. *Symbols Sonata*™, from http://www.htcinfo.com/.
30. Buisson, J., Richard, P. Y., & Cormerais, H. (2000). Ideal versus non-ideal approaches in bond graph modelling of switching devices: a comparison based on singular perturbation theory. In *Proceedings of the 4th International Confrence on Automation of Mixed Processes: Hybrid Dynamic Systems (ADPM' 2000)* (pp. 257–264).
31. Buisson, J. (1993). Analysis and characterisation of hybrid systems with bond-graphs. In J. J. Granda & F. E. Cellier (Eds.), *International Conference on Bond Graph Modeling, ICBGM'93, Proceedings of the 1993 Western Simulation Multiconference* (Vol. 25, No. 2, pp. 264–269). Simulation Series. San Diego: SCS Publishing. ISBN: 1-56555-019-6.
32. Campbell, S. L., & Gear, C. W. (1995). The index of general nonlinear DAEs. *Numerische Mathematik*, *72*, 173–196.
33. Gear, C. W. (1971). Simultaneous numerical solution of differential/algebraic equations. *IEEE Transactions on Circuits Theory*, *CT–18*(1), 89–95.

34. Hairer, E., & Wanner, G. (1996). *Solving ordinary differential equations II, stiff and differential-algebraic problems* (2nd ed.). Berlin: Springer.
35. Brenan, K. E., Campbell, S. L., & Petzold, L. R. (1996). *Numerical solution of initial-value problems in differential-algebraic equations*. Philadelphia: SIAM.
36. Pantelides, C. C. (1988). The consistent initialization of differential-algebraic systems. *SIAM Journal of Scientific and Statistical Computation*, *9*, 213–231.
37. Mattsson, S., & Söderlind, G. (1993). Index reduction in differential algebraic equations using dummy derivatives. *SIAM Journal on Scientific Computing*, *14*(3), 677–692.
38. Vieira, R. C., & Biscaia, E. C, Jr. (2001). Direct methods for consistent initialization of DAE systems. *Computers and Chemical Engineering*, *25*, 1299–1311.
39. Frasca, R., Camlibel, M. K., Goknar, I. C., Iannelli, I., & Vasca, F. (2010). Linear passive networks with ideal switches: Consistent initial conditions and state discontinuities. *IEEE Transactions on Circuits and Systems*, *57*(12), 3138–3151.
40. Lundvall, H., Fritzson, P., & Bachmann, B. (2008). *Event handling in the openmodelica compiler and runtime system*. Series: Technical Reports in Computer and Information Science. Linkping University, Department of Computer and Information Science, PELAB—Programming Environment Laboratory. ISSN 1654–7233; 2.
41. OpenModelica Consortium. *OpenModelica*, from https://www.openmodelica.org/.
42. van Dijk, J. (1994). On the role of bond graph causality in modelling mechatronic systems. *Ph.D. Thesis*. The Netherlands: University of Twente. Enschede.
43. Li, S. Petzold, L. (1999). Design of new DASPK for sensitivity analysis. *UCSB*, from http://www.engineering.ucsb.edu/~cse/.
44. DASPK, from http://www.cs.ucsb.edu/~cse/software.html.
45. Cash, J. (2003). Efficient numerical methods for the solution of stiff initial-value problems and differential algebraic equations. *Proceedings of the Royal Society*, *459*, 797–815.
46. Maplesoft, from http://www.maplesoft.com/.
47. Zeigler, B. (1976). *Theory of modeling and simulation*. New York: Wiley.
48. Zeigler, B., Tag Gon, K., & Praehofer, H. (2000). *Theory of modeling and simulation* (2nd ed.). New York: Academic Press.
49. Naturo, J. (2011). *Building software for simulation: Theory and algorithms, with applications in C++*. New York: Wiley.
50. adevs—A simulation library for discrete event and hybrid dynamics systems, from http://sourceforge.net/projects/adevs/.
51. Bergero, F., & Kofman, E. (1229). PowerDEVS. *A tool for hybrid system modeling and real time simulation*. Rosario: Laboratorio de Sistemas Dinámicos. FCEIA - UNR. CIFASISCONICET, from http://www.fceia.unr.edu.ar/kofman/files/bergero_kofman_pd.pdf.
52. PowerDEVS, from http://sourceforge.net/projects/powerdevs/.
53. Kofman, E. (2003). Discrete event based simulation and control of continuous systems. *Ph.D. Thesis*. Facultad de Ciencias Exactas, Ingeniería y Agrimensura, Rosario: Universidad Nacional de Rosario.
54. Kofman, E. (2004). Discrete event simulation of hybrid systems. *SIAM Journal on Scientific Computing*, *25*(5), 1771–1797, from http://dx.doi.org/10.1137/S1064827502418379.
55. Kofman, E. (2002). A second order approximation for DEVS simulation of continuous systems. *Simulation*, *78*(2), 76–89.
56. Kofman, E., & Junco, S. (2001). Quantized bond graphs: An approach for discrete event simulation of physical systems. In G. Dauphin-Tanguy & J. J. Granda (Eds.), *Proceedings of the 2001 International Conference on Bond Graph Modeling and Simulation (ICBGM'01)* (Vol. 33, No. 1, pp. 369–374). Simulation Series. San Diego: SCS.
57. Naamane, A., Damiba, A., & Giambiasi, N. (2001). Bond ggraphs and gdevs. In G. Dauphin-Tanguy & J. J. Granda (Eds.), *Proceedings of the 2001 International Conference on Bond Graph Modeling and Simulation (ICBGM'01)* (Vol. 33, No. 1, pp. 375–380). Simulation Series. San Diego: SCS.
58. D'Abreu, M. C., & Wainer, G. A. (2003). Models for continous and hybrid system simulation. In S. Chick, P. J. Sánchez, D. Ferrin, & D. J. Morrice (Eds.), *Proceedings of the 2003 Winter Simulation Conference* (pp. 641–649). New Orleans: IEEE Press.

Chapter 3
Structural Control Properties of Switched LTI Systems

It is obvious that for FDI a system must be observable and in order to allow for fault accommodation it must be controllable. Therefore, it is important to analyse the structural properties of a model. If it turns out that, for a given set of sensors, a system is structurally not observable then it is numerically not observable. In a hybrid system model, switches disconnect or reconnect model parts at the advent of discrete events. That is, the model structure is system mode dependent. Also, the set of model equations that holds for a system mode differs from the one for another system mode with regard to its form and its properties.

This chapter addresses structural control properties of linear hybrid system models as well as mathematical properties. If the underlying mathematical model is of the form of a DAE system then its index is an information relevant with regard to its symbolic and numerical processing.

3.1 Structural Observability

Suppose that the dynamic behaviour of a real system can be described by means of a linear time-invariant (LTI) system given by the state space model

$$\dot{\boldsymbol{x}}(t) = \mathbf{A}\boldsymbol{x}(t) + \mathbf{B}\boldsymbol{u}(t)\,, \quad \boldsymbol{x}(0) = \boldsymbol{x}_0 \tag{3.1a}$$

$$\boldsymbol{y}(t) = \mathbf{C}\boldsymbol{x}(t) + \mathbf{D}\boldsymbol{u}(t) \tag{3.1b}$$

where $\boldsymbol{x}$ denotes the vector of states, $\boldsymbol{u}$ the vector of inputs, and where $\mathbf{A}$, $\mathbf{B}$, $\mathbf{C}$, $\mathbf{D}$ are matrices of appropriate dimensions characterising the system.

Let n be the number of states. Then the system is *numerically* completely state observable according to Kalman's criterion if and only if the rank of the observability matrix

W. Borutzky, *Bond Graph Model-based Fault Diagnosis of Hybrid Systems*,
DOI 10.1007/978-3-319-11860-4_3

$$\mathbf{O} = \begin{bmatrix} \mathbf{C} \\ \mathbf{CA} \\ \mathbf{CA}^2 \\ \vdots \\ \mathbf{CA}^{n-1} \end{bmatrix} \tag{3.2}$$

is equal to the number of states.

The disadvantage of this criterion is that the rank of the observability matrix depends on the numerical values of the system parameters, and that it does not give an indication for which eigenvalue of $\mathbf{A}$ the associated transient behaviour is not observable in the case $rank(\mathbf{O}) < n$. In other words, the criterion does not identify the unobservable modes. The Hautus test gives an answer to the latter question but still relies on numerical values of the system parameters. According to the Hautus test, an nth order LTI system $(\mathbf{A}, \mathbf{C})$ is completely observable if and only if

$$\text{rank}\left(\begin{bmatrix} s\mathbf{I} - \mathbf{A} \\ \mathbf{C} \end{bmatrix}\right) = n \quad \forall s \in \mathbb{C} \tag{3.3}$$

3.1.1 Structural Observability Matrix

In order to check whether an nth order LTI system $(\mathbf{A}, \mathbf{C})$ is completely observable independent of any parameter values, i.e. whether it is *structurally* completely observable, one may replace its matrices $\mathbf{A}, \mathbf{C}$ in the observability matrix $\mathbf{O}$ by the structurally equivalent interconnection matrices $\mathbf{A}^*$, $\mathbf{C}^*$ and determine the rank of the resulting structural observability matrix $\mathbf{O}^*$. The structural equivalent matrices are obtained by replacing each non-zero entry in $\mathbf{A}, \mathbf{C}$ by the value 1. Every zero entry in $\mathbf{A}, \mathbf{C}$ remains a zero entry in $\mathbf{A}^*$, $\mathbf{C}^*$. An nth order LTI system $(\mathbf{A}, \mathbf{C})$ is then said to be structurally observable if the structural observability matrix $\mathbf{O}^*$ has full rank, i.e.

$$\text{rank}(\mathbf{O}^*) = \text{rank}\left(\begin{bmatrix} \mathbf{C}^* \\ \mathbf{C}^*\mathbf{A}^* \\ \mathbf{C}^*(\mathbf{A}^*)^2 \\ \vdots \\ \mathbf{C}^*(\mathbf{A}^*)^{n-1} \end{bmatrix}\right) = n \tag{3.4}$$

Note that from $\text{rank}(\mathbf{O}^*) < n$ it cannot be concluded that $\mathbf{O}$ does not have full rank, i.e. that the system is not observable.

3.1.1.1 Bond Graph-based Construction of Interconnection Matrices

Given a causal bond graph of the system, there is no need to derive state-space equations in order to get the matrices of an LTI system and to replace their non-zero matrix entries by the value 1. The entries of the interconnection matrices may be obtained directly from a causal bond graph by following causal paths [1, 2]. If there is a causal path between two storage elements associated with the jth and the ith state variable that contains the least number of storage elements that does not pass through any other storage element in integral causality and if none of its bonds is activated, then $a^*_{ij} = a^*_{ji} = 1$. Likewise, $b_{ij} = 1$ if there is a causal path from the jth input to the ith state that does not pass through any other storage element, and $c_{ij} = 1$ if there is a causal path from the ith state to the jth output that does not pass through any other storage element. Such causal paths are termed direct paths. In classical control theory, structural dependencies between inputs, states, and outputs are often represented by a digraph obtained from the equations of a state space model.

3.1.1.2 Matrix-based Analysis of Structural Observability of Switched LTI Systems

A slight modification of this construction of interconnection matrices for continuous time LTI systems may also be applied to bond graphs of hybrid system models if the behaviour between two discrete events can be described by an LTI system, in other words, if the system behaviour can be described by a switched LTI system. If non-ideal switches are used and represented by a switching MTF in conjunction with an ON resistor in conductance causality and if a causal path from the storage element of state j to the storage element of state i that passes through no other storage element in integral causality but through a switch or a sequence of switches then entry a^*_{ij} is set to the modulus of the switched MTF associated with the one switch on that path or by the product of the moduli of the switched MFTs involved in the traversal of that causal path. Likewise, if a direct path from the storage element of state i to a detector of the jth output passes through a switch or a sequence of switches then c^*_{ij} is set to the modulus of the switched MTF associated with the one switch on that path or by the product of the moduli of the switched MFTs encountered in the traversal of that causal path.

3.1.1.3 Example: Electrical Network with Independent Switches

As an example, consider the network with two independent switches in Fig. 2.17. Figure 2.18 displays the bond graph of the network with two switches.

Let the inductor current $i_L = f_6$ be the first state variable x_1 and the current across the capacitor $u_C = e_8$ the second state variable x_2. As there is a direct causal path from the capacitor $\mathrm{C} : C$ to the inductor $\mathrm{I} : L$

$$e_8 \rightarrow e_7 \rightarrow e_6 \rightarrow \mathrm{I} : L \rightarrow f_6 \rightarrow f_7 \rightarrow f_8$$

that does not pass a switch, $a^*_{12} = a^*_{21} = 1$. Furthermore, there is a direct causal path

$$f_6 \rightarrow f_5 \rightarrow f_3 \rightarrow f_2 \rightarrow \mathrm{R} : R_1 \rightarrow e_2 \rightarrow e_3 \rightarrow e_5 \rightarrow e_6$$

that starts from $\mathrm{I} : L$, does not pass through another storage element and relates input and output of the I-element. Hence, $a^*_{11} = 1$. Finally, there is a direct causal path from the capacitor $\mathrm{C} : C$ to the switch $\mathrm{Sw} : m_2$

$$e_8 \rightarrow e_{12} \rightarrow e_{11} \rightarrow \mathrm{Sw} : m_2 \rightarrow f_{11} \rightarrow f_{12} \rightarrow f_8$$

Accordingly, $a^*_{22} = m_2$. As a result,

$$\mathbf{A}^* = \begin{bmatrix} 1 & 1 \\ 1 & m_2 \end{bmatrix} \tag{3.5}$$

Furthermore, there is a direct causal path from the C-element to the detector $\mathrm{De} : y$ but no direct causal path from the I-element to the sensed output variable $y = u_C$. Thus,

$$\mathbf{C}^* = \begin{bmatrix} 0 & 1 \end{bmatrix} \tag{3.6}$$

With these two interconnection matrices the structural observability matrix reads

$$\mathbf{O}^* = \begin{bmatrix} \mathbf{C}^* \\ \mathbf{C}^*\mathbf{A}^* \end{bmatrix} = \begin{bmatrix} 0 & 1 \\ 1 & m_2 \end{bmatrix} \tag{3.7}$$

Its rank equals two. That is, with this sensor, the circuit is observable in all system modes as would be expected.

3.1.2 Bond Graph-based Analysis of Structural Observability

Another test for structural complete observability of continuous time LTI systems that is solely based on an inspection of a causal bond graph has been given by Sueur and Dauphin-Tanguy [3–5]. Their purely graphical test has the advantage that it also answers the question in which locations which type of sensor is needed to make a model completely observable with a minimum number of sensors. The necessary and sufficient criterion requires that for a bond graph model with a given set of sensors the following two conditions are met.

1. If integral causality is applied to the energy storage elements as the preferred causality then every storage element in integral causality must have at least one causal path to either an effort sensor De or to a flow sensor (Df). This reachability requirement is a necessary condition.
2. If derivative causality is the preferred causality, then all storage elements in integral causality must be able to take derivative causality. If integral causality cannot be turned into derivative causality for some storage elements, then this must be possible after some sensors in appropriate locations have been dualised, i.e. a flow sensor (Df) has been replaced by an effort sensor (De) or vice versa. This requirement is a sufficient condition.

It both conditions are met, then the model is structurally completely observable with the given set of sensors and the rank of the observability matrix $\mathbf{O}$ equals the number of states n. If $\mathrm{rank}(\mathbf{A}) = n$ then one single sensor is sufficient to assure complete state observability. Its type and position is to be chosen so that both conditions are fulfilled.

If the second condition cannot be satisfied so that $k := n - q$ storage elements remain in integral causality in the bond graph with preferred derivative causality (BGD), then k additional sensors are needed to assure complete observability. Their type and their position is to be chosen so that both conditions are satisfied.[1]

3.1.2.1 Bond Graph-based Analysis of Structural Observability of Switched LTI Systems

The graphical criterion given by Sueur and Dauphin-Tanguy can also be used for systems modelled as a switched LTI system.

Its application to the example circuit in Fig. 2.17 confirms that the circuit with the one voltage sensor is completely observable for all system modes. In the bond graph with preferred integral causality (BGI) (Fig. 2.18), there is a direct causal path from the C-element to the detector and a causal path from the I-element through the C-element to the detector so that condition 1 is satisfied.

If derivative causality is assigned to the bond graph as preferred causality, (Fig. 3.1), then both storage elements take derivative causality. The resulting causal conflict at the right-hand 0-junction is resolved by turning the effort sensor into a flow sensor.

In bond graphs of hybrid system models, some storage elements in integral causality may be connected to a detector only via a causal path through a switch model Sw : m or even through several switch models. Clearly, the necessary reachability condition is only satisfied when the switch involved in the causal path is closed. That is, a model is not fully state-observable in those system modes in which a switch being part of the only causal path from a storage element to a detector is off, i.e. $m = 0$ for the switched MTF of that switch model.

---

[1] The number q of storage elements that take derivative causality when derivative causality is assigned to a bond graph as preferred causality is termed the *bond graph-rank* of the state space matrix $\mathbf{A}$ and is equal to $\mathrm{rank}(\mathbf{A})$ while k equals the number of null modes of $\mathbf{A}$ [3].

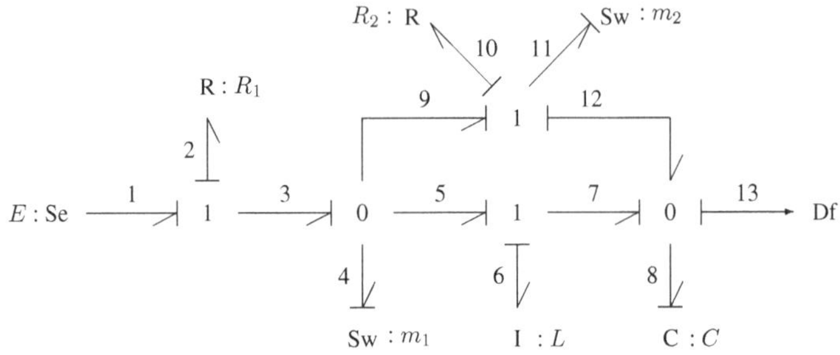

Fig. 3.1 Bond graph of the electrical circuit in Fig. 2.17 in preferred derivative causality

Moreover, if an inductor and a switch are in series, it is the switch that determines the current through the series connection forcing the inductor into derivative causality. If the switch is off, i.e. the switch in conductance causality delivers a current equal to zero, then the switch and the inductor can be considered non-existent. If the non-ideal switch is on, its conductance causality can be changed into resistive causality and the causality of the inductor can be changed into integral causality. Then, for this mode, it has to be checked whether there is a causal path from the inductance to a detector and if for this mode, derivative causality can be assigned to all storage elements as the preferred causality. If both conditions are fulfilled, the model is structurally completely observable for that mode. That is, in bond graphs of hybrid models it may become necessary not only to dualise detectors but also to invert the causality of switches. If the fixed conductance causality at switch ports forces some storage elements into derivative causality in the bond graph with preferred integral causality (BGI), then this indicates that the reachability condition is not satisfied for some switch state combinations. Hence, the model is not structurally completely observable for these switch state combinations. Some of them may indicate a system mode that is physically not feasible.

If there are n_d storage elements with derivative causality in a bond graph with preferred integral causality and their causality cannot be changed by inverting the causality of some switches then they do not contribute a state variable and do not need to be considered in a test for structural observability. The number of state variables equals the number of storage elements in integral causality.

3.1.2.2 Example: DC-DC Buck Converter

For illustration, consider the example of a buck converter displayed in Fig. 2.19. As can be seen, in this bond graph in preferred integral causality, the inductor I : L is in derivative causality. The bond graph captures the physical feasible system mode in which the two switches are off ($m_2 = 0 \wedge m_1 = 0$). However, derivative causality at the I-element indicates that the inductor current is not observable in this mode. If one of the two switches is closed, its causality can be changed into resistive causality. As

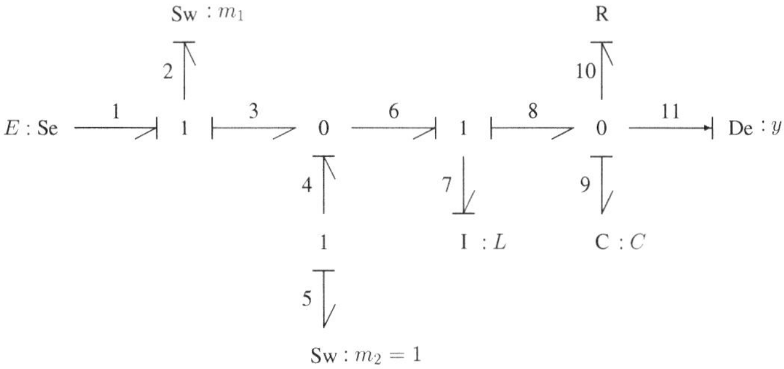

Fig. 3.2 Bond graph in preferred integral causality of the buck converter in the case Sw : m_2 is on

a result, integral causality can be assigned to the inductance. Figure 3.2 shows the bond graph for the case that Sw : m_2 is on, i.e. $m_2 = 1$. (In this example, the two switches operate in a complementary fashion, hence $m_1 = 0$.)

The switch state combination $m_2 = 1 \wedge m_1 = 0$ indicates a physically feasible system mode. In this system mode, the two storage elements are in integral causality. As there is a causal path from each of them to the detector, the reachability is satisfied in this system mode.

$$\text{C} \rightarrow \text{De} : e_8 \rightarrow e_{10} \rightarrow \text{De}$$
$$\text{I} \rightarrow \text{De} : f_6 \rightarrow f_8 \rightarrow f_9 \rightarrow \text{C} \rightarrow e_{10} \rightarrow e_{11} \rightarrow \text{De}$$

The second condition is also satisfied. The two storage elements can take derivative causality as the preferred causality as indicated in Fig. 3.3.

That is, in system mode ($m_2 = 1 \wedge m_1 = 0$), the system is structurally completely state observable with the one given voltage sensor as would be expected.

Likewise, the switch state combination $m_1 = 1 \wedge m_2 = 0$ also indicates a physically feasible system mode. In this case, the conductance causality at Sw : m_1

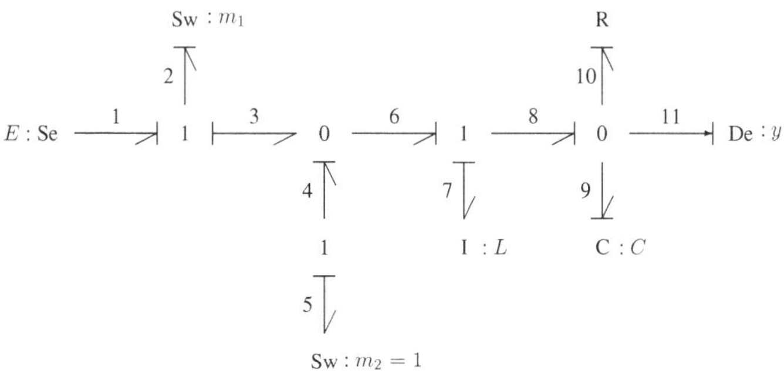

Fig. 3.3 Bond graph of the buck converter in preferred derivative causality in the case Sw : m_2 is on

can be inverted and derivative causality at the inductor I : L can be turned into integral causality in the BGI of Fig. 2.20. In this case, there is a causal path from the I-element in integral causality to the detector and in the BGD, the two storage elements take derivative causality even without dualisation of the detector. That is, the model is structurally completely state observable in this system mode. Clearly, there is a current through the inductor that can be observed if one of the two switches is closed. Hence, it is sufficient to assume one of the switches to be on and to invert its causality in order to remove the derivative causality at the I-element in the BGI of Fig. 2.20.

If one disregards that the switches in this example commutate in a complementary fashion, then $m_1 = 1 \wedge m_2 = 1$ would be another possible switch state combination that denotes a physically feasible system mode. There would be no short circuit because of the ON resistance of the switches. A bond graph with one of the two switches in resistive causality would also capture this case.

3.1.2.3 Example: Electrical Network with Two Independent Switches and One Storage Element

For illustration of the bond graph-based test for structural observability consider the switched circuit in Fig. 3.4. Figure 3.5 displays the bond graph of the switched network.

If there is no voltage detector De : u_C then one of the other two detectors is only reachable via a switch.

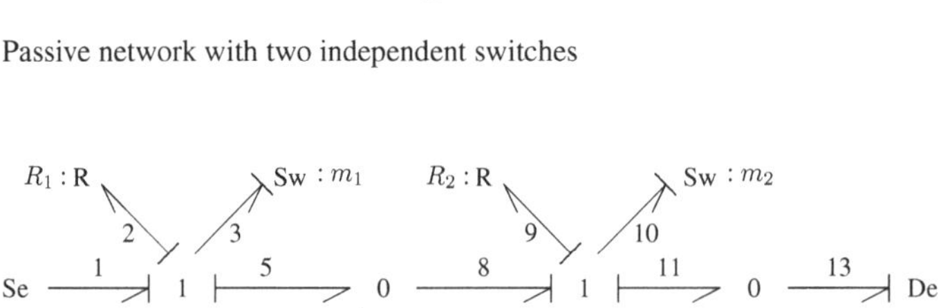

Fig. 3.4 Passive network with two independent switches

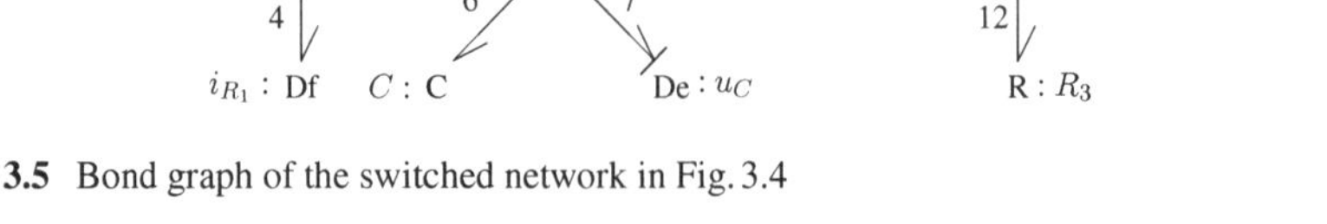

Fig. 3.5 Bond graph of the switched network in Fig. 3.4

Causal path from C : C *to* De : u_{R_3} :

$$e_6 \rightarrow e_8 \rightarrow e_{10} \rightarrow \boxed{\mathrm{Sw} : m_2} \rightarrow f_{10} \rightarrow f_{11} \rightarrow f_{12} \rightarrow \mathrm{R} : R_3 \rightarrow e_{12} \rightarrow e_{13} \rightarrow \mathrm{De} : u_{R_3}$$

Causal path from C : C *to* Df : i_{R_1} :

$$e_6 \rightarrow e_5 \rightarrow e_3 \rightarrow \boxed{\mathrm{Sw} : m_1} \rightarrow f_3 \rightarrow f_4 \rightarrow \mathrm{Df} : i_{R_1}$$

The capacitor C : C can take preferred derivative causality if one of the switches is on and if causality at the switch is inverted. For these modes, the model is structurally observable with the sensor Df : i_{R_1} or the sensor De : u_{R_3} respectively. If sensor De : u_C is available, the model is observable independently of the switch states.

For large hybrid system models with many switches and sensors (controlled sources), the bond graph-based test for structurally complete observability (structurally complete controllability) may become laboriously. The sufficient condition of the test for observability (controllability) requires that for all those system modes in which the reachability condition is satisfied, all storage elements in integral causality in the BGI must take derivative causality if preferred derivative causality is assigned. It may be difficult to decide which sensors (controlled sources) must be dualised to meet this requirement. For continuous time models, Samantaray and Bouamama propose an approach that allows for software support. All storage elements are dualised, i.e. a C-element becomes an I-element and vica versa. Moreover, sensors (controlled sources) are replaced by linear R-elements. Bond graph software then assigns preferred integral causality to the storage elements of the modified bond graph. If integral causality cannot be assigned to some storage elements, then their duals in the original bond graph are those storage element that cannot be assigned derivative causality [7]. This software support can also be used for hybrid system models in which switching devices are represented by a switched MTF and an ON resistance in conductance causality.

For illustration, Fig. 3.6 shows the revised bond graph of Fig. 3.3 with dualised storage elements and the voltage sensor De : y replaced by a linear resistor. Both storage elements in the revised bond graph in Fig. 3.6 are in integral causality. That is, corresponding storage elements in the original bond graph can be assigned derivative causality if derivative causality is the preferred causality. There are no storage elements remaining in integral causality. Hence, the second sufficient condition of the structural observability test is satisfied.

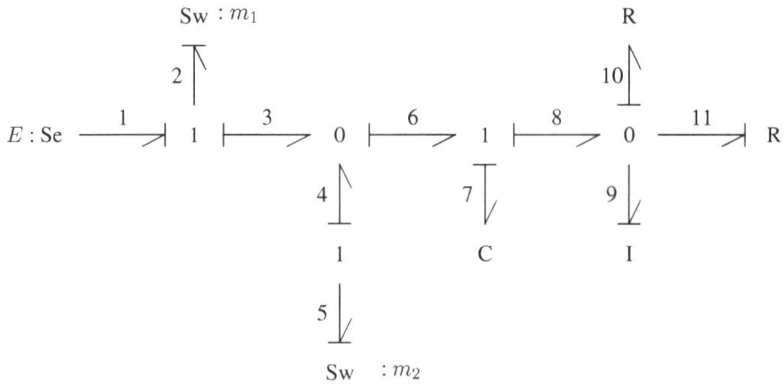

Fig. 3.6 Revised bond graph of the buck converter with dualised storage elements and the sensor replaced by a resistor

3.2 Structural Controllability

As to the controllability of a switched LTI system, tests similar to the ones for observability considered in the previous section, are known. According to Kalman's criterion, a continuous time LTI system is numerically completely state controllable if and only if the rank of the controllability matrix **S** equals the number n of states.

$$\text{rank}(\mathbf{S}) := \text{rank}\left[\mathbf{B}|\mathbf{AB}|\ldots|\mathbf{A}^{n-1}\mathbf{B}\right] = n \tag{3.8}$$

This test has the same disadvantage as the observability criterion. The rank of the controllability matrix depends on the actual numerical values of the system parameters. If the test fails, it does not give any indication for which eigenvalue of **A** the associated transient behaviour is not controllable in case $\text{rank}(\mathbf{S}) < n$. That is, the criterion does not identify the uncontrollable modes. The latter ones can be identified by means of the Hautus test which is also based on numerical values of the system parameters and not on structural properties. According to the Hautus test, an nth order LTI system $(\mathbf{A}, \mathbf{B})$ is completely controllable if and only if for all eigenvalues λ_i $(i = 1, 2, \ldots, n)$ of the state space matrix **A**

$$\text{rank}([\lambda_i \mathbf{I} - \mathbf{A})|\mathbf{B}]) = n \tag{3.9}$$

A structurally equivalent controllability matrix $\mathbf{S}^*$ is obtained by replacing the matrices in (3.8) by their structurally equivalent matrices.

$$\mathbf{S}^* := \left[\mathbf{B}^* \mid \mathbf{A}^*\mathbf{B}^* \mid \ldots \mid (\mathbf{A}^*)^{n-1}\mathbf{B}^*\right] \tag{3.10}$$

An nth order LTI system $(\mathbf{A}, \mathbf{B})$ is then said to be structurally complete state controllable if

$$\text{rank}(\mathbf{S}^*) = n \tag{3.11}$$

Note that if $\text{rank}(\mathbf{S}^*) < n$, it cannot be concluded that the system with the given sources is not controllable.

3.2.1 *Matrix-based Analysis of Structural Controllability of Switched LTI Systems*

If there are switches encountered on the traversal of a direct causal path from the jth controlled source to the storage element of state i, the entry b^*_{ij} of the interconnection matrix $\mathbf{B}^*$ is set to the product of the moduli of the switched MTFs associated with the switches. If a storage element of index i has got derivative causality in the bond graph with preferred causality and if the conductance causality of a switch $\text{Sw} : m$ has to be changed into resistive causality in order to enable a direct causal path from a controlled source of index j through that switch to the storage element of index i with its causality changed into integral causality, then b^*_{ij} is set to the modulus of the switched MTF associated with that switch.

For illustration, the example of the buck converter is considered once again. As can be seen from the bond in preferred integral causality in Fig. 2.20, there is a causal path from the effort source to the C-element. However, the I-element is in derivative causality. It can be brought into integral causality by inverting the causality at one of the two switches. If the causality at switch $\text{Sw} : m_2$ is inverted (cf. Fig. 3.2) then there is a direct causal path from the effort source to the I-element that passes through both switches.

Causal path from $\text{Se} : E$ *to* $\text{I} : L$:

$e_1 \rightarrow e_2 \rightarrow \boxed{\text{Sw} : m_1} \rightarrow f_2 \rightarrow f_3 \rightarrow f_4 \rightarrow f_5 \rightarrow \boxed{\text{Sw} : m_2} \rightarrow e_5 \rightarrow e_4 \rightarrow e_6 \rightarrow e_7 \rightarrow \text{I} : L$

The C-element, however, is not reachable by a direct causal path starting from the effort source. Hence,

$$\mathbf{B}^*_1 = \begin{bmatrix} 0 \\ m_1 m_2 \end{bmatrix} \tag{3.12}$$

and the structural controllability matrix reads

$$\mathbf{S}^*_1 = [\mathbf{B}^*_1 \,|\, \mathbf{A}^*\mathbf{B}^*_1] = m_1 m_2 \begin{bmatrix} 0 & 1 \\ 1 & 1 \end{bmatrix} \tag{3.13}$$

Its has full rank only if $m_1 = 1 \wedge m_2 = 1$. However, in this example circuit, the switches commutate conversely. That is, $\text{rank}(\mathbf{S}_1^*) = 0$ for system mode $m_1 = 0 \wedge m_2 = 1$. From a look at the circuit schematic it is clear that the system is not controllable by the effort source when switch Sw : m_1 is open. However, this cannot be concluded by just evaluating the rank of the structural controllability matrix $\mathbf{S}_1^*$.

If switch Sw : m_1 is closed, its causality can be inverted. Consequently, the I-element gets integral causality. As a result, there is a direct causal path from the effort source to the I-element that does not pass through any of the two switches. Also, as in the previous case, there is no direct causal path from the effort source to the C-element. Hence,

$$\mathbf{B}_2^* = \begin{bmatrix} 0 \\ m_1 \end{bmatrix} \tag{3.14}$$

In this case, the structural controllability matrix reads

$$\mathbf{S}_2^* = [\mathbf{B}_2^* \mid \mathbf{A}^* \mathbf{B}_2^*] = m_1 \begin{bmatrix} 0 & 1 \\ 1 & m_2 \end{bmatrix} \tag{3.15}$$

This structural controllability matrix has full rank for switch state combination $m_1 = 1 \wedge m_2 = 0$. That is, the system is structurally completely controllable by the one effort source in system mode $m_1 = 1 \wedge m_2 = 0$ as would be expected.

3.2.2 *Bond Graph-based Analysis of Structural Controllability of Switched LTI Systems*

In [3–5], Sueur and Dauphin-Tanguy have also given a purely graphical test for the structural complete controllability of continuous time LTI systems that can also be used for systems modelled as a switched LTI system. For a continuous time LTI system to be structurally completely controllable with a given set of controlled sources a necessary and a sufficient condition are to be satisfied that are similar to the conditions for structural complete observability.

1. The necessary reachability condition requires that for every storage element in integral causality, there must be at least one causal path from a controlled source.
2. All storage elements in integral causality in the bond graph with preferred integral causality must take derivative causality if derivative causality is the preferred causality. If some storage elements cannot be assigned derivative causality then this must be possible after dualisation of some or all controlled sources. Uncontrolled sources must not be dualised.

If both conditions are met, the rank of the controllability matrix $\mathbf{S}$ equals the number n of states. If $\text{rank}(\mathbf{A}) = n$, then one single actuator is sufficient to assure complete state controllability. Its type and position is to be chosen so that both conditions are fulfilled.

If the second condition cannot be satisfied so that $k := n - q$ storage elements remain in integral causality in the bond graph with preferred derivative causality (BGD) then k additional actuators are needed to assure complete state controllability. Their type and their position is to be chosen so that both conditions are satisfied.

Inspection of the bond graph in Fig. 2.18 shows that there is a direct causal path from the effort source to the I-element and a causal path from the effort source through the I-element to the capacitor. That is, both storage elements are reachable from the effort source independently of the switch states. The sufficient condition is also satisfied. Both storage elements take derivative causality in the bond graph with preferred derivative causality in Fig. 3.1. Hence, the model of the circuit in Fig. 2.17 with two independent switches is structurally completely state controllable with the one effort source for all four system modes.

Application of the purely graphical test to the buck converter example confirms the results obtained by means of the structurally controllability matrix test in the previous section. If switch $\mathrm{Sw} : m_1$ is assumed to be on, its causality can be inverted and the I-element gets integral causality. Then there is a direct causal path from the effort source to the I-element and a causal path from the effort source through the I-element to the C-element. That is, the reachability condition is satisfied. In addition, both storage elements can be assigned derivative causality in the bond graph with preferred derivative causality. Thus, also the sufficient condition is met. Hence, in system mode $m_1 = 1 \wedge m_2 = 0$, the model is structurally completely controllable with the one effort source.

The I-element may also be brought into integral causality by inverting the causality at switch $\mathrm{Sw} : m_2$. In that case, there is a direct causal path from the effort source to the I-element that passes through both switches and a causal path from the effort source to the C-element that passes through both switches and the I-element. Clearly, for the storage elements to be reachable from the effort source, both switches need to be closed. If preferred derivative causality is assigned, both storage elements get derivative causality. That is, the model would be structurally complete controllable by the one effort source in the case that both switches would be closed. However, as both switches commutate in a complementary fashion, system mode $m_1 = 1 \wedge m_2 = 1$ does not happen in the healthy system.

As a result, if there are storage elements that take derivative causality in a bond graph of a switched LTI system when preferred integral causality is assigned, the necessary reachability condition is not satisfied for all system modes. If some of the these storage elements can be brought into integral causality by assuming that some switches are closed and by inverting their conductance causality then this means considering specific system modes and to test for structural controllability in these system modes. The selection of specific system modes is guided by the aim to switch causality at storage elements in derivative causality and is not unique. For a selected specific system mode, the reachability condition may be met for all storage elements in integral causality, but the sufficient condition may fail. Or both conditions may be satisfied, but there is no admissible system mode corresponding to a combination of switch states.

The bond graph based test for structural observability (structural controllability) can also be applied to models in which instantaneous structural changes are accounted for by residual sinks that are switched on and off at discrete events. As an example consider the bond graph of clutch in Fig. 2.15. A check of the causal bond graph confirms the expectation that the clutch model is structurally observable with a sensor for the angular velocities on the drive side and on the load side. Furthermore, the model is structurally controllable in the case when the clutch is engaged as to be expected. In that case, there is a causal path from the motor moment M_1 to both inertia elements through the residual sink MSe : M modulated by the difference of the angular velocities. The two inertia elements can take derivative causality by dualising the flow sensors or the modulated sources MSe : M_1 and MSe : M_2 respectively.

3.3 Summary

A condition for model-based FDI is that the model is structurally observable. A design of online fault accommodation requires that the model used is structurally controllable. This chapter has recalled matrix based as well as pure graphical tests for LTI systems. They can be applied to systems of which the dynamic behaviour between two consecutive discrete events can be described by an LTI system. As the example of the network with two switches in Fig. 2.17 shows, structurally complete observability (controllability) may be given for all system modes. For other systems, assignment of preferred integral causality and conductance causality to the ON resistors of the switches may force some storage elements into derivative causality. If derivative causality at some storage elements is caused by the conductance causality at the non-ideal switch models, structurally complete observability (structurally complete controllability) is not possible for all system modes. Assuming that an appropriately chosen switch is on, its causality can be inverted and derivative causality at storage element can be removed. This means that a subsequent check for structural observability (structural controllability) only applies for all those system modes in which the selected switch is on.

References

1. Zhang, X. (2009). *Modeling, control, fault detection and isolation of chemical processes using a bond graph framework* [PhD thesis]. Texas Tech University.
2. Merzouki, R., Samantaray, A., Pathak, P., & Ould Bouamama, B. (2013). *Intelligent mechatronic systems*. Springer.
3. Dauphin-Tanguy, G., Rahmani, A., & Sueur, C. (1999). Bond graph aided design of controlled systems. *Simulation Practice Theory*, *7*(5–6), 493–513.

4. Sueur, C., Dauphin-Tanguy, G. (1991). Bond graph approach for structural analysis of MIMO linear systems. *Journal Franklin Institute*, *328*(1), 55–70.
5. Sueur, C. (1989). Structural controllability/observability of linear systems represented by bond graphs. *Journal Franklin Institute*, *326*(6), 869–883.
6. Wang, D., Yu, M., Low, C., & Arogeti, S. (2013). *Model-based health monitoring of hybrid systems*. New York: Springer.
7. Samantaray, A. K., Ould Bouamama B. (2008). *Model-based process supervision - a bond graph approach. advances in industrial control*. London: Springer.

Chapter 4
Bond Graph Model-based Quantitative FDI in Hybrid Systems

The representation of a hybrid system model by means of a bond graph with system mode independent causalities has the advantage that a unique set of equations can be derived from the bond graph that holds for all system modes. Discrete switch state variables in these equations account for the system modes. In this chapter, this bond graph representation is used to derive analytical redundancy relations (ARRs) from the bond graph. The result of their numerical evaluation called residuals can serve as fault indicator. Analysis of the structure of ARRs reveals which system components, sensors, actuators or controllers contribute to a residual if faults in these devices happen. This information is usually expressed in a so-called *structural fault signature matrix* (FSM). As ARRs derived from the bond graph of a hybrid system model contain discrete switch state variables, the entries in a FSM are mode dependent. Moreover, the FSM is used to decide if a fault has occurred and whether it can unequivocally be attributed to a component. Finally, the chapter discusses the numerical computation of ARRs.

4.1 ARR Residuals as Fault Indicators

An essential step in FDI is the evaluation of the time history of residuals serving as fault indicators. To that end, one approach is to establish ARRs from a model of a system (cf. Fig. 1.3). These relations are algebraic or dynamic constraints between known continuous variables, i.e. system inputs $\boldsymbol{u}$ and measured output variables $\boldsymbol{y}$ that include known model parameters $\boldsymbol{\Theta}$. ARRs derived from a bond graph of a hybrid system model also depend on discrete switch state variables. For a healthy system, the time evolution of ARRs should ideally be identical to zero in all system modes. In practice, residuals will be within certain small error bounds due to measurement noise, parameter uncertainties, and numerical inaccuracies. If, however, faults in some system components occur, then the values of some residuals will be outside given thresholds and can serve as fault indicators. The structure of ARRs expressed in the FSM will then indicate whether faults can be isolated. Let $\boldsymbol{m}(t)$ denote the vector

W. Borutzky, *Bond Graph Model-based Fault Diagnosis of Hybrid Systems*,
DOI 10.1007/978-3-319-11860-4_4

of discrete switch state values at time instant t, let the ARRs be expressed by real-valued functions f_k and let ε_k^j be the error bound for the absolut values of residuals r_k in system mode j where a system mode is given by a feasible combination of switch states. Then

$$|f_k(\boldsymbol{u}(t), \boldsymbol{y}(t), \boldsymbol{\Theta}, \boldsymbol{m}(t))| < \varepsilon_k^j \quad \forall\, k \quad \forall\, t > 0 \tag{4.1}$$

indicates a healthy system. In the case of a single fault or multiple simultaneous faults or a temporary disturbance at time instant t, however, the absolute values of some residuals exceed given thresholds thr_i^j.

$$\exists\, i \;\; \exists\, D_i^j \subset \mathbb{R}^+ \text{ such that} \, |f_i(\boldsymbol{u}(t), \boldsymbol{y}(t), \boldsymbol{\Theta}, \boldsymbol{m}(t))| = |r_i(t)| \geq thr_i^j \quad \forall\, t \in D_i^j \tag{4.2}$$

Clearly, chosen thresholds must not be too big in order to avoid that true faults are not detected, nor must they be too small in order to avoid the report of faults that did not happen.

In order to get a clear indication for a fault by numerical evaluation of ARRs, these constraints should be insensitive to disturbances while sensitive to true faults. Moreover, analysis of their structure should allow to locate the cause of a fault. For a specific fault, only a subset of ARR residuals should be outside given thresholds. Ideally each residual is sensitive to only one fault while all other residuals are insensitive to that same fault. In that case, ARRs are said to be *structured*. However, if the number of faults taken into account exceeds the number of sensors or if the latter are not appropriately located then a set of structured ARRs cannot be achieved. If a residual is affected by a set of faults that does not affect any other residual then this residual is called *structurally independent*. Such ARRs cannot be obtained by algebraic manipulation of other ARRs. The number of structurally independent ARRs that can be derived from a model equals the number of sensors used in the model [1].

Before the next section addresses the question how ARRs can be derived from a causal bond graph it has to be noted that a bond graph is the result of a modelling process based on physical first principles. Modelling means that assumptions are made, that effects are conceptually captured in an idealised manner while others are neglected at all. Because of such modelling assumptions, because of simplifications such as a linearisation of model equations or a reduction of the model order and because of parameter uncertainties, the numerical evaluation of some ARRs may give non-zero residuals even if no fault has happened and in the first place it may not be clear which assumptions or simplifications in the course of the modelling process may have given rise to a non-zero residual in the case of no fault. The effect of parameter uncertainties on ARR residuals is addressed in Chap. 5.

4.2 Generation of ARRs from a Bond Graph of a Hybrid Model

Given a causal bond graph of a hybrid model, the question is how ARRs containing only known system inputs, outputs, system parameters, and information about the system mode can be derived. This section considers three methods that have been reported in the literature. One approach is the so-called *causality inversion method* [2–4]. It has been introduced for causal bond graphs of continuous time models but can also be applied to bond graphs of hybrid models [5, Chap. 7].

4.2.1 Causality Inversion Method

For the purpose of online FDI, first, two steps are to be carried out.

1. Causalities at bond graph sensor element ports (detector ports) are inverted. In online FDI, outputs of a real process or a real system are measured by real sensors. For a model of the system used for the determination of ARRs, these measurements are known inputs.
2. For the determination of the dynamic behaviour of a system, a bond graph model in preferred integral causality is used. If ARRs are to be derived from such a model, initial values would be needed for their numerical solution, which in turn would require a state estimation. Therefore, for online FDI, preferred derivative causality is assigned to the bond graph. However, as it is well known, differentiation of measurements amplifies the noise carried by a signal so that appropriate signal processing is needed. Moreover, differentiation with respect to time of inputs with discontinuities results in pulses at the time instances of the discontinuities.

Remark 4.1 Such a bond graph in preferred derivative causalities and inverted causalities at sensor elements has been termed *diagnostic* bond graph (DBG) [1, 6], while a bond graph in preferred integral causality used for the determination of the dynamic system behaviour is called a *behavioural* bond graph. Bond graph sensor elements for which causality cannot be inverted turn out to be redundant. Their measured value can be obtained by means of other sensors. Moreover, it may be possible that not all storage elements take preferred derivative causality. As a result, some ARRs derived from a diagnostic bond graph will depend on the initial values of some state variables so that it will be necessary to differentiate these ARRs. □

Derivation of ARRs from a diagnostic bond graph then starts by summing power variables at all those junctions that have a BG sensor element in inverted causality attached to it. At first, these balances of power variables will contain unknown variables. They may be eliminated by following causal paths and by using constitutive equations of bond graph elements. The result may be a set of ARRs in closed symbolic form [cf. (4.2)] if nonlinear constitutive element equations permit necessary eliminations. For closed loop feedback systems, further ARRs are obtained by adding the reformulated constitutive equations of modulated sources representing actuators and of controllers.

Remark 4.2 This approach to the derivation of ARRs in closed form has been implemented in the software ModelBuilder [7]. Also, a module of the software Symbols™ [8] can automatically generate ARRs. If ARRs in closed form are not possible, i.e. unknowns cannot be eliminated then ARR residuals are to be computed numerically simultaneously with the model equations. □

For a bond graph of a hybrid system model with system mode independent causalities, model equations derived from the bond graph contain discrete switch state variables and thus hold for all system modes. The same is true for ARRs.

4.2.2 Example: Network with a Semiconductor Switch

The derivation of ARRs from a bond graph of a hybrid system model that holds for all system modes is illustrated by means of a simple network with one switch and elements with a linear constitutive equation displayed in Fig. 4.1.

Figure 4.2 shows a diagnostic bond graph (DBG) of the switched network in Fig. 4.1 with the semiconductor switch explicitly modelled by an element Sw : b, i.e. an MTF-R pair.

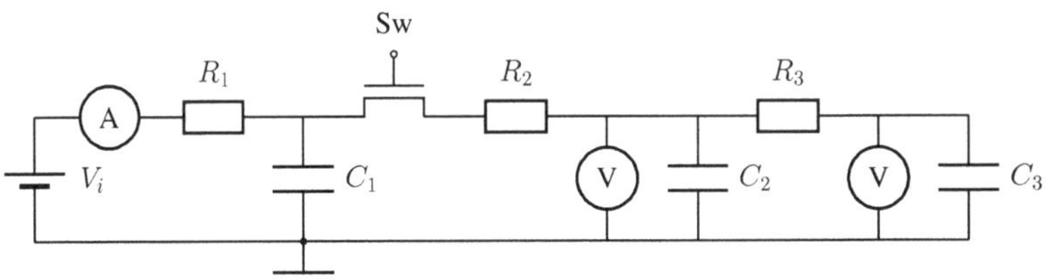

Fig. 4.1 Electrical network with a semiconductor switch

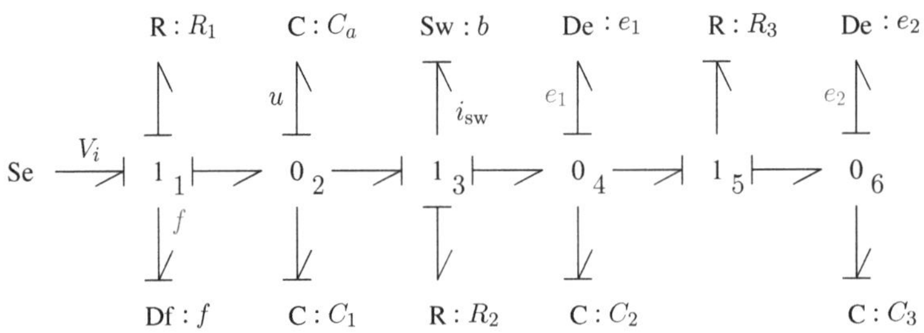

Fig. 4.2 Invariant causality diagnostic bond graph of the switched network in Fig. 4.1

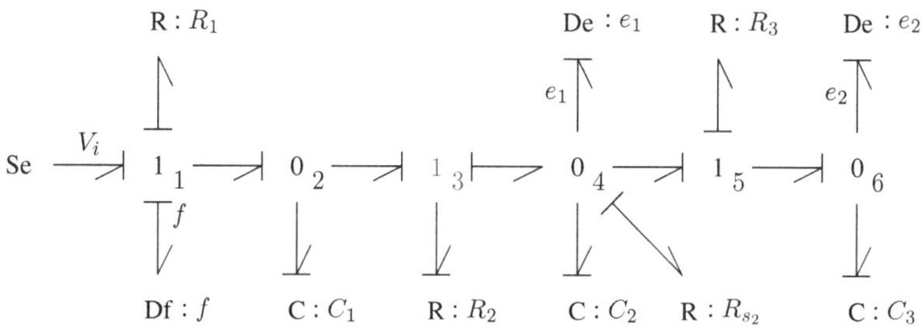

Fig. 4.3 Diagnostic bond graph of the switched network in Fig. 4.1 according to [9]

Notice that the fixed conductance causality of the switch model and the inverted causality at the flow sensor Df : f cause a causal conflict at junction 0_2 that has been resolved by attaching an auxiliary C-element C : C_a in integral causality. However, that does not mean that the ARRs derived from this bond graph depend on the initial condition of this C-element. In the process of ARR formulation, its capacitance parameter C_a is set to zero. Instead of an auxiliary C-element a resistor R : R_a with $R_a \rightarrow \infty$ could also be used. Alternatively, the C element C : C_1 could be assigned integral causality, or the causality at the flow sensor Df : f is not inverted. Clearly, the causal conflict vanishes if an additional effort sensor is used that replaces the auxiliary C-element C : C_a.

Furthermore, there is a causal path between resistor R : R_2 and the ON resistor of the switch R_{on}, which means that there is an algebraic loop for the current i_{sw}. As both resistors are linear, this algebraic loop can be solved symbolically.

In comparison to the BG in Fig. 4.2, the BG in Fig. 4.3 reproduced from [9] does not reflect the series connection of a resistor and a semiconductor switch. The latter one is not explicitly visible in the BG. Additional information must be known that junction 1_3 is a controlled junction. Moreover, detectors are not in inverted causality although sensors deliver a measured and thus known variable into a DBG. Also note that an auxiliary resistor R : R_{s_2} is needed to determine the effort at junction 0_4.

The sum of power variables at the junctions in the bond graph of Fig. 4.2 read

$$1_1: \quad 0 = V_i - R_1 f - u \tag{4.3a}$$

$$0_2: \quad 0 = f - C_1 \dot{u} - i_{sw} \tag{4.3b}$$

$$1_3: \quad i_{sw} = \frac{b}{R_{on}}(u - R_2 i_{sw} - e_1) \tag{4.3c}$$

$$0_4: \quad 0 = i_{sw} - C_2 \dot{e}_1 - \frac{1}{R_3}(e_1 - e_2) \tag{4.3d}$$

$$0_6: \quad 0 = \frac{1}{R_3}(e_1 - e_2) - C_3 \dot{e}_2 \tag{4.3e}$$

By eliminating unknowns the following mode dependent ARRs are obtained.

$$1_1: \quad 0 = \dot{V}_i - R_1\dot{f} - \frac{1}{C_1}\left[f - \frac{b}{R_{\text{on}} + bR_2}(V_i - R_1 f - e_1)\right] \tag{4.4a}$$

$$0_4: \quad 0 = \frac{b}{R_{\text{on}} + bR_2}(V_i - R_1 f - e_1) - C_2\dot{e}_1 - \frac{1}{R_3}(e_1 - e_2) \tag{4.4b}$$

$$0_6: \quad 0 = \frac{1}{R_3}(e_1 - e_2) - C_3\dot{e}_2 \tag{4.4c}$$

As $b(t) \in \{0, 1\} \; \forall t \geq 0$, the expression for the switch current i_{sw} can be simplified to

$$i_{\text{sw}} = \frac{b}{R_{\text{on}} + R_2}(u - e_1) \tag{4.5}$$

and the ARRs can be reformulated.

$$0_2: \quad 0 = f - C_1(\dot{V}_i - R_1\dot{f}) - \frac{b}{R_{\text{on}} + R_2}(V_i - R_1 f - e_1) \tag{4.6}$$

$$0_4: \quad 0 = \frac{b}{R_{\text{on}} + R_2}(V_i - R_1 f - e_1) - C_2\dot{e}_1 - \frac{1}{R_3}(e_1 - e_2) \tag{4.7}$$

$$0_6: \quad 0 = e_1 - e_2 - R_3C_3\dot{e}_2 \tag{4.8}$$

In the reformulated ARRs, the ON resistance of the switch can be set to zero turning the non-ideal switch into an ideal switch. The resulting ARRs are then the same Wang et al. obtained in [9] by a completely different approach that uses controlled junctions, and a version of the SCAP adapted for the purpose of FDI. Wang et al. call ARRs that hold for all system modes global analytical redundancy relations (GARRs). A disadvantage of controlled junctions with regard to FDI is that they provide a balance of power variables only if they are switched on. That is, if controlled junctions are used to derive ARRs then the number of ARRs is mode dependent and may be less than the number of sensors for some system modes.

The causality inversion method obtains N ARRs for N sensors by summing power variables at those junctions to which a sensor element has been attached and by eliminating unknowns in the resulting balances. In the example, summing power variables at junctions $1_1, 0_4, 0_6$ results in three ARRs for the three sensors. An alternative set of N ARRs may be obtained by starting with the sum of power variables at other junctions and by eliminating unknowns. That is, a set of ARRs for a given number of sensors is not unique. However, for a switched LTI system, one set of ARRs can be transformed into another. Actually, (4.6) is obtained by summing flows at junction 0_2. ARR (4.8) is the result of summing up efforts at junction 1_5.

4.2.3 Covering Path Method

ARRs could also be derived from a bond graph in preferred integral causality and without causality inversion at sensor ports. The covering path method [3, 10, 11] starts by summing the power variables at an arbitrary junction, replaces the unknowns by considering causal paths in the bond graph and then proceeds to the next junction. A resulting ARR is checked against the ones obtained in previous steps whether it contains the same known variables and parameters as one of the ARRs already derived. In that case, the obtained ARR is said to have the same *fault signature* as another ARR. It is not structurally independent and is discarded. Given N sensors, the procedure stops when N ARRs have been found even when not all junctions have been considered. However, as the authors of reference [3] admit, the procedure is computationally costly. Moreover, the replacement of unknowns involves the causal inversion of constitutive equations of storage elements and the inversion of sensor causalities not expressed in the bond graph. The causality inversion method reduces the computational effort by derivation of ARRs from junctions with a sensor attached to it.

4.2.4 Extended Covering Path Method

The previously outlined covering path method and the causality inversion method start the generation of ARRs by summing efforts or flows respectively at junctions and by replacing unknowns following causal paths and by using constitutive element equations. In contrast, it is also possible to start with the constitutive relations of an element and to substitute unknown variables by known ones. In [9], Wang et al. propose a procedure that starts from a causal bond graph in preferred derivative causality with *controlled* junctions and N sensors with *non-inverted* causality, chooses one of the sensors and looks for an element whose output variable is measurable by that sensor. Its constitutive relation is then turned into an ARR by eliminating all unknowns following causal paths. These steps are repeated with the next sensor until all N sensors have been considered. Wang et al. call this ARR generation method *extended covering path method.*

For comparison with the causality inversion method, this procedure is used to generate again ARRs for the switched network in Fig. 4.1. ARRs are derived from the bond graph in Fig. 4.4.

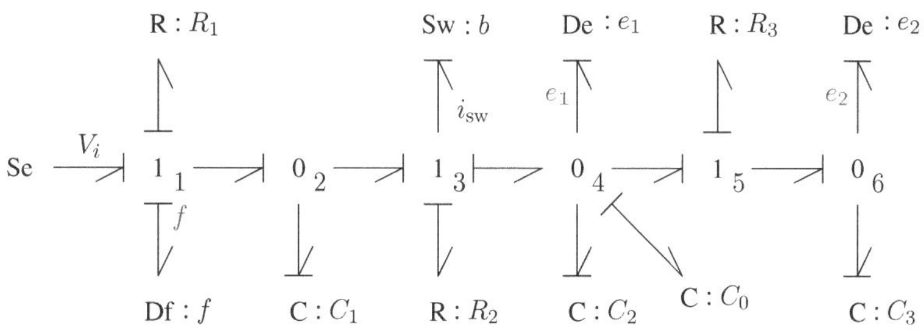

Fig. 4.4 Bond graph of the switched network in Fig. 4.1 with non-inverted causality at the sensors

Note that the C elements in derivative causality, the sensors with non-inverted causality and the switch in conductance causality lead to a causal conflict at junction 0_4 that has been resolved by adding an auxiliary capacitor C : C_0. Its capacitance is set to zero in the formulation of ARRs.

First, the flow sensor Df : f is chosen. Junction 0_2 is the element whose output variable is measured by the flow sensor. Its constitutive relation reads

$$f = C_1(\dot{V}_i - R_1\dot{f}) + i_{sw} \tag{4.9}$$

The constitutive equation of the switch Sw is

$$i_{sw} = \frac{b}{R_{on}}[(V_i - R_1 f) - R_2 i_{sw} - e_1] \tag{4.10}$$

Solving (4.10) for i_{sw} and substituting the result into (4.9) yields the ARR

$$0 = f - C_1(\dot{V}_i - R_1\dot{f}) - \frac{b}{R_{on} + bR_2}(V_i - R_1 f - e_1) \tag{4.11}$$

equivalent to ARR (4.4a)

Next, the effort sensor De : e_1 is considered. The auxiliary capacitor is the element whose output variable is measured by this effort sensor. Its constitutive relation reads

$$e_1 = \frac{1}{C_0}\int_0^t \left[i_{sw} - C_2\dot{e}_1 - \frac{1}{R_3}(e_1 - e_2) \right] d\tau + e_1(0) \tag{4.12}$$

Reformulation yields the ARR

$$\underbrace{C_0}_{=0}\ \dot{e}_1 = i_{sw} - C_2\dot{e}_1 - C_3\dot{e}_2 \tag{4.13}$$

$$0 = \frac{b}{R_{on} + bR_2}(V_i - R_1 f - e_1) - C_2\dot{e}_1 - \frac{1}{R_3}(e_1 - e_2) \tag{4.14}$$

The output variable e_2 of the last remaining sensor De : e_2 equals the output of junction 1_5. Its output is

$$e_2 = e_1 - R_3C_3\dot{e}_2 \tag{4.15}$$

and gives the ARR

$$0 = \frac{1}{R_3}(e_1 - e_2) - C_3\dot{e}_2 \tag{4.16}$$

As a result, for this example, the extended covering path method gives the same ARRs as the inverted causality method.

Remark 4.3 If no auxiliary C element C : C_0 is attached to junction 0_4, then C element C : C_2 must take integral causality. In that case, an ARR can also be derived for sensor De : e_1. Its output e_1 equals the output of C : C_2. The constitutive equation of C : C_2 is

$$\dot{e}_1 = \frac{1}{C_2}\left[i_{\mathrm{sw}} - \frac{1}{R_3}(e_1 - e_2)\right] \tag{4.17}$$

which, in fact, is equivalent to the sum of flows at junction 0_4. Substitution of i_{sw} gives ARR (4.14).

4.3 All-Mode Structural Fault Signature Matrix

ARRs derived from a bond graph of a hybrid system model contain discrete switch state variables. Depending on their values, i.e. depending on the system mode, some parts of the ARRs are disabled. An inspection of ARRs derived from a bond graph of a hybrid system model thus shows which component parameters affect which residuals in which system mode. This information may be expressed in an *all-mode* structural fault signature matrix (FSM) $\mathbf{S} = (S_{ij})$, $i = 1, \ldots, p$, $j = 1, \ldots, N$, where p is the number of component parameters Θ_i and N is the number of ARR residuals r_j equal to the number of sensors. If parameter Θ_i affects residual r_j in all system modes, then $S_{ij} = 1$. If Θ_i does not contribute at all to r_j, then $S_{ij} = 0$. If its contribution depends on a function $f_{ij}()$ of the discrete switch states and if the value of $f_{ij}()$ is m_{ij} then $S_{ij} = m_{ij}$. This all-mode FSM is an extension of a FSM Tagina et al. used back in 1995 to express the structural dependencies of ARRs from component parameters for bond graphs of continuous time models [11].

Let $\boldsymbol{u}(t)$ denote the known inputs into a system, $\boldsymbol{y}(t)$ known measured outputs, $\boldsymbol{m}(t)$ the vector of all discrete switch states at time t and $\boldsymbol{\Theta}$ the vector of component parameters. Each ARR$_j$, $j = 1, \ldots, N$, may then be expressed by means of function $g_j()$ that maps the vector $\boldsymbol{p} := (\boldsymbol{u}^T(t) \quad \boldsymbol{y}^T(t) \quad \boldsymbol{\Theta}^T \quad \boldsymbol{m}^T(t))^T$ onto a residual $r_j(t) \in \mathbb{R}$.

$$r_j(t) = g_j(\boldsymbol{u}(t), \boldsymbol{y}(t), \boldsymbol{\Theta}, \boldsymbol{m}(t)) \qquad j = 1, \ldots, N \tag{4.18}$$

An all-mode FSM $\mathbf{S} = (S_{ij})$, $i = 1, \ldots, \dim \boldsymbol{p}$, $j = 1, \ldots, N$ may be considered a structural Jacobian of a vector valued function $\boldsymbol{g} : \boldsymbol{p} \mapsto \boldsymbol{r}$ with entries

$$S_{ij} = \begin{cases} 0 & : \quad \dfrac{\partial g_j}{\partial p_i} = 0 \\ m_{ij} & : \quad \dfrac{\partial g_j}{\partial p_i} = m_{ij}\tilde{g}_i(\boldsymbol{p}) \\ 1 & : \quad \text{otherwise} \end{cases} \tag{4.19}$$

If the inputs into the system, the measured outputs and the discrete switch states may be considered faultless, $g_j()$ becomes a function of the component parameter vector, their rows may be removed from the FSM and the FSM can be considered a structural parameter sensitivity matrix for ARR residuals.

Faults in a system component can be related to deviations of parameter values from those of the healthy component. For instance, if a hydraulic check valve that autonomously switches on and off is modelled as a non-ideal switch, then a stuck-open fault can be captured by a permanent change in its switch state. Thus, a matrix entry S_{ij} that is non-zero for some system modes means that in these system modes, residual r_j is structurally sensitive to faults in the ith component. A FSM thus relates discrepancies in components to changes in residuals. The columns of a FSM indicate the fault signatures of the ARRs of the residuals. Structurally independent ARRs, i.e. ARRs that cannot be algebraically constructed from other ARRs have a unique fault signature. The rows of a FSM are called component fault signatures. An all-mode FSM will be termed *hybrid* FSM (HFSM).

Detection and Isolation of Faults

Clearly, a fault in a component can be detected if in some system mode, at least one entry in the component fault signature does not vanish, i.e. at least one residual in one system mode is sensitive to it. If, moreover, the component fault signature differs from all others in that system mode, then the fault can be isolated. If a fault in the ith component can be detected in all modes, then this is captured by an entry equal to one in the ith row of an extra column with the heading D_b added to the FSM. For a component fault signature that depends on switch states, the detectability of the fault is given by the logical OR of the switch states.

If a fault in the ith component can be isolated in a system mode, then this is usually expressed by adding another column with the heading I_b to the FSM for that system mode and by inserting an entry equal to one in its ith row. Parameters of components that are considered less sensitive to possible faults than others or that can be considered free from faults may be omitted from the FSM.

Note that it depends on the number of sensors and where they are placed whether parametric faults can be isolated. Under the assumption of a single fault hypothesis and N given sensors, the maximum number of parameter faults that can be isolated is equal to $2^N - 1$. However, often, the number of sensors, N, is less than the number of component parameters, p, so that the FSM is not quadratic. Some component parameters may have the same component fault signature (in some modes) so that faults in these parameters cannot be isolated by inspection of the all-mode FSM. Clearly, additional sensors may improve the isolability of faults. But detectors can be placed in a model only for those variables that are accessible by real sensors in the real system. Even if quantities can be measured, cost considerations may suggest to limit the number of real sensors. For illustration, inspection of ARRs (4.6)–(4.8) yields the FSM in Table 4.1.

Table 4.1 All-mode FSM for the switched network in Fig. 4.1

Component	Parameter	r_1	r_2	r_3	D_b	I_b
Switch control	b	1	1	0	1	0
Switch Sw	R_{on}	b	b	0	b	0
R : R_1	R_1	1	b	0	1	0
C : C_1	C_1	1	0	0	1	b
R : R_2	R_2	b	b	0	b	0
C : C_2	C_2	0	1	0	1	1
R : R_3	R_3	0	1	1	1	1
C : C_3	C_3	0	0	1	1	1
Se	V_i	1	b	0	1	0
Sensor Df : f	f	1	b	0	1	0
Sensor De : e_1	e_1	b	1	1	1	b
Sensor De : e_2	e_2	0	1	1	1	0

The component providing a signal $b(t)$ that externally controls the pass transistor modelled as a switch is denoted as switch control. Its output signal $b(t)$ is an input to the switched network and may be subject to disturbances.

If the voltage supply V_i is considered robust, and the sensors are considered faultless, the last four rows of the FSM in Table 4.1 may be omitted.

4.4 Construction of an All-Mode FSM by Inspection of Causal Paths

The structural information contained in the ARRs, i.e. the information on which ARR depends on which component parameters can be obtained directly by inspection of causal paths in a diagnostic bond graph [1]. There is no need to derive equations and to eliminate unknowns in order to set up a mode-dependent FSM. To that end, causal paths from model inputs to inputs of sensor elements are considered. Elements that are traversed on these causal paths contribute to the ARR of a residual related to a sensor element. An output of a source or an element that is followed directly or indirectly by switches on the causal path to a sensor element provides an entry in the FSM equal to the product of the switch states.

For illustration, Fig. 4.5 reproduces the diagnostic bond graph of the switched network in Fig. 4.1.

Notice that element C : C_1 in derivative causality and the switch Sw : b in conductance causality do not allow a flow sensor Df : f in inverted causality attached to junction 1_1. The non-inverted flow sensor Df : f has been replaced by a virtual detector Df* : r_1 for residual r_1 and a modulated flow source MSf : f. Virtual detectors are distinguished from detectors of power variables by an asterisk. The modulated flow source delivers a measured flow f.

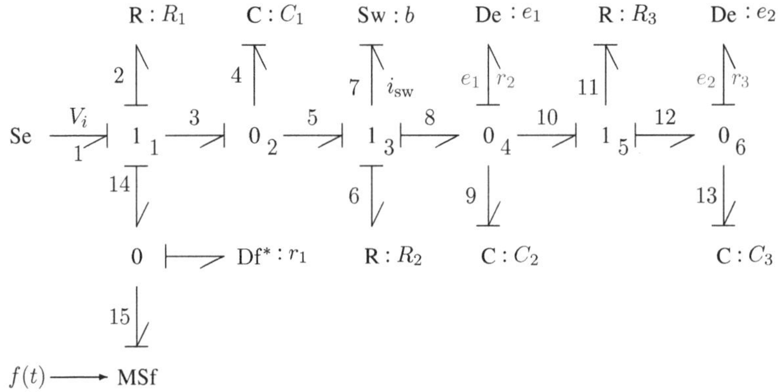

Fig. 4.5 Diagnostic bond graph of the switched network in Fig. 4.1

Causals paths from model inputs to detector $\mathrm{Df}^* : r_1$:

$f \rightarrow \mathrm{MSf} \rightarrow f_{16} \rightarrow r_1$
$V_i \rightarrow e_3 \rightarrow e_4 \rightarrow \boxed{\mathrm{C} : C_1} \rightarrow f_4 \rightarrow f_3 \rightarrow f_{15} \rightarrow r_1$
$e_1 \rightarrow e_9 \rightarrow e_7 \rightarrow \boxed{\mathrm{Sw} : b} \rightarrow f_7 \rightarrow f_5 \rightarrow f_3 \rightarrow f_{15} \rightarrow r_1$
$f \rightarrow f_2 \rightarrow \boxed{\mathrm{R} : R_1} \rightarrow e_2 \rightarrow e_3 \rightarrow e_5 \rightarrow e_7 \rightarrow \boxed{\mathrm{Sw} : b} \rightarrow f_7 \rightarrow f_6 \rightarrow$
$\boxed{\mathrm{R} : R_2} \rightarrow e_6 \rightarrow e_7 \rightarrow \boxed{\mathrm{Sw} : b} \rightarrow f_7 \rightarrow f_5 \rightarrow f_3 \rightarrow f_{15} \rightarrow r_1$

Traversing the switch model $\mathrm{Sw} : b$ means traversing its internal ON resistor $\mathrm{R} : R_{\mathrm{on}}$.

The causal paths from inputs to the detector of residual r_1 show that elements $\mathrm{Se} : V_i$, $\mathrm{R} : R_1$ and $\mathrm{C} : C_1$ contribute to residual r_1 *independently* of the switch state b, while elements $\mathrm{R} : R_{\mathrm{on}}$ and $\mathrm{R} : R_2$ contribute to residual r_1 only if the switch is on, i.e. $b = 1$ in accordance with ARR (4.6). Thus, the latter two elements provide an entry b to the column of ARR residual r_1 in their row respectively, while the first three ones have an entry equal to one in the column of r_1 in agreement with Table 4.1.

Causals paths from model inputs to detector $\mathrm{De} : e_1$:

$f \rightarrow \mathrm{MSf} \rightarrow f_{16} \rightarrow f_{15} \rightarrow f_2 \rightarrow \boxed{\mathrm{R} : R_1} \rightarrow e_2 \rightarrow e_3 \rightarrow e_5 \rightarrow e_7 \rightarrow$
$\boxed{\mathrm{Sw} : b} \rightarrow f_7 \rightarrow f_9 \rightarrow r_2$
$V_i \rightarrow e_3 \rightarrow e_5 \rightarrow e_7 \rightarrow \boxed{\mathrm{Sw} : b} \rightarrow f_7 \rightarrow f_9 \rightarrow r_2$
$e_1 \rightarrow e_{10} \rightarrow \boxed{\mathrm{C} : C_2} \rightarrow f_{10} \rightarrow r_2$
$e_2 \rightarrow e_{13} \rightarrow e_{12} \rightarrow \boxed{\mathrm{R} : R_3} \rightarrow f_{12} \rightarrow f_{11} \rightarrow r_2$

The elements in these causal paths are the elements involved in ARR (4.7).

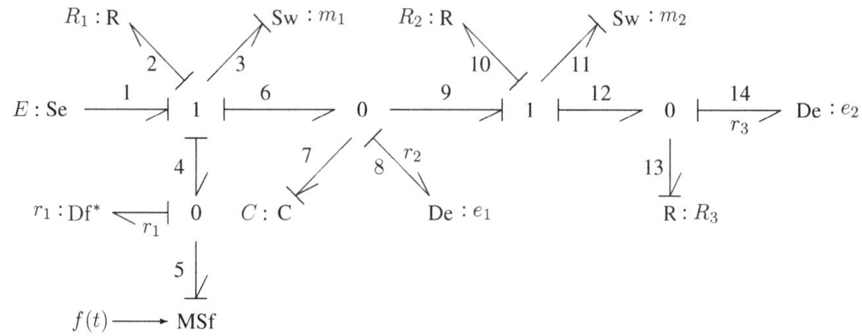

Fig. 4.6 Diagnostic bond graph of the switched network in Fig. 3.4

Causals paths from model inputs to detector De : e_2 :

$$e_2 \to e_{14} \to \boxed{\mathrm{C} : C_3} \to f_{14} \to r_3$$
$$e_2 \to e_{13} \to e_{12} \to \boxed{\mathrm{R} : R_3} \to f_{12} \to f_{13} \to r_3$$

The elements in the last two causal paths are the elements involved in ARR (4.8). As a result, the elements involved in the considered causal paths lead to the entries in the FSM in Table 4.1.

In the same manner, mode-dependent ARRs can be generated from diagnostic bond graphs of hybrid system models with more than one switch. As an example, Fig. 4.6 depicts the diagnostic bond graph of the switched network in Fig. 3.4.

The DBG in Fig. 4.6 contains the following causal paths from inputs to residuals. Causals paths from model inputs to detector Df* : r_1 :

$$f \to \mathrm{MSf} \to f_5 \to r_1$$
$$E \to e_3 \to \boxed{\mathrm{Sw} : m_1} \to f_3 \to f_4 \to r_1$$
$$e_1 \to e_5 \to e_3 \to \boxed{\mathrm{Sw} : m_1} \to f_3 \to f_4 \to r_1$$
$$E \to e_3 \to \boxed{\mathrm{Sw} : m_1} \to f_3 \to f_2 \to \boxed{\mathrm{R} : R_1} \to e_2 \to e_3 \to \boxed{\mathrm{Sw} : m_1} \to f_3 \to f_4 \to r_1$$

Causals paths from model inputs to detector De : e_1 :

$$f \to \mathrm{MSf} \to f_5 \to f_4 \to f_6 \to r_2$$
$$e_1 \to e_6 \to \boxed{\mathrm{C} : C} \to f_6 \to r_2$$
$$e_1 \to e_9 \to e_{11} \to \boxed{\mathrm{Sw} : m_2} \to f_{11} \to f_{10} \to \boxed{\mathrm{R} : R_2} \to e_{10} \to e_{11} \to \boxed{\mathrm{Sw} : m_2} \to f_{11} \to f_9 \to r_2$$
$$e_2 \to e_{12} \to e_{11} \to \mathrm{Sw} : m_2 \to f_{11} \to f_9 \to r_2$$

Causals paths from model inputs to detector De : e_2 :

$$e_2 \to e_{13} \to \boxed{\mathrm{R} : R_3} \to f_{13} \to r_3$$

Table 4.2 All-mode FSM for the switched network in Fig. 3.4

Component	Parameter/ output	r_1	r_2	r_3	$\mathrm{D_b}$	$\mathrm{I_b}$
Switch control m_1	m_1	1	0	0	1	0
Switch control m_2	m_2	0	1	1	1	0
Switch Sw_1	R^1_{on}	m_1	0	0	m_1	0
Switch Sw_2	R^2_{on}	0	m_2	m_2	m_2	0
$\mathrm{R} : R_1$	R_1	m_1	0	0	m_1	0
$\mathrm{C} : C$	C_1	0	1	0	1	1
$\mathrm{R} : R_2$	R_2	0	m_2	m_2	m_2	0
$\mathrm{R} : R_3$	R_3	0	0	1	1	1
Voltage source	E	m_1	0	0	m_1	0
Sensor Df : f	f	1	1	0	1	0
Sensor De : e_1	e_1	m_1	m_1	m_2	$m_1 \vee m_2$	$m_1 \wedge m_2$
Sensor De : e_2	e_2	0	m_2	m_2	m_2	0

$e_2 \rightarrow e_{12} \rightarrow e_{11} \rightarrow \boxed{\mathrm{Sw} : m_2} \rightarrow f_{11} \rightarrow f_{10} \rightarrow \boxed{\mathrm{R} : R_2} \rightarrow e_{10} \rightarrow e_{11} \rightarrow \boxed{\mathrm{Sw} : m_2} \rightarrow f_{11} \rightarrow f_{12} \rightarrow r_3$

From these causal paths the mode-dependent FSM in Table 4.2 is obtained.

If the voltage source Se : E and the sensors can be considered free of faults, then their rows in the FSM may be omitted.

In accordance with the FSM in Table 4.2, the following ARRs can be derived from the DBG in Fig. 4.6.

$$0 = f - \frac{m_1}{R^1_{\mathrm{on}} + R_1}(E - e_1) \tag{4.20}$$

$$0 = f - C\dot{e}_1 - \frac{m_2}{R^2_{\mathrm{on}} + R_2}(e_1 - e_2) \tag{4.21}$$

$$0 = \frac{m_2}{R^2_{\mathrm{on}} + R_2}(e_1 - e_2) - \frac{e_2}{R_3} \tag{4.22}$$

In these ARRs, the On resistance of the switches can be set to zero turning them into ideal switches.

4.5 Coherence Vector and Decision Procedures

In online model-based FDI, sampled real sensor measurements and known inputs are used for evaluation of ARRs. As measurements carry noise proper pre-filtering is necessary to reduce the effect of noise. This is of importance because ARRs contain derivatives of measured variables. Moreover, model and parameter uncertainties and numerical inaccuracies affect the evaluation of ARRs. However, as long as the result, i.e. the residuals of all ARRs are close to zero, the monitored system can be considered healthy. In order to decide whether faults have occurred between two sampling points in time, ARR residuals are passed to a decision procedure. In this procedure, a function Φ_i is applied to residual r_i, $i = 1, \ldots, N$ that maps the residual to a value c_i either equal to one or to zero where a zero value indicates normal behaviour and a value equal to one abnormal behaviour. The results are usually collected into a *coherence vector* $\boldsymbol{c} = [c_1 \ldots c_N]$. This vector is computed at every sampling step. According to its definition, the system is considered healthy as long as $\boldsymbol{c} = \mathbf{0}$. A non-null coherence vector means that at least one residual has been mapped to a value one, that is, faults in at least one component have occurred.

In order to isolate the faulty component the coherence vector is matched with the rows of the FSM, i.e with the component fault signatures. Given a hybrid system model, there is a FSM for each system mode. That is, in order to use the correct FSM for comparison, it is important to know in which mode the monitored system is at the present time point. Chapter 7 shows that ARRs derived from a diagnostic bond graph can also be used for system mode identification.

An important goal of any decision procedure is to reliably detect true faults and to minimise misdetections. A common strategy is to test the absolute value of a residual r_i against a threshold thr_i.

$$c_i(t) = \Phi_i(r_i(t)) = \begin{cases} 1 \text{ if } |r_i(t)| > thr_i(t) \\ 0 \text{ otherwise} \end{cases} \qquad \text{for i} = 1, \ldots, N \tag{4.23}$$

In the simplest case, one single mapping $\Phi : \mathbb{R} \to \mathbb{R}$, $r_i \mapsto c_i$ for all residuals and fixed values on the basis of experience can be used for the thresholds thr_i. However, given a hybrid system model, the dynamic system behaviour in one mode may be quite different from the one in another mode. Hence, in order to avoid conservative thresholds that may lead to a non-detection of faults, appropriate thresholds thr_i^j must be chosen for each system mode j. Furthermore, in each mode, modelling and parameter uncertainties may affect ARR residuals more or less. Therefore, more generally, adaptive thresholds may be defined that are a function of time, inputs and measurements. Chapter 5 presents a bond graph approach to the introduction of adaptive thresholds that are insensitive to parameter uncertainties.

4.6 Fault Isolation

In online FDI, the coherence vector is computed at every sampling step. If it is not a null vector, a fault is detected and an alarm is raised. Clearly, detectability is a necessary condition for a fault to be isolated. In order to simplify the task of isolating the fault, often a single fault hypothesis is adopted. It is assumed that more than one fault have not occurred simultaneously, that only one single fault may occur at a time.

This also means that faults do not cancel each other in their effect on an ARR residual. Given a single fault hypothesis, the faulty component is identified by comparing the coherence vector against the rows of the FSM, i.e. the component fault signatures. If this comparison results in a match, the faulty component is isolated. However, there may be no match, or more than one match may be obtained. That is, the faulty component cannot be isolated. In the case of multiple simultaneous faults, FDI can be performed e.g. by means of parameter estimation as is discussed in Chap. 6.

Clearly, a single fault can be isolated, if the FSM contains as many residuals as faults are considered and if the residuals are structured, i.e. each residual is sensitive to one and only one fault. That is, the FSM should be diagonal. With this goal in mind, the question is where to place which type of sensor. As the entries of a FSM can be obtained by inspection of causal paths in a diagnostic bond graph, assessing the impact of a sensor placement is of moderate effort. In addition, software such as ModelBuilder [7] can take the task of building a FSM. While in a model sensor elements may be placed where suitable, this may not always be possible for the real system. Some variables may not be accessible for direct monitoring and need to be constructed by means of observers, other variables such as an enthalpy flow cannot be measured at all.

If the number of fault candidates exceeds the number of sensors, structured residuals cannot be obtained. In that case, the FSM is non-square, rows, i.e. some component fault signatures are identical and not all faults can be isolated. More faults may be isolated by adding more detectors to the model if the real system permits to attach more real sensors.

Moreover, as to hybrid system models, the previously considered examples show that fault detectability and isolability is mode dependent. As can be seen from the FSM in Table 4.2, faults in some components may be detectable in all modes, e.g. a faulty capacitance in the switched network of Fig. 3.4, while other faults can only be detected in some modes, such as faulty resistances R_1 and R_2. (Wang et al. call such faults weakly detectable [9].) It may also be possible that a fault cannot be detected in none of the system modes. Similar observations can be made with regard to fault isolation bearing in mind that a necessary condition for a fault to be isolated is that it can be detected.

On the basis of a single fault hypothesis, fault isolation is performed by comparing the periodically updated coherence vector with the rows of the FSM. However, for a hybrid system model, the entries of a FSM are mode dependent. For a model with n_s switches, $n_f \leq 2^{n_s}$ physical feasible switch state combinations, i.e. n_f system modes are to be considered. The FSM holding for all modes provides a specific FSM for each mode. To make sure that the coherence vector is compared with the component fault signatures in the right FSM, the current system mode of operation must be identified from measured system or process outputs. Figure 4.7 depicts a flowchart of a bond graph model-based FDI process.

The diagram clearly indicates the fundamental role of generating ARRs. A DBG model with system inputs $\boldsymbol{u}(t)$ and measurements $\tilde{\boldsymbol{y}}(t)$ from the real system can serve this purpose. In the decision procedure, ARR residuals are assessed and mapped onto a coherence vector. Faults that have happened are indicated by components of this

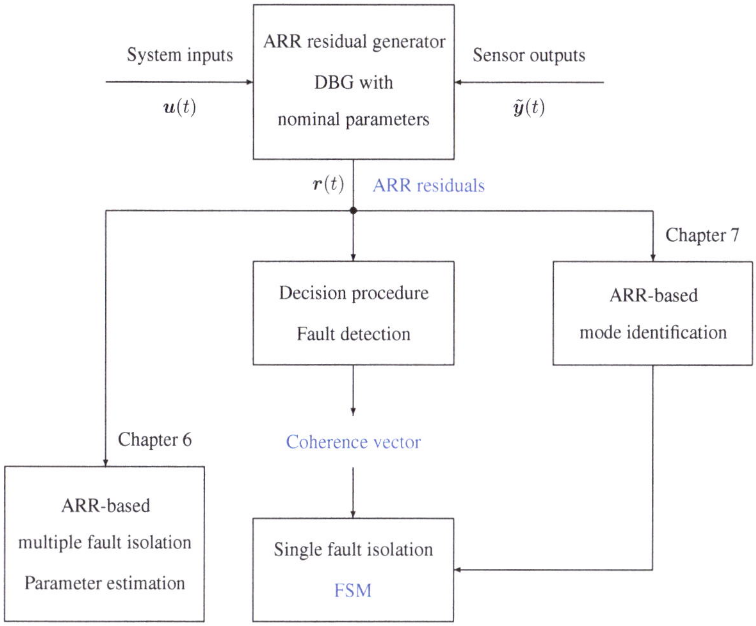

Fig. 4.7 Flowchart of a model-based FDI process

vector equal to one. Single fault isolation takes place by comparing a regularly updated coherence vector with the rows of a FSM. As ARRs are mode-dependent, there is a single all-mode FSM and a FSM for each system mode. To isolate a fault that has happened in the current system mode, the latter one must be known in order to use the FSM pertinent to the present mode. Chapter 7 addresses the use of ARRs for system mode identification. If the pattern of the actual coherence vector matches more than one component fault signature then the fault cannot be isolated by inspection of the FSM for the current system mode. In the case of multiple simultaneous faults, parameter estimation can be used to isolate them. This is addressed in Chap. 6.

4.7 Parameter Sensitivities of ARRs

The structural fault signature matrix considered so far indicates which component parameters are contained in which ARR. As faults in a component can be related to unwanted changes of parameter values, a structural FSM displays which ARRs are *structurally* sensitive to which faults. This matrix does not capture that variations in some of the parameters contained in an ARR may have only little effect on the residual of an ARR and may be overshadowed by the affect of other parameter variations in an ARR. As to hybrid models, the sensitivity of an ARR residual with respect to a parameter may be different in different system modes. If the parameter variations predominant on the sensitivity of an ARR are identified, then a FSM can be simplified by turning some non-zero entries into zeros. A FSM that takes the magnitude of relative parameter sensitivities of ARRs into account is sometimes called a practical fault signature matrix [3].

For ARRs in closed symbolic form, parameter sensitivities of ARR residuals can be obtained by symbolic differentiation. In case an explicit formulation of ARRs is not achievable, e.g. due to nonlinear algebraic loops, parameter sensitivities of ARR residuals can be numerically computed by using a *sensitivity* bond graph, in which bonds carry sensitivities of power variables [12–14], or by using *incremental* bond graphs, in which bonds carry variations of power variables [5]. In Chap. 5, incremental bond graphs are used for the determination of adaptive fault thresholds.

In a recent paper [15], Levy, Arogeti and Wang introduce a FSM $\mathbf{S} = (S_{ij})$ with more specific entries. If residual r_j is a function of parameter p_i then they define entry S_{ij} as $-\text{sign}(\partial r_j/\partial p_i)(t_n)$ or $\text{sign}(\partial r_j/\partial p_i)(t_n)$ depending on whether p_i is expected to increase or to decrease due to a fault. If r_j does not depend on p_i then $S_{ij} = 0$. Parameters that can increase as well as decrease contribute two rows to the FSM. The entries of a FSM extended in this way are instantaneous values $S_{ij} \in \{0, +1, -1\}$ and are to be computed online at each sampling point.

A more specific decision procedure assessing ARR residuals is designed accordingly. It does not produce a coherence vector $\boldsymbol{c} = [c_1, c_2, \ldots, c_N]$ with entries that are either zero or one but with entries $c_i \in \{0, 1, -1\}$ depending on whether residual r_i is within given bounds ($c_i = 0$), has exceeded an upper positive threshold ($c_i = +1$), or has crossed a lower negative threshold ($c_i = -1$). This discrimination of increasing and decreasing parameters and of rising and falling residuals can help to narrow the set of fault candidates when the coherence vector is compared with the component fault signatures or can even isolate faults that cannot be identified in a standard FSM with entries $S_{ij} \in \{0, 1\}$ because rows share the same component fault signature.

4.8 Numerical Computation of ARR Residuals

For a diagnostic bond graph model with nominal parameters used in FDI, inputs into the real system or process and measured outputs are model inputs. Outputs of a DBG are time evolutions of ARR residuals. DBGs are thus an ARR evaluation block. Introducing one fault at a time into the real system leads to changed measured system outputs that affect the evaluation of ARRs. As a result, some ARR residuals change. A parameter change $\Delta\Theta_i$ results in a variation of some residuals. Hence, an approximation of the sensitivity of these residuals with respect to parameter change $\Delta\Theta_i$ can be computed and can be used to decide whether the effect of some parameter changes on an ARR residual can be neglected in comparison to other parameter changes. This way, a practical FSM can be build up.

4.8.1 Coupling of a Behavioural and a Diagnostic Bond Graph Model

However, introducing faults into a real system or process often is either not possible or too hazardous. Some components do not allow to introduce a fault, e.g. if a tank has not got an additional outlet valve, leakage from that tank cannot be introduced. If a tank contains a poisonous gas, introduction of a leakage for test purposes is not an option. Even if the introduction of faults into a real system is feasible, controllers may compensate for them so that it is difficult to detect them, or the introduced faults may lead to instability of the system. For these reason, an option is to replace the real system or process by a behavioural model, to consider its numerically computed output as 'measured' and to feed them into the ARR evaluation block [1]. All kinds of faults can then be deliberately introduced into the behavioural model at no risk by adding controlled sources and/or by altering parameters, while the diagnostic model remains unchanged. The introduced faults may be of different type, i.e. they may be abrupt or progressive. The results of such an off-line simulation may be used to classify faults into weakly or strongly detectable faults and to define proper thresholds for ARR residuals. To make the off-line simulation more realistic, the outputs of the behavioural model may be superimposed with noise. However, as ARRs contain derivatives of inputs, noise can lead to significant instantaneous changes in residuals that must be filtered by a post processing of residuals, or inputs into the diagnostic model are filtered before they are used in the evaluation of ARRs. Figure 4.8 displays the basic framework of this approach to an off-line numerical computation of ARR residuals.

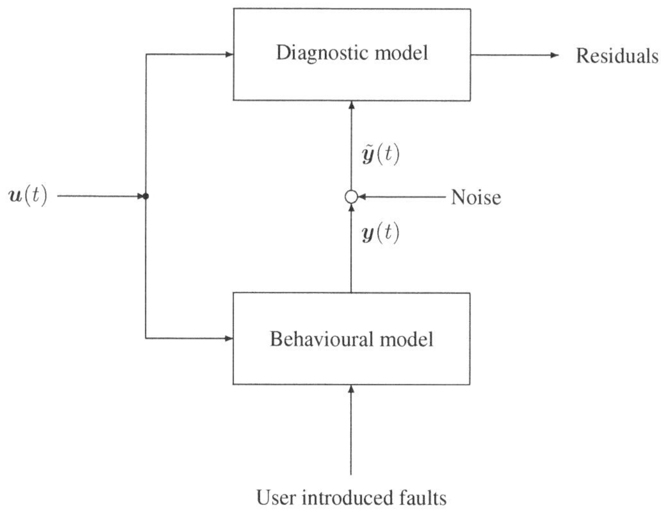

Fig. 4.8 Basic framework of coupling a behavioural model and a diagnostic model

Essential advantages of this framework are that ARRs in symbolic form are not required. The diagnostic model is numerically computed by calculating derivatives of its inputs in discrete time. The result is the time evolution of ARR residuals. Finally, in the case of slow real processes, off-line simulation using a behavioural model reveals the effects of parametric faults much faster than a real-time computation of ARRs that uses measured output from the real process.

4.8.1.1 Example: Network with a Semiconductor Switch

The basic framework for an off-line numerical computation of ARR residuals is illustrated by means of the network in Fig. 4.1. The coupling of its behavioural BG and its diagnostic BG is displayed in Fig. 4.9.

From the coupled causal BGs a set of equations can be easily derived and coded, e.g. in the Scilab script language [16]. In this case, the script can be simplified a bit because ARRs in symbolic form can be obtained from the diagnostic bond graph. That is, let the equations of the behavioural model be collected in a script file named `f`, then a simulation run can be performed by calling the Scilab function `ode()`, i.e.

```
y = ode(y0,t0,t,f);
```

where `y0` denotes the vector of initial conditions, `t0` the start time of the simulation, and `t` the vector of discrete time points in the interval from `t0` to `tf`, where `tf` denotes the final time point of the simulation run. The values of the model outputs, i.e. flow $f(t)$ and efforts $e_1(t)$, $e_2(t)$ together with the voltage supply V_i can be used

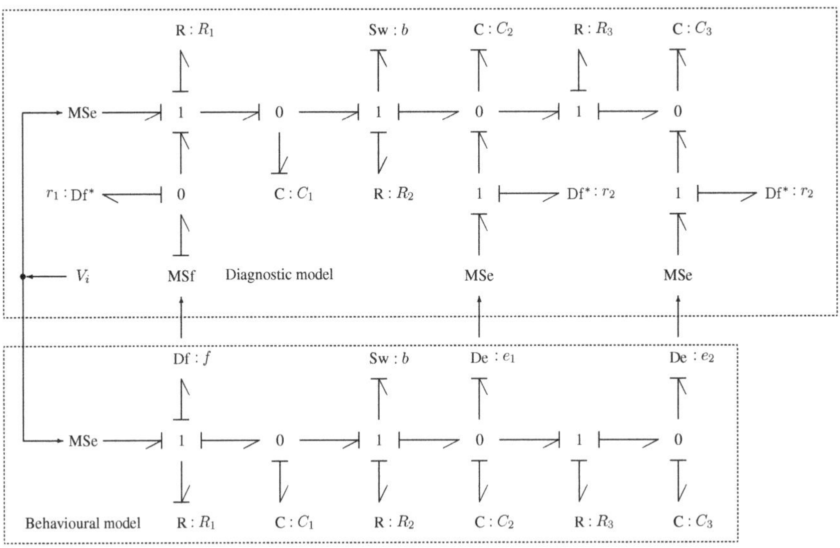

Fig. 4.9 Coupling a behavioural model and a diagnostic model of the network in Fig. 4.1

Table 4.3 Parameters of the switched network in Fig. 4.1

Parameter	Value	Units
V_i	5	V
R_1	100	kΩ
C_1	10	μF
R_2	100	kΩ
C_2	10	μF
R_3	100	kΩ
$\varDelta R_3$	100	kΩ
C_3	10	μF

in another script for evaluation of the ARRs. The derivatives of inputs required by the ARRs can be computed in discrete time by the Scilab function `diff()`.

The simulation of the dynamic behaviour of the faultless switched network uses the parameters in Table 4.3.

The ON resistance of the switch has been chosen to be $R_{\text{on}} = 0.1\ \Omega$. The pass transistor in the network of Fig. 4.1 is externally switched on and off periodically every 0.5 s by a signal b. Its discrete values $b(t) \in \{0, 1\}$ denote the two system modes in which the circuit operates. The hybrid FSM in Table 4.1 shows that parametric faults in the elements C : C_2, R : R_3, and C : C_3 can be detected and isolated in all modes.

4.8.1.2 Fault Scenario: The Value of Resistance R_3 Is Changed for Some Time Interval

For validation of the ARRs in symbolic form obtained from the BG in Fig. 4.2 it is assumed that R : R_3 is a resistor of variable resistance R_3 that can be abruptly doubled at $t = 2.5$ s and abruptly reduced to its original value R_{30} at $t = 10.5$ s.

$$R_3(t) = R_{30} + \varDelta R_3 \cdot \text{pulse}(t, 2.5, 10.5) \qquad (4.24)$$

where `pulse(t,2.5,10.5)` denotes a unit pulse lasting from 2.5 s to 10.5 s.

During this time interval, the system mode changes every 0.5 s. According to the HFSM, residual r_3 should clearly indicate this fault in parameter R_3 independent of the system mode, while residual r_1 should be insensitive to this parametric fault.

Figure 4.10 indicates that due to an increase of resistance R_3 the values of current i_3 are in a more narrow envelope during that time interval.

For simplicity, a single mode-independent fixed threshold $thr = 0.02$ is adopted for all three residuals. As to be expected, Fig. 4.11 shows that the values of residual r_3 in the time interval [2.5 s, 10.5 s] are significantly above this threshold in both system modes.

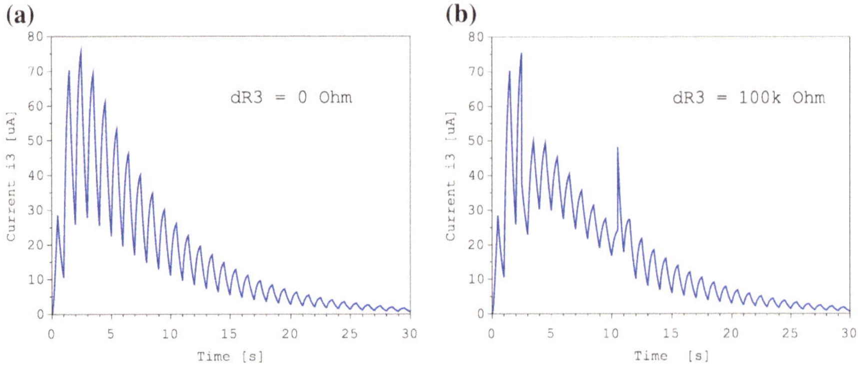

Fig. 4.10 Time evolution of current i_3 in the circuit of Fig. 4.1. **a** Current i_3: resistence R_3 is not increased. **b** Current i_3: increased resistance R_3 in the time interval [2.5 s, 10.5s]

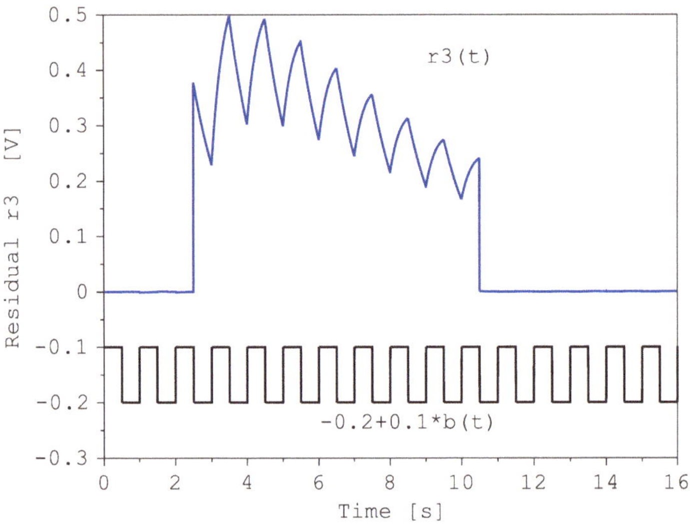

Fig. 4.11 Residual r_3 due to an increase of resistance R_3 in the time interval [2.5 s, 10.5 s]

On the contrary, residual r_1 in fact, is clearly insensitive to a change in resistance R_3 (Fig. 4.12).

4.8.2 Coupling of a Real System Model to a Non-faulty Model by Means of Residual Sinks

If a real engineering system or process is replaced by a behavioural model that may be subject to deliberately injected faults, then simulation of the real system dynamics

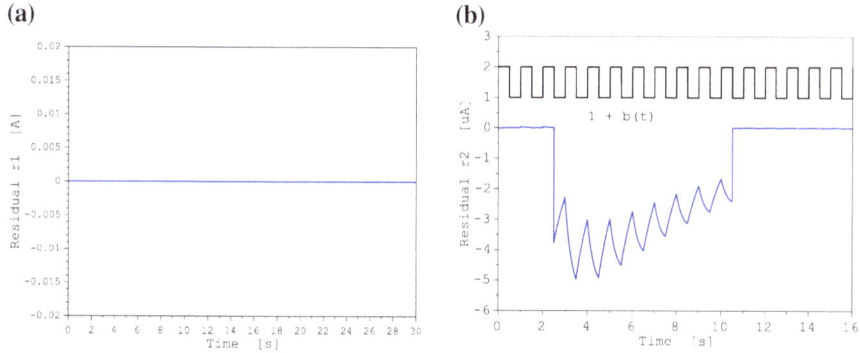

Fig. 4.12 Time evolution of residuals r_1 and r_2. **a** Time history of residual r_1. **b** Time history of residual r_2

needs initial conditions which can also be used for the numerical evaluation of a non-faulty system model with fixed nominal parameters. Both bond graph models can be in preferred integral causality. The introduction of faults into the real system model by switching on sources or by changing a component parameter results in an output vector $\tilde{\boldsymbol{y}}$ that may be considered a substitute of measured variables and that differs from the output vector $\boldsymbol{y}$ of the non-faulty system model.

The 'measured' variables $\tilde{y}_i$ obtained from the faulty system model and the outputs y_i of the non-faulty system model may be coupled by feeding the differences $\tilde{y}_i - y_i$ into modulated sinks that deliver an output r_i so that their input becomes zero. These sinks are termed *residual sinks*. The output r_i of a residual sink is a power variable and is inserted into the balance of power variables at that junction in the non-faulty system model from which its co-variable y_i is obtained. The outputs of the residual sinks are additional inputs into the non-faulty system model that force the faultless system to alter its behaviour so that it becomes identical to the one of the faulty system. If no fault is introduced into the real system model, then there are no differences and all values $r_i(t)$ are close to zero. Differences, however, lead to values $r_i(t)$ that remain distinguishably different from zero as long as a fault is effective, i.e. is not repaired. That is, the outputs of the residual sinks can serve as fault indicators. The balance of power variables at a junctions in the non-faulty system model connected to a residual sink becomes an ARR when unknowns have been eliminated and the output of the residual sink becomes the residual of that ARR.

In some way, this coupling of two models by residual sinks may be compared to the temporary coupling of two bodies such as the plates of a clutch by a residual sink (cf. Fig. 2.15). As long as the clutch is disengaged, there is no force acting between the two plates. If, however, the clutch engages, a torque acts on both of them such that the difference of their angular velocities is zero. The load side is forced to adapt to drive side. This approach has been applied for the numerical computation of ARR residuals from continuous time models [14, 17] and recently also to systems described by a hybrid model [18].

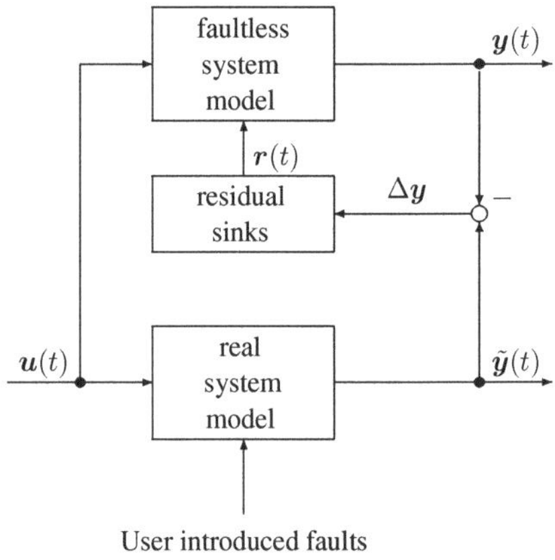

Fig. 4.13 Coupling a real engineering system model to a non-faulty system model by means of residual sinks for off-line numerical evaluation of ARR residuals

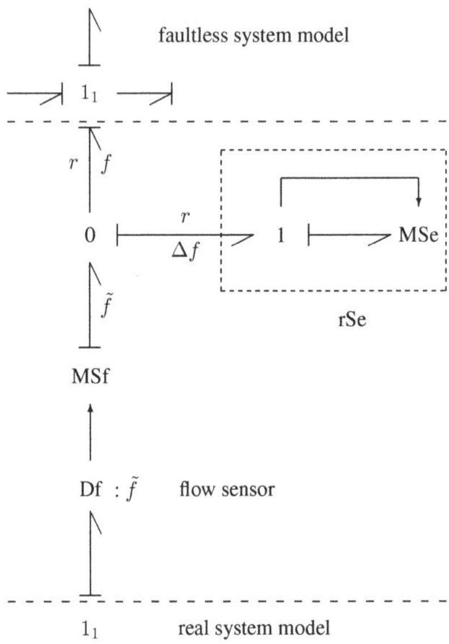

Fig. 4.14 A residual effort sink, rSe, forcing the difference of flows $\tilde{f} - f$ to zero and contributing an effort r to the upper 1-junction

Figure 4.13 indicates the basic idea of this approach to a numerical off-line computation of ARR residuals.

The coupling of the real system model to the faultless model is depicted in the BG fragment of Fig. 4.14 for the case of a residual effort source. The flow $\tilde{f}_1$ obtained by numerical computation of the real system model may be considered a substitute of a measured flow. For brevity, modulated residual sinks are denoted by rSe or rSf respectively.

Remark 4.4 Derivation of equations from the bond graphs coupled by residual sinks results in a DAE system as there are no time derivatives of the outputs of the coupling residual sinks. Let some elements of the switched models have non-linear

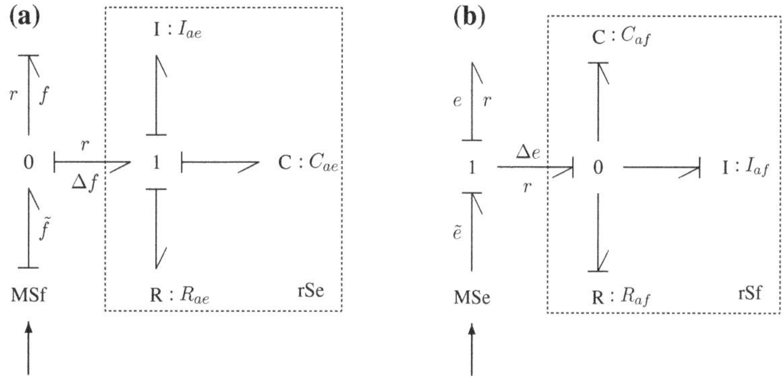

Fig. 4.15 Bond graph models of residual sinks. **a** Model of a residual effort sink. **b** Model of a residual flow sink

characteristics and assume that all storage elements take preferred integral causality in all modes, then the equations derived from the coupled BGs form a semi-explicit DAE system

$$\dot{\boldsymbol{x}}(t) = \boldsymbol{f}_1(t, \boldsymbol{x}(t), \boldsymbol{r}(t), \boldsymbol{u}(t)) \tag{4.25a}$$
$$\boldsymbol{0} = \boldsymbol{f}_2(t, \boldsymbol{x}(t)) \tag{4.25b}$$

where $\boldsymbol{x}$ denotes the vector of state variables and $\boldsymbol{r}$ the vector of residuals. The residuals are *numerically* computed as components of the descriptor vector $[\boldsymbol{x}^T \; \boldsymbol{r}^T]^T$. There is no need to set up sums of power variables at junctions to which a sensor has been attached and to eliminate unknown variables in order to obtain ARRs in *symbolic* form. The DAE system is of index 2 if

$$\det\left(\frac{\partial \boldsymbol{f}_2}{\partial \boldsymbol{x}} \frac{\partial \boldsymbol{f}_1}{\partial \boldsymbol{r}}\right) \neq 0\,. \tag{4.26}$$

Equations 4.25a, 4.25b are known as Hessenberg index-2 form. Codes suitable for numerical solution of such DAE systems are, e.g. DASPK 3.1 [19, 20] (based on the BDF method) and Radau5 [21] that uses an implicit Runge-Kutta method of order 5.

If the engagement of a clutch is considered instantaneously and is modelled by means of a residual effort sink that provides a moment M as long as the plates of the clutch are in contact, then the resulting model is also of Hessenberg index 2 form. □

Remark 4.5 The index 2 DAE system may be turned into a stiff ODE system if the residual sinks are replaced by an artificial storage element in conjunction with an artificial resistor as shown in Fig. 4.15.

The bond graph model for a residual effort sink in Fig. 4.15a yields for the residual r

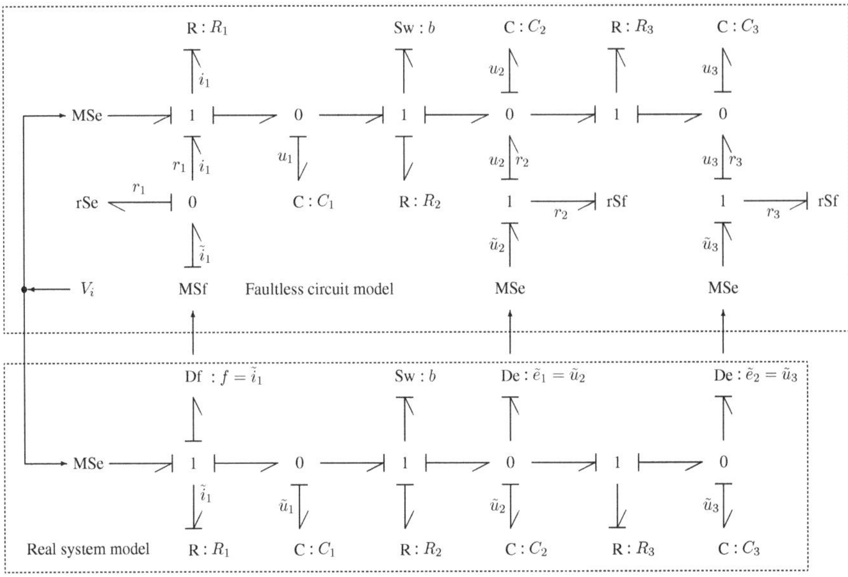

Fig. 4.16 Coupling of a real system model and a faultless model with nominal parameters for the network in Fig. 4.1

$$r(t) = R_{ae} \cdot (\Delta f)(t) + \frac{1}{C_{ae}} \int_0^t (\Delta f)(\tau) \mathrm{d}\tau + I_{ae}(\Delta \dot{f})(t) \tag{4.27}$$

or

$$\mathscr{L}r(s) = R_{ae} \left[1 + \frac{1}{(R_{ae} C_{ae}) s} + \frac{I_{ae}}{R_{ae}} s \right] \mathscr{L}(\Delta f)(s) \tag{4.28}$$

Equation 4.28 can be considered the equation of a PID controller which is known to force its input, i.e. the error to zero. □

Example: network with a semiconductor switch

For illustration, Fig. 4.16 displays the coupling of a real system model and a faultless model by residual sinks for the switched circuit in Fig. 4.1.

Fault scenario: resistance R_3 abruptly changes at $t = 2\,s$

For $0 \leq t < 5$ s the switch is on. At $t = 5$ s it is switched off. At $t = 2\,s$, resistance R_3 in the behavioural model of the real circuit abruptly doubles its value and remains at

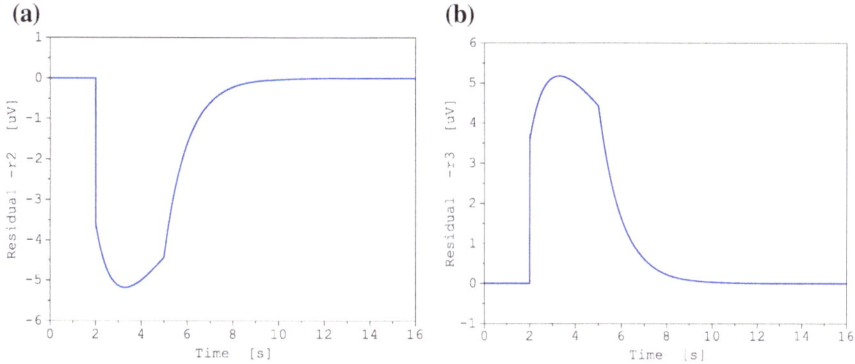

Fig. 4.17 Time evolution of residuals r_2 and r_3 due to an abrupt increase of resistance R_3 at $2s$ (switch Sw is open for $t \geq 5s$). **a** Time evolution of residuals r_2. **b** Time evolution of residuals r_3

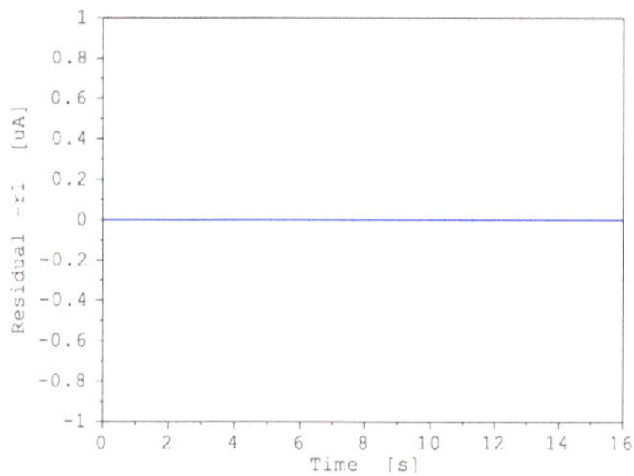

Fig. 4.18 Residual r_1 insensitive to a change in resistance R_3

that increased value while resistance R_3 in the non-faulty model remains at its initial value.

Simulation results in Fig. 4.17 obtained by numerical evaluation of the coupled bond graph models in Fig. 4.16 confirm that residuals r_2 and r_3 indicate this change in the behavioural model of the real circuit in accordance with the FSM in Table 4.1.

In contrast, Fig. 4.18 shows that residual r_1 is not sensitive to this change of resistance R_3 as to be expected.

For this simulation, the residual sinks have been replaced by bond graph models depicted in Fig. 4.15. The simulation run performed with the dassl solver has used the following values for the artificial elements $C_{ae} = C_{af} = 10^{-5}$, $R_{ae} = 10^{-3}$, $I_{ae} = I_{af} = 10^{-5}$, $R_{af} = 10^{+3}$.

4.8.3 Residual Sinks Accounting for Uncertainties

In the previous section, a behavioural model of the faulty system and a non-faulty system model with nominal parameters have been coupled by residual sinks that are modulated by the difference between a 'measured' and a computed variable. They contribute a power variable to a junction of the non-faulty system model that forces the non-faulty system model with nominal parameters to adapt its dynamic behaviour to the one of the faulty system so that their inputs, i.e. the difference between a 'measured' and a computed variable becomes zero. Their output added to the sum of power variables at a junction of the non-faulty system model is an ARR residual. The difference between a 'measured' and a computed variable, however, is unlikely to be identical zero due to model and parameter uncertainties. Moreover, real measurements provided by sensors carry noise so that the residual sinks will produce non-zero residuals even when no fault has happened. In order to avoid incorrect fault detections and false alarms reported to a supervision system differences between a 'measured' variable and its corresponding computed variable must be allowed to vary within given bounds. This can be achieved by a slight modification in the use of the residual sinks. The difference between 'measured' and computed variable is first input into a decision procedure that decided whether the signal modulating a residual sink is set to zero or not. Consider the residual effort sink rSe in Fig. 4.19.

As long as the absolute value $|\Delta f_i(t)| := |\tilde{f}_i(t) - f_i(t)|$ is below a given threshold thr_i, the signal s_i modulating the residual sink rSe_i is set to zero and the sink remains

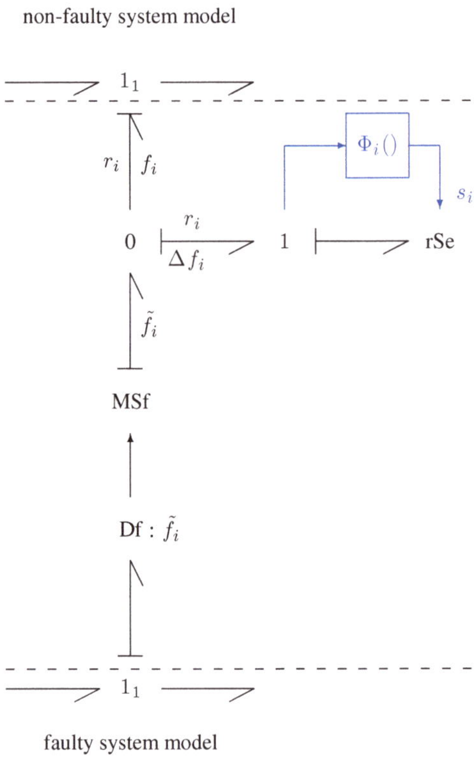

Fig. 4.19 A residual effort sink rSe allowing its input to vary within bounds

inactive. Otherwise, the modulating signal is equal to $|\Delta f_i(t)|$ having the residual sink force its input to zero.

$$s_i(t) = \Phi_i(|\Delta f_i(t)|) = \begin{cases} 0 & \text{if} \quad |\Delta f_i(t)| \leq thr_i \\ |\Delta f_i(t)| & \text{if} \quad |\Delta f_i(t)| \geq thr_i \end{cases} \tag{4.29}$$

The threshold thr_i may be a constant chosen on the basis of experience for each system mode or a function of variables and thus may be time dependent.

4.9 Analytical Determination of ARR Residuals for Switched LTI Systems

The coupling of two bond graph models by means of residual sinks and the numerical evaluation of the integrated model may be complemented by some analytical formulations in case the real system along with its sensors and actuators being subject to parametric faults can be described sufficiently accurate by a switched LTI model and represented by a bond graph in preferred integral causality for all modes. Let the state variables, the outputs and the matrices of the model representing the faulty system be distinguished from their counterparts in the switched LTI model of the non-faulty system by a tilde. Moreover, let all disturbances be composed into a vector $\boldsymbol{d}$. Then the state space model of the faulty system reads

$$\dot{\tilde{\boldsymbol{x}}}(t) = \tilde{\mathbf{A}}\tilde{\boldsymbol{x}}(t) + \tilde{\mathbf{B}}_1\boldsymbol{u}(t) + \tilde{\mathbf{B}}_2\boldsymbol{d}(t) \,, \quad \tilde{\boldsymbol{x}}(0) = \tilde{\boldsymbol{x}}_0 \tag{4.30a}$$

$$\tilde{\boldsymbol{y}}(t) = \tilde{\mathbf{C}}\tilde{\boldsymbol{x}}(t) + \tilde{\mathbf{D}}_2\boldsymbol{d}(t) \tag{4.30b}$$

where $\tilde{\mathbf{A}}$, $\tilde{\mathbf{B}}_1$, $\tilde{\mathbf{B}}_2$, $\tilde{\mathbf{C}}$, $\tilde{\mathbf{D}}_2$ are matrices of appropriate dimensions with coefficients that depend on switch states and on the events of parametric changes. They are constant between two discrete events. At the event of a discrete system mode change or an abrupt parametric fault the values of some coefficients jump to new levels.

The state space model of the non-faulty model reads

$$\dot{\boldsymbol{x}}(t) = \mathbf{A}\boldsymbol{x}(t) + \mathbf{B}\boldsymbol{u}(t) \,, \quad \boldsymbol{x}(0) = \boldsymbol{x}_0 \tag{4.31a}$$

$$\boldsymbol{y}(t) = \mathbf{C}\boldsymbol{x}(t) \tag{4.31b}$$

where the coefficients of the matrices $\mathbf{A}$, $\mathbf{B}$ and $\mathbf{C}$ depend on switch states. They are constant for the duration of a system mode. At the event of a system mode change they jump to new values.

If the behavioural model of the real system being subject to faults is coupled to the model of the non-faulty system, then the residual sinks force the difference $\tilde{\boldsymbol{y}} - \boldsymbol{y}$ to zero by introducing their outputs $\boldsymbol{r}$ into the model of the non-faulty system. As a result, the model of the non-faulty system with original unmodified parameters adapts to the behaviour of the simulated faulty system. Hence, $\boldsymbol{x}(t) = \tilde{\boldsymbol{x}}(t)$, $\boldsymbol{y}(t) = \tilde{\boldsymbol{y}}(t)$

and

$$\dot{\tilde{\boldsymbol{x}}}(t) = \mathbf{A}\tilde{\boldsymbol{x}}(t) + \mathbf{B}\boldsymbol{u}(t) + \mathbf{E}_1\boldsymbol{r}(t)\,, \quad \tilde{\boldsymbol{x}}(0) = \tilde{\boldsymbol{x}}_0 \tag{4.32a}$$

$$\tilde{\boldsymbol{y}}(t) = \mathbf{C}\tilde{\boldsymbol{x}}(t) + \mathbf{E}_2\boldsymbol{r}(t) \tag{4.32b}$$

Let t_1, t_2 denote the time instances of two discrete events and let (t_1, t_2) be a time interval, in which no discrete mode change and no new parametric fault takes place, i.e. the coefficients of the matrices are constant for $t \in [t_1, t_2)$. It is assumed that some parametric faults that happened for $t \leq t_1$ still last for $t > t_1$ so that some residuals do not vanish. Laplace transformation then gives a linear system of algebraic equations for $\mathscr{L}\boldsymbol{r}$.

$$\begin{aligned}[\mathbf{C}(s\mathbf{I} - \mathbf{A})^{-1}\mathbf{E}_1 + \mathbf{E}_2]\mathscr{L}\boldsymbol{r} &= \mathscr{L}\tilde{\boldsymbol{y}} - \mathbf{C}(s\mathbf{I} - \mathbf{A})^{-1}\mathbf{B}\mathscr{L}\boldsymbol{u} \\ &\quad -\mathbf{C}(s\mathbf{I} - \mathbf{A})^{-1}\tilde{\boldsymbol{x}}(t_1)\end{aligned} \tag{4.33}$$

The state vector $\tilde{\boldsymbol{x}}$ of the faulty system model differs from the state vector $\boldsymbol{x}_n$ of the non-faulty system model with nominal parameters due to parametric faults. The same holds for the 'measured' outputs $\tilde{\boldsymbol{y}}$, i.e. $\tilde{\boldsymbol{x}} = \boldsymbol{x}_n + \Delta\boldsymbol{x}$ and $\tilde{\boldsymbol{y}} = \boldsymbol{y}_n + \Delta\boldsymbol{y}$. Substituting the decomposed vectors into (4.32a and 4.32b) yields

$$\Delta\dot{\boldsymbol{x}}(t) = \boldsymbol{A}\Delta\boldsymbol{x}(t) + \boldsymbol{E}_1\boldsymbol{r}(t) \tag{4.34a}$$

$$\Delta\boldsymbol{y}(t) = \boldsymbol{C}\Delta\boldsymbol{x}(t) + \boldsymbol{E}_2\boldsymbol{r}(t) \tag{4.34b}$$

Laplace transformation yields another equation that determines the vector $\mathscr{L}\boldsymbol{r}$ of Laplace transformed residuals.

$$[\mathbf{C}(s\mathbf{I} - \mathbf{A})^{-1}\mathbf{E}_1 + \mathbf{E}_2]\mathscr{L}\boldsymbol{r} = \mathscr{L}\Delta\boldsymbol{y} - \mathbf{C}(s\mathbf{I} - \mathbf{A})^{-1}\Delta\boldsymbol{x}(t_1) \tag{4.35}$$

Substituting (4.30a) into (4.32a) gives

$$\mathbf{E}_1\boldsymbol{r}(t) = (\Delta\mathbf{A})\tilde{\boldsymbol{x}}(t) + (\Delta\mathbf{B})\boldsymbol{u}(t) + \tilde{\mathbf{B}}_2\boldsymbol{d}(t) \tag{4.36}$$

where $\Delta\mathbf{A} := \tilde{\mathbf{A}} - \mathbf{A}$ and $\Delta\mathbf{B} := \tilde{\mathbf{B}}_1 - \mathbf{B}$ account for parametric faults. If there are none, then $\boldsymbol{r}$ is only caused by possible disturbances.

$$\mathbf{E}_1\boldsymbol{r}(t) = \mathbf{B}_2\boldsymbol{d}(t) \tag{4.37}$$

The decomposition of the system matrix $\tilde{\mathbf{A}}$ into a matrix $\mathbf{A}$ with nominal parameters and a matrix $\Delta\mathbf{A}$ accounting for parameter changes is illustrated for the example network with one semiconductor switch and the previously considered abrupt change of resistance R_3 at $t = 2$ s.

From the bond graph of the circuit with one switch (Fig. 4.16), the following state equations can be derived.

$$\dot{u}_1 = \frac{1}{C_1}\left[\frac{1}{R_1}(V_i + r_1 - u_1) - \frac{b}{R_{on} + R_2}(u_1 - u_2)\right] \tag{4.38a}$$

$$\dot{u}_2 = \frac{1}{C_2}\left[\frac{b}{R_{on} + R_2}(u_1 - u_2) + r_2 - \frac{1}{R_3}(u_2 - u_3)\right] \tag{4.38b}$$

$$\dot{u}_3 = \frac{1}{C_3}\left[\frac{1}{R_3}(u_2 - u_3) + r_3\right] \tag{4.38c}$$

Writing (4.38a, 4.38b and 4.38c) as an equation for the state vector $[u_1\, u_2\, u_3]^T$ yields the matrices **A**, **B** and $\mathbf{E}_1$ in (4.32a).

$$\begin{bmatrix}\dot{u}_1\\ \dot{u}_2\\ \dot{u}_3\end{bmatrix} = \underbrace{\begin{bmatrix} -\frac{1}{C_1R_1} - \frac{b}{C_1(R_{on}+R_2)} & \frac{b}{C_1(R_{on}+R_2)} & 0 \\ \frac{b}{C_2(R_{on}+R_2)} & -\frac{b}{C_2(R_{on}+R_2)} - \frac{1}{C_2R_3} & \frac{1}{C_2R_3} \\ 0 & \frac{1}{C_3R_3} & -\frac{1}{C_3R_3}\end{bmatrix}}_{\mathbf{A}} \begin{bmatrix}u_1\\ u_2\\ u_3\end{bmatrix}$$

$$+ \underbrace{\begin{bmatrix}\frac{1}{C_1R_1}\\ 0\\ 0\end{bmatrix}}_{\mathbf{B}}[V_i] + \underbrace{\begin{bmatrix}\frac{1}{C_1R_1} & 0 & \\ 0 & \frac{1}{C_2} & 0\\ 0 & 0 & \frac{1}{C_3}\end{bmatrix}}_{\mathbf{E}_1}\begin{bmatrix}r_1\\ r_2\\ r_3\end{bmatrix} \tag{4.39}$$

The considered fault scenario assumes that the value of resistance R_3 increases by ΔR_3 at time instant $t_s = 2\,\text{s}$. This change affects the coefficients a_{22}, a_{23}, a_{32} and a_{33} of matrix **A**. That is, matrix $\tilde{\mathbf{A}}$ differs from the system matrix **A** of the non-faulty system only with regard to these coefficients. All of them include the factor $1/R_3$. For $t > t_s$ this factor can be decomposed into the form

$$\frac{1}{R_3} = \frac{1}{R_3^n} - \frac{\Delta R_3}{R_3^n(R_3^n + \Delta R_3)} \tag{4.40}$$

where R_3^n denotes the original nominal parameter.

Equation 4.40 may be extended to account for the time instant t_s at which R_3 abruptly changes.

$$\frac{1}{R_3(t)} = \frac{1}{R_3^n} - \frac{\Delta R_3}{R_3^n(R_3^n + \Delta R_3)}\texttt{step}(t, t_s) \tag{4.41}$$

where $\texttt{step}(t, t_s)$ denotes the step function that jumps at $t = t_s$.

That is, the system matrix $\tilde{\mathbf{A}}$ of the faulty system model can be decomposed into a matrix $\mathbf{A}$ with nominal parameters and a matrix $\Delta\mathbf{A}$ with coefficients depending on the parameter change ΔR_3.

$$\tilde{\mathbf{A}} = \mathbf{A} + \Delta\mathbf{A} \tag{4.42}$$

4.10 Summary

As to continuous time models, it is well known that ARR residuals can serve as fault indicators. If element characteristics allow for the elimination of unknown variables, as many ARRs as sensors used can be derived from a bond graph in closed symbolic form by either applying the causality inversion method or the extended covering path method. The structure of the obtained ARRs, i.e. the information which parameters or variables affect which ARR can be expressed in a structural fault signature matrix. In this book, hybrid system models are represented by causal bond graphs that hold for all system modes. Therefore, the entries in the FSM depend on switch states. Accordingly, fault detectability and fault isolation become mode dependent. That is, faults in some components may be detectable in all modes, while other faults can only be detected in some modes, or not at all. The result is an all-mode FSM. It is shown that the entries in a structural all-mode FSM can be obtained by considering causal paths from sources to detectors as in the case of a continuous time model.

In online fault detection, ARR residuals are close to zero for a healthy system. Generally, they are not identical to zero for various reasons such as modelling uncertainties, uncertain parameters, noise, or numerical inaccuracies. For correct online fault detection it is important that true faults are reliably detected and false alarms are avoided. To that end, residuals are fed into a fault decision procedure. The result is a coherence vector. If this vector is a null vector, then the system is healthy, no fault has happened. If some of its entries are non-zero, then the coherence vector is compared with the rows of the structural FSM. Given a single fault hypothesis, the fault is isolated if there is a match with one row of the FSM. If there is more than one match then the detected fault cannot be isolated. Also, if the number of fault candidates exceeds the number of sensors, not all faults can be isolated. Isolation of multiple simultaneous faults by means of parameter estimation is considered in Chap. 6.

As the dynamic behaviour of systems described by a hybrid model can be quite different in different system modes, mode dependent thresholds for the absolute values of ARR residuals should be chosen. Also, modelling and parameter uncertainties may affect ARR residuals in one mode more than in another one. Therefore, even more generally, adaptive thresholds may be defined that are a function of time, inputs and measurements. The next chapter presents a bond graph approach to the introduction of adaptive thresholds that are insensitive to parameter uncertainties.

Furthermore, an all-mode structural FSM holds for all modes. That is, from this FSM one FSM for each mode can be obtained. In online FDI, the current system mode must be identified from measured system outputs in order to use the correct FSM. Chapter 7 presents an ARR-based approach to system mode identification.

A structural FSM does not capture parameter sensitivities of ARRs in a quantitative manner. Some parameter sensitivities of an ARR may be negligible (in some modes) compared to others. A structural FSM then may be simplified by setting corresponding entries to zero. For ARRs in closed symbolic form, parameter sensitivities of ARR residuals can be obtained by symbolic differentiation. In case this is not possible, parameter sensitivities of ARR residuals can be numerically computed by using sensitivity bond graphs or incremental bond graphs. The latter are considered in the next chapter.

As to the numerical computation of ARR residuals, two possible bond graph based approaches have been presented. One approach suited for online as well as for offline FDI is to use a diagnostic bond graph in which storage elements are in preferred derivative causality in order to be independent of initial conditions that are difficult to be obtained in online FDI. Moreover, sensors are in inverted causality.

For an offline FDI, the real system may be replaced by a behavioural model that allows to introduce deliberately all kinds of fault and to study their effect with no risk. The equations derived from a model of the faulty system are integrated with respect to time by means of an appropriate solver starting from initial conditions, while the derivatives of inputs into the diagnostic bond graph model are computed in discrete time. Evaluation of the faulty system model and computation of the ARR residuals can be performed by means of Scilab script files.

Alternatively, for offline FDI, a faulty system bond graph model can be coupled to a model of the non-faulty system by means of residual sinks. In this approach, storage elements in both models are in preferred integral causality as in offline simulation initial values are available and are the same for both models. The equations derived from the coupled bond graph models are of Hessenberg index 2 form. Replacing the residual sinks by a bond graph model with artificial elements turns the DAE system into a set of ODEs. The advantage of both methods is that the computation of ARR residuals does not require ARRs in closed symbolic form. If a faulty system model is coupled to a non-faulty system model by means of residual sinks, then ARR residuals are components of the descriptor vector and are computed simultaneously with state variables of both models.

Finally, the coupling of two bond graph models in preferred integral causality by means of residual sinks can be complemented by some analytical formulation in case the behaviour of the real system can be sufficiently accurately modelled by means of a switched LTI system and if the bond graph is in preferred integral causality for all system modes.

References

1. Samantaray, A. K., Medjaher, K., Ould Bouamama, B., Staroswiecki, M., & Dauphin-Tanguy, G. (2006). Diagnostic bond graphs for online fault detection and isolation. *Simulation Modelling Practice and Theory*, *14*(3), 237–262.
2. Merzouki, R., Samantaray, A., & Pathak, P. (2013). *Intelligent mechatronic systems*. New York: Springer.
3. Samantaray, A. K., & Ould Bouamama, B. (2008). *Model-based Process Supervision - A Bond Graph Approach. Advances in Industrial Control*. London: Springer.
4. Mukherjee, A., Karmakar, R., & Samantaray, A. K. (2006). *Bond graph in modeling, simulation and fault identification*. New Delhi, India: I.K. International Publishing House.
5. Borutzky, W. (Ed.). (2011). *Bond graph modelling of engineering systems—Theory, applications and software support*. New York: Springer-Verlag.
6. Medjaher, K., Samantaray, A.K., Ould Bouamama, B. (2005) Diagnostic bond graphs for direct residual evaluation. In J. J. Granda, & F. E. Cellier (Eds.), *Proceedings of the International Conference on Bond Graph Modeling, ICBGM'05* (vol. 37(1), pp. 307–312). New Orleans, Louisiana, USA: SCS Publishing. Simulation Series. ISBN: 1-56555-287-3.
7. Ould Bouamama, B., Samantaray, A. K., Medjaher, K., Staroswiecki, M., & Dauphin-Tanguy, G. (2005). Model builder using functional and bond graph tools for FDI design. *Control Engineering Practice*, *13*(7), 875–891.
8. HighTec consultants. SYMBOLS Sonata™. http://www.htcinfo.com/.
9. Wang, D., Yu, M., Low, C., & Arogeti, S. (2013). *Model-based health monitoring of hybrid systems*. Heidelberg: Springer.
10. Ould Bouamama, B., Medjaher, K., Bayart, M., Samantaray, A., & Conrard, B. (2005). Fault detection and isolation of smart actuators using bond graphs and external models. *Control Engineering Practice*, *13*(2), 159–175.
11. Tagina, M., Cassar, J. P., Dauphin-Tanguy, G., Staroswiecki, M. (1995). Monitoring of systems modelled by bond-graphs. In: F. E. Cellier & J. J. Granda (Eds.), *International Conference On Bond Graph Modeling And Simulation, ICBGM'95*. (vol. 27(1), of Simulation Series, pp. 275–280). Las Vegas, Nevada, USA: SCS Publishing.
12. Gawthrop, P. J. (2000). Sensitivity bond graphs. *Journal of the Franklin Institute*, *337*, 907–922.
13. Samantaray, A. K., & Ghoshal, S. K. (2007). Sensitivity bond graph approach to multiple fault isolation through parameter estimation. *Proceedings of the Institution of Mechanical Engineers Part I: Journal of Systems and Control Engineering*, *221*(4), 577–587.
14. Borutzky, W. (2009). Bond graph model-based fault detection using residual sinks. *Proceedings of the Institution of Mechanical Engineers Part I Journal of Systems and Control Engineering*, *223*(3), 337–352.
15. Levy, R., Arogeti, S., Wang, D. (2013). Improving mode-change and fault isolation of hybrid systems using instantaneous sensitivity matrices. In *Proceedings of the 1st International and 16th National Conference on Machines and Mechanisms, (iNaCoMM2013)*, (pp. 116–123).
16. Scilab Enterprises, Scilab. 78000 Versailles, France. http://www.scilab.org/.
17. Borutzky, W. (2010). *Bond graph methodology—Development and analysis of multidisciplinary dynamic system models*. UK: Springer.
18. Borutzky, W. (2012). Bond-graph-based fault detection and isolation for hybrid system models. *Proceedings of IMechE Part I: J Systems and Control Engineering*, *226*(6), 742–760.
19. Li, S., & Petzold, L. (1999). Design of new DASPK for sensitivity analysis, UCSB. http://www.engineering.ucsb.edu/cse/.
20. DASPK. http://www.cs.ucsb.edu/cse/software.html.
21. Radau5. http://www.unige.ch/hairer/software.html.

Chapter 5
Parameter Uncertainties

This chapter uses an *incremental* bond graph approach in order to determine parameter sensitivities of ARR residuals as well as adaptive mode-dependent ARR thresholds for systems described by a hybrid model. To that end, first, the underlying idea and some basics of incremental bond graphs are briefly recalled.

5.1 Introduction

In order to avoid that false alarms are reported to a supervisor system or that true faults are not detected, ARR residuals should be significantly sensitive to true faults but little sensitive to parameter variations given uncertain system parameter values. Parameter sensitivities of ARR residuals can be singled out by defining appropriate thresholds. As the dynamic behaviour of a real system described by a hybrid model can be quite different in different system modes, thresholds should be adapted to system modes.

If ARRs can be obtained in closed symbolic form, parameter sensitivities can be determined by symbolic differentiation with respect to parameters. If this is not possible, parameter sensitivities of ARRs can be computed numerically by using either a *sensitivity* bond graph [1–4] or an *incremental* bond graph [5, 6]. Incremental bond graphs were initially introduced for the purpose of frequency domain sensitivity analysis of LTI models. Furthermore, they have also proven useful for the determination of parameter sensitivities of state variables and output variables, transfer functions of the direct model as well as of the inverse model, and for the determination of ARR residuals from continuous time models [7, Chap. 4]. In this chapter, the incremental bond graph approach is applied to systems described by switched LTI systems.

As to FDI robust with regard to parameter uncertainties, an approach based on so-called uncertain bond graphs in linear fractional transformation form (LFT) has been reported in the literature [8–10] for time-continuous models. In an uncertain bond graph, bonds carry power variables uncertain with regard to parameter variations

W. Borutzky, *Bond Graph Model-based Fault Diagnosis of Hybrid Systems*,
DOI 10.1007/978-3-319-11860-4_5

and bond graph elements are decomposed into a part with nominal parameters and another one with uncertain parameters.

In contrast, in an incremental bond graph, bonds carry variations of power variables, and variations of ARR residuals are outputs of special interest. In the case of switched LTI systems with uncertain parameters, variations of ARR residuals are a weighted sum of variations of power variables. The weighting factors depend on power variables of the original bond graph. The weighted sum suggests to apply the triangle inequality to obtain adaptive bounds for the part of an ARR residual that is uncertain due to parameter variations. The nominal part of an ARR residual is obtained from the original bond graph with nominal parameters.

An incremental bond graph can be constructed in a systematic manner from the original bond graph of a switched LTI system by replacing an element that is due to parameter variations by its incremental element model. Equations for variations of power variables can be automatically derived in the same way as they are derived from an initial bond graph with nominal parameters.

5.2 Incremental Bond Graphs for Switched LTI Systems

Switched LTI systems are just LTI systems for the time intervals betweens between discrete mode changes. Therefore, first, the incremental bond graph approach is recalled for LTI systems. In a second step, an incremental model for switches is proposed.

5.2.1 Incremental Models of Bond Graph Elements

In order to simplify the presentation, it is assumed that energy sources, energy storage elements and resistors are linear 1-port elements and that transformers and gyrators have got two ports.

The underlying idea of incremental bond graphs is that if a parameter Θ of a component model varies, then both power variables at its port are perturbed due to its interaction with the ports of other elements in the model. That is, an effort $e_n(t)$ in a bond graph with nominal parameters becomes $e(t) = e_n(t) + \Delta e(t)$. The same holds for the conjugate power variable $f(t)$. In incremental bond graphs, bonds carry the increments $\Delta e(t)$, $\Delta f(t)$ of power variables. In other words, they represent energy flows carrying the amount of power $\Delta e(t) \cdot \Delta f(t)$.

An incremental bond graph of an LTI system is constructed from an initial bond graph with nominal parameters by replacing those bond graph elements by their incremental model for which a parameter variation has taken place. Clearly, sources that do not depend on a parameter become null sources. Furthermore, the incremental model of a 1(0)-junction remains a 1(0)-junction. As to the incremental models of storage elements, resistors, transformers and gyrators, it turns out that they differ from the respective element just by modulated sinks added to junctions. The sinks are modulated by a power variable of the initial bond graph. That is, the structure of the initial bond graph is retained by the incremental bond graph. Structurally, both bond graphs only differ with respect to sources and sinks and their location. As a result, existing software can be used to set up a state space model for the variations of the state variables.

In order to see how an incremental bond graph model for a bond graph element is obtained, a linear 1-port C-element with the nominal capacitance C_n is considered. In the following, an index n indicates a parameter or a variable of the non-faulty bond graph model with nominal parameters. In the case of a time constant parameter variation ΔC the linear constitutive equation of a 1-port C element in derivative causality takes the form

$$f_C(t) = f_{C_n}(t) + (\Delta f_C)(t) = (C_n + \Delta C)(\dot{e}_{C_n} + (\Delta \dot{e}_C))(t) \tag{5.1}$$

Expanding the right hand side yields

$$\begin{aligned}(\Delta f_C)(t) &= C_n(\Delta \dot{e}_C)(t) + (\Delta C)\dot{e}_C(t) \\ &= C_n(\Delta \dot{e}_C)(t) + \underbrace{\frac{\Delta C}{C}}_{=:\,\delta_C} f_C(t)\end{aligned} \tag{5.2}$$

Figure 5.1 shows a bond graph representation of (5.2). In essence, the incremental model of a C element is again a C element. A modulated sink MSe : $\delta_C f_C$ contributes to its output Δf_C.

The power variable f_C controlling the modulated source is an output variable of the original bond graph model. If f_C has been obtained by measurements of the real system, then the contribution to the output of the incremental bond graph model of the C element may contain sensor noise. In any case, the outputs of the incremental bond graph of a bond graph element indicate a parameter variation.

Recall that in FDI, energy storage elements are in preferred derivative causality because initial conditions are not known or difficult to obtain in real time FDI. In offline simulation, the storage elements in the non-faulty model may be in preferred integral causality. Reformulation of (5.2) gives

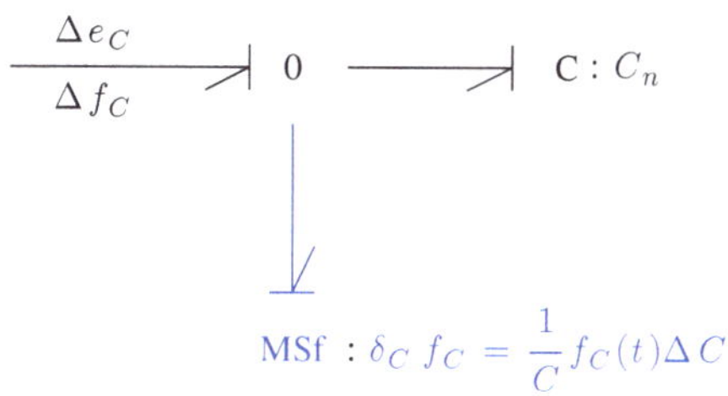

Fig. 5.1 Incremental bond graph model of a linear 1-port C element in preferred derivative causality

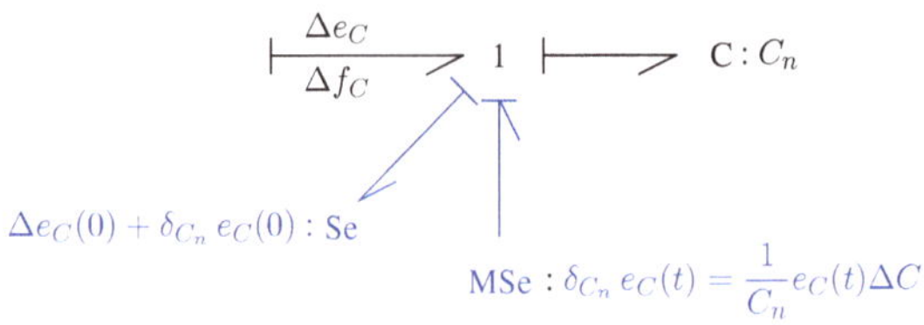

Fig. 5.2 Incremental bond graph model of a linear 1-port C element in preferred integral causality

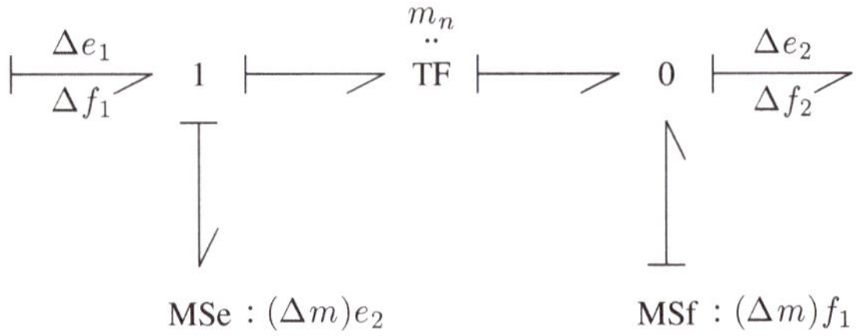

Fig. 5.3 Incremental bond graph model of a 2-port transformer

$$\Delta e_C(t) = \frac{1}{C_n} \int_0^t (\Delta f)(\tau) d\tau - \delta_{C_n}(e_C(t) - e_c(0)) + \Delta e_C(0) \tag{5.3}$$

where $\delta_{C_n} := \Delta C / C_n$. Equation 5.3 may be represented by the bond graph model in Fig. 5.2.

In the same manner, incremental models for the other bond graph elements may be obtained. As an example, Fig. 5.3 depicts the incremental bond graph model of a transformer TF : m_n.

5.2.2 *Incremental Models of Nonlinear Bond Graph Elements*

The incremental bond graph approach is not limited to linear 1-port elements with one parameter [5]. The constitutive equation of the incremental model of a bond graph element can be easily obtained by taking the total differential of the constitutive equation of the bond graph element.

For instance, let

$$q(t) = \Phi(e(t), \theta) \tag{5.4}$$

be the constitutive equation of a nonlinear 1-port C element, $\theta = \Theta_n + \Delta\Theta$ its parameter and $\Delta\Theta$ a time constant deviation from the nominal value Θ_n. Then

$$f(t) = \dot{q}(t) = \frac{\partial \Phi(e(t), \theta)}{\partial e} \cdot \dot{e}(t) = f(e(t), \dot{e}(t), \theta) \tag{5.5}$$

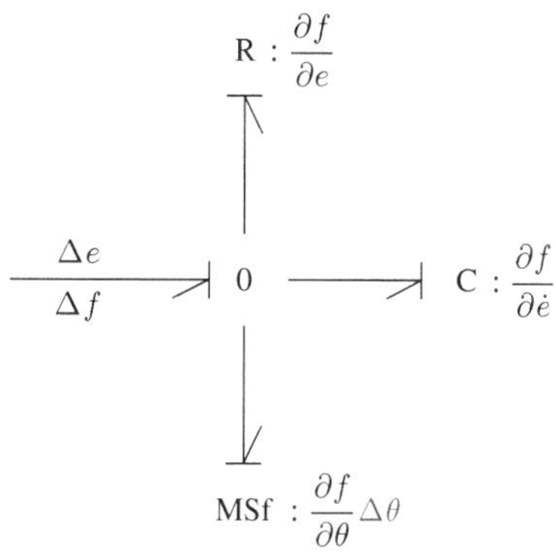

Fig. 5.4 Incremental bond graph model of a nonlinear 1-port C element in preferred derivative causality

and

$$\Delta f = \frac{\partial f}{\partial e}\Delta e + \frac{\partial f}{\partial \dot{e}}\Delta \dot{e} + \frac{\partial f}{\partial \theta}\Delta\theta \tag{5.6}$$

Equation 5.6 may be represented by an incremental bond graph model with nonlinear elements as depicted in Fig. 5.4.
In Fig. 5.4, the partial derivatives of f may be complex expressions depending on the nonlinear function Φ.

5.2.3 An Incremental Model of the Non-ideal Switch

For the sake of a unified bond graph representation of hybrid models that hold for all system modes, in this book, non-ideal switches are used and are represented by means of an MFT : $m_{\text{sw}}(t)$ with $m_{\text{sw}}(t) \in \{0, 1\}\ \forall\, t \geq 0$ in conjunction with a resistor R : R_{on} in fixed conductance causality (Fig. 2.12b).

In ON-mode, the switch model simply reduces to a resistor with the small nominal ON-resistance R^n_{on}. The constitutive equation of a linear 1-port resistor in conductance causality then leads to an equation for the increments of the power variables.

$$f_{\text{sw}}(t) \;=\; f^n_{\text{sw}}(t) + (\Delta f_{\text{sw}})(t) = \frac{1}{R^n_{\text{on}} + \Delta R_{\text{on}}}\,(e^n_{\text{sw}}(t) + \Delta e_{\text{sw}}(t)) \tag{5.7}$$

Reformulation yields

$$(\Delta f_{\text{sw}})(t) = \frac{1}{R^n_{\text{on}}}(\Delta e_{\text{sw}})(t) - \underbrace{\frac{\Delta R_{\text{on}}}{\text{R}^n_{\text{on}}}}_{=:\,\delta_{R_{\text{on}}}} f_{\text{sw}}(t) \tag{5.8}$$

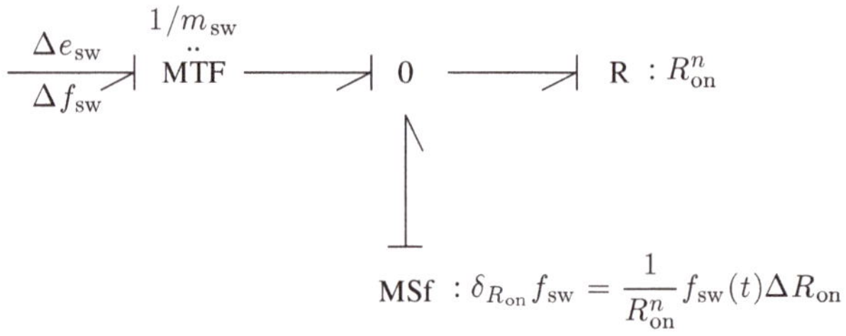

Fig. 5.5 Incremental bond graph model of a non-ideal switch

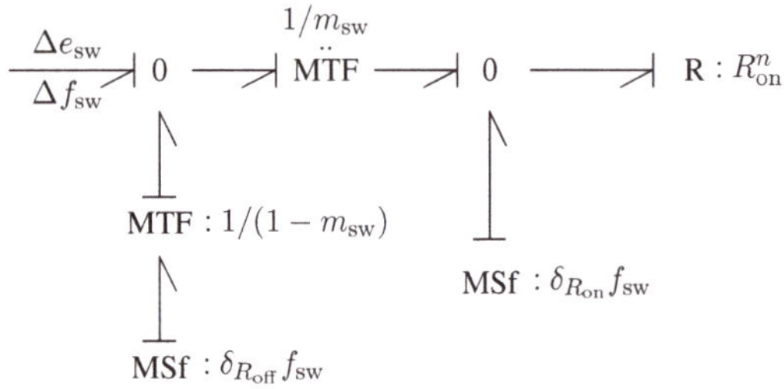

Fig. 5.6 Extended incremental bond graph model of a non-ideal switch accounting for a non-zero flow in OFF-mode

In OFF-mode, $f_{\text{sw}} = 0$. Hence, $\Delta f_{\text{sw}} = 0$. Figure 5.5 depicts an incremental bond graph model of a non-ideal switch that captures both switch states.

If the flow close to zero in OFF-mode is not neglected, then the switch in OFF-mode can be considered a resistor with a very high OFF resistance R_{off}. For this resistor, an incremental model can be developed in the same manner as for the resistor R : R_{on}. In the resulting model, the resistor R : R_{off} can be neglected as the nominal value of $1/R_{\text{off}}$ is very small. Figure 5.6 depicts an extended incremental bond graph model of a non-ideal switch that accounts for a non-zero flow in OFF-mode.

The incremental bond graph model of a linear 1-port resistor R : R_n in resistance causality is obtained by solving (5.8) for Δe.

$$\Delta e_R(t) = R_n \Delta f_R(t) + R_n \delta_{R_n} f_R(t) \tag{5.9}$$

in accordance with Fig. 5.7.

Fig. 5.7 Incremental bond graph model of a linear 1-port resistor in resistance causality

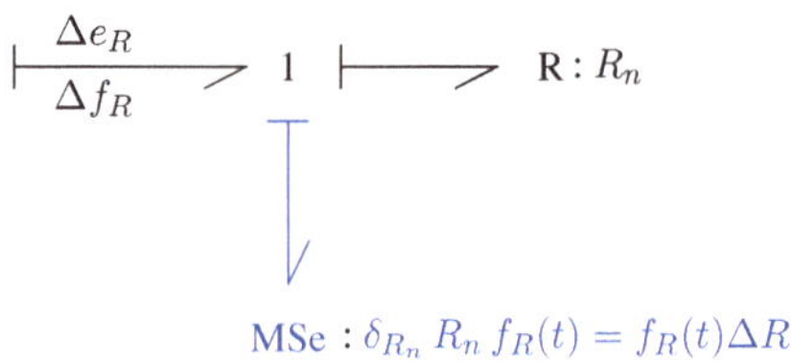

5.3 Incremental Bond Graphs for FDI

Given incremental models for all bond graph elements, an incremental bond graph (incBG) can be systematically constructed from a bond graph of the non-faulty system with nominal parameters by replacing sources of constant value by sources of value zero and by replacing elements with varying parameters by their incremental bond graph model. As the incremental model of a bond graph element differs from the element only by modulated sinks representing parameter variations, the incremental bond graph retains the structure of the original bond graph. The incremental DBG differs from the initial DBG only with respect to the sources and the additional sinks modulated by a power variable of the original BG. The number of these additional sinks equals the number of varying parameters. If the original model is given by a switched LTI system so is the incremental bond graph model. Hence, the incremental model and the original one have the same system matrix **A** with nominal parameters. Existing software can be used to derive equations from the incremental DBG and to set up the matrices of a switched LTI system in symbolic form in the same manner as for the original DBG.

Let $\boldsymbol{u}$ denote the vector of inputs into the original bond graph model with nominal parameters, $\boldsymbol{x}$ its state vector composed of all inputs into storage elements in preferred derivative causality and let an index n indicate a dependency from nominal parameter values. The state space model derived from the original bond graph then reads

$$\dot{\boldsymbol{x}}(t) = \mathbf{A}_n \boldsymbol{x}(t) + \mathbf{B}_n \boldsymbol{u}(t), \quad \boldsymbol{x}(0) = \boldsymbol{x}_0 \tag{5.10a}$$

$$\boldsymbol{y}(t) = \mathbf{C}_n \boldsymbol{x}(t) \tag{5.10b}$$

Furthermore, let $\boldsymbol{\Theta}$ denote the vector of all component parameters. The state space model derived from the incremental bond graph then is of the form

$$\Delta\dot{\boldsymbol{x}}(t) = \mathbf{A}_n \Delta\boldsymbol{x}(t) + \mathbf{B}^*(\boldsymbol{\Theta}_n)\boldsymbol{w}(t) \tag{5.11a}$$

$$\Delta\boldsymbol{y}(t) = \mathbf{C}_n \Delta\boldsymbol{x}(t) \tag{5.11b}$$

The matrices $\mathbf{A}_n$ and $\mathbf{C}_n$ are obtained deriving equations from the original bond graph with nominal parameters. Matrix $\mathbf{B}^*$ can be automatically set up from the incremental bond graph. The vector $\boldsymbol{w}$ denotes the outputs of the modulated sinks in the incremental bond graph that represent parameter variations. These outputs are of the form

$$\boldsymbol{w}(t) = \text{diag}(\delta_i z_i(t))\Delta\boldsymbol{\Theta} \tag{5.12}$$

where diag() is a diagonal matrix and z_i a power variable from the original bond graph controlling the ith modulated sink that represents a parameter variation. The coefficients δ_i depend on the type of the bond graph element that has been replaced by its incremental model. For a capacitor for instance, $\delta_i = 1/C^i$ and $z_i = f_C^i$

(cf. Fig. 5.1). For a linear resistor in conductance causality with nominal resistance R_n, $\delta_i = 1/R_n$ and $z_i = f_R^i$ (cf. Fig. 5.5).

5.3.1 Parameter Sensitivities of ARR Residuals

Inputs to an incremental bond graph are the outputs of the sinks representing a parameter variation. They are modulated by a power variable from the original bond graph. Outputs of the incremental bond graph model with respect to FDI are variations Δr_i of ARR residuals r_i. In the case of a switched LTI system, they can be expressed as a weighted sum of the inputs and their time derivatives into the incremental bond graph. As the weighting factors may include transformer moduli $m_i(t) \in \{0, 1\}$, variations of ARR residuals are system mode dependent. According to (5.11a, 5.11b) and (5.12), there is a matrix $\mathbf{C}_n$ such that the Laplace transform of the variation of the ith residual r_i reads

$$\mathscr{L}\Delta r_i = \sum_{j=1}^{m} F_{ij}^* \underbrace{\delta_j\,(\mathscr{L}z_j)\,\Delta\Theta_j}_{\mathscr{L}w_j} \tag{5.13}$$

where

$$(F_{ij}^*) = \mathbf{C}_n(s\mathbf{I} - \mathbf{A}_n)^{-1}\mathbf{B}^* \tag{5.14}$$

Hence, the parameter sensitivity of ARR residual r_i with respect to the jth parameter Θ_j is

$$\mathscr{L}\frac{\partial r_i}{\partial \Theta_j} = F_{ij}^* \delta_j \mathscr{L}z_j \tag{5.15}$$

The power variable z_j is an output variable of the original bond graph model with nominal parameters and as such it is a weighted sum of the inputs $u_k(t)$ into the original bond graph in the case of a switched LTI system.

$$\mathscr{L}z_j = \sum_{k=1}^{n} F_{jk}\,\mathscr{L}u_k \tag{5.16}$$

where

$$(F_{jk}) = \mathbf{C}(s\mathbf{I} - \mathbf{A})^{-1}\mathbf{B} \tag{5.17}$$

As a result, parameter sensitivities of ARR residuals r_i with respect to parameter Θ_j can be obtained by constructing a matrix $\mathbf{F}$ from the matrices of the original bond graph and a matrix $\mathbf{F}^*$ from the matrices of the incremental bond graph and by multiplying the ith row of matrix $\mathbf{F}$ by the factor $F^*_{ij}\delta_j$.

$$\mathscr{L}\frac{\partial r_i}{\partial \Theta_j} = F^*_{ij}\delta_j \sum_{k=1}^{n} F_{jk}\,\mathscr{L}u_k \tag{5.18}$$

These operations can be hardly manually performed, even for models of small size. However, a bond graph preprocessor such as CAMPG [11] can automatically derive the equations from the original as well as from the incremental bond graph. MATLAB®[12] or Scilab [13] script files can then generate the matrices $\mathbf{F}$ and $\mathbf{F}^*$ in symbolic form and can perform the multiplication of a row of $\mathbf{F}$ by the factor $F^*_{ij}\delta_j$ for each requested parameter sensitivity of an ARR residual.

For small switched LTI systems, variations of ARR residuals can be manually derived from an incremental bond graph by applying the principle of superposition. That is, only one bond graph element at a time is assumed to have an uncertain parameter. It is replaced by its incremental model. Detectors are replaced by a dual virtual detector for the variation of an ARR residual. Summing variations of flows or efforts, respectively, at these junctions and eliminating unknowns yields variations of residuals of ARRs as a weighted sum of the inputs supplied by those modulated sinks that represent parameter variations. The weighting factors in these sums are the sensitivities to be determined.

Example: Network with a Semiconductor Switch

As an example, the circuit with one switch in Fig. 4.1, is considered. To keep the illustration of the procedure short and simple it is assumed that only one parameter is uncertain. Accordingly, the incremental bond graph is obtained by replacing the element by its incremental model and by replacing the constant voltage source Se : V_i by an effort source of value zero and by replacing detectors by dual virtual detectors for the variations of ARR residuals.

Parameter R_1 Is Uncertain

Figure 5.8 displays the corresponding incremental bond graph. Again, the purpose of the auxiliary storage element C : C_a is just to resolve the causal conflict at junction 0_2. In the process of equation formulation, the capacitance C_a is set to zero.
Summing variations of power variables at junctions 1_1, 0_2, and i_3 yields

$$1_1: \quad \Delta\tilde{r}_1 = 0 - f\,\Delta R_1 - \Delta u \tag{5.19a}$$

$$0_2: \quad 0 = C^n_1\,\Delta\dot{u}_{C_1} + \underbrace{C_a}_{\approx 0}\,\Delta\dot{u} + \Delta i_{sw} \tag{5.19b}$$

$$1_3: \quad \Delta i_{sw} = \frac{b}{R^n_{on}}(\Delta u_{C_1} - R^n_2\,\Delta i_{sw} - 0) \tag{5.19c}$$

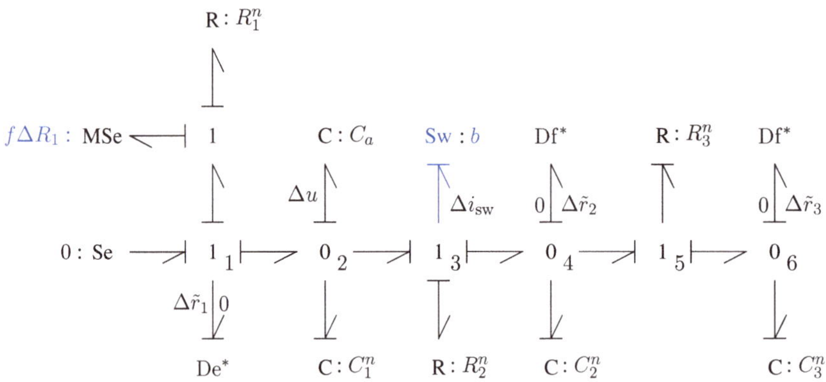

Fig. 5.8 Incremental bond graph of the switched circuit in Fig. 4.1 in the case of an uncertain parameter R_1

Combining (5.19a)–(5.19c) gives

$$\Delta r_1 := -C_1^n \Delta\dot{\tilde{r}}_1 - \frac{b}{R_{on}^n + R_2^n}\Delta\tilde{r}_1 = C_1^n \dot{f} \Delta R_1 + \frac{b}{R_{on}^n + R_2^n} f \Delta R_1 \quad (5.20)$$

Hence,

$$\frac{\partial r_1}{\partial R_1} = C_1^n \dot{f} + \frac{b}{R_{on}^n + R_2^n} f \quad (5.21)$$

in accordance with (4.6).

Sensitivity $\partial r_2 / \partial R_1$ is likewise obtained. Summation of flow variations at junction 0_4 yields

$$\Delta\tilde{r}_2 = \Delta i_{sw} \quad (5.22)$$

Reformulation gives

$$\Delta r_2 := \Delta\tilde{r}_2 + \frac{b}{R_{on}^n + R_2^n}\Delta\tilde{r}_1 = -\frac{b}{R_{on}^n + R_2^n} f \Delta R_1 \quad (5.23)$$

Thus,

$$\frac{\partial r_2}{\partial R_1} = -\frac{b}{R_{on}^n + R_2^n} f \quad (5.24)$$

in accordance with (4.7).

Parameter R_3 Is Uncertain

Figure 5.9 displays the corresponding incremental bond graph. Summation of flow variations at junction 0_6 gives

$$\Delta r_3 = -\frac{1}{R_3^n}\, i_3\, \Delta R_3 \;=\; -\frac{1}{R_3^n}\, C_3^n\, \dot{e}_2\, \Delta R_3 \tag{5.25}$$

and

$$\Delta r_3 := R_3^n\, \Delta\tilde{r}_3 \;=\; -\, C_3^n\, \dot{e}_2\, \Delta R_3 \tag{5.26}$$

Thus

$$\frac{\partial r_3}{\partial R_3} = -\, C_3^n\, \dot{e}_2 \tag{5.27}$$

in accordance with (4.8).
Likewise,

$$\Delta r_2 := \Delta\tilde{r_2} \;=\; \Delta i_{\mathrm{sw}} + \frac{1}{R_3^n}\, i_3\, \Delta R_3 \;=\; \Delta i_{\mathrm{sw}} + \frac{e_1 - e_2}{(R_3^n)^2}\, \Delta R_3 \tag{5.28}$$

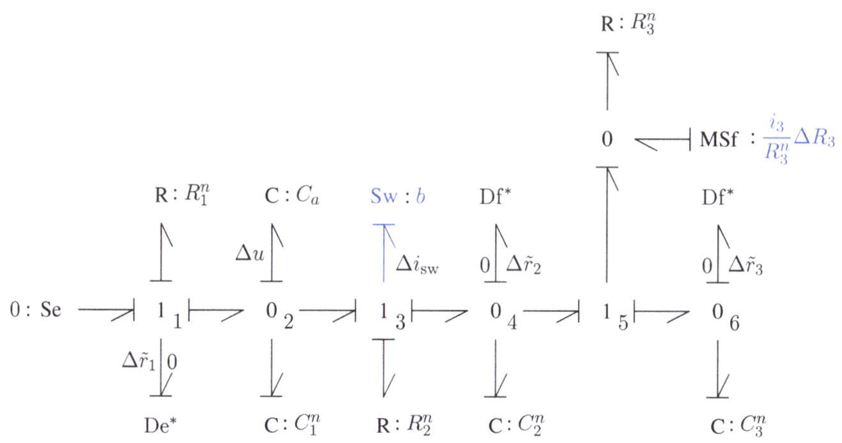

Fig. 5.9 Incremental bond graph of the switched circuit in Fig. 4.1 in the case of an uncertain parameter R_3

and

$$\frac{\partial r_2}{\partial R_3} = \frac{1}{(R_3^n)^2}(e_1 - e_2) \tag{5.29}$$

in accordance with (4.7).

5.3.2 *Adaptive Mode-Dependent Thresholds for Parameter Variations of ARR Residuals*

According to (5.13), the Laplace transform of the variation of an ARR residual is a weighted sum of the inputs into the incremental bond graph. The weighting factors are transfer functions

$$F_{ij}^*(s) = \frac{N_{ij}^*(s)}{D^*(s)} \tag{5.30}$$

where $N_{ij}^*(s)$ and $D^*(s)$ are polynomials and $s \in \mathbb{C}$.

Let $(n_{ij}^*)_\kappa$ be the κth coefficient of the polynomial $N_{ij}^*(s)$ and $\delta_{\Theta_j} := \delta_j \Delta\Theta_j$. Then

$$\begin{aligned} \mathscr{L}\Delta r_i := D^* \mathscr{L}\Delta\tilde{r}_i &= \sum_j \delta_{\Theta_j} N_{ij}^*(s) \mathscr{L} z_j \\ &= \sum_j \delta_{\Theta_j} \sum_{\kappa=0}^{k_j} (n_{ij}^*)_\kappa \, s^\kappa \, \mathscr{L} z_j \end{aligned} \tag{5.31}$$

or in the time domain

$$\Delta r_i(t) = \sum_j \delta_{\Theta_j} \sum_{\kappa=0}^{k_j} (n_{ij}^*)_\kappa \, z_j^{(\kappa)} \tag{5.32}$$

The sum in (5.32) suggests in a natural way to apply the triangle inequality in order to define an adaptive upper bound for the absolut values of parameter variations of ARR residuals.

$$|\Delta r_i(t)| \leq \sum_j \sum_k |\delta_{\Theta_j} (n_{ij}^*)_k \, z_j^{(k)}| =: thr_i(t) \tag{5.33}$$

No fault is reported as along as variations of ARR residuals due to uncertain parameters are within the adaptive bounds $\pm thr_i(t)$.

As a result, ARR residuals as fault indicators may be obtained by evaluating ARRs derived from a diagnostic bond graph with nominal parameters. In order to assess the effect of uncertain parameters on ARR residuals, parameter variations of ARR residuals may be derived from an incremental bond graph. Application of the triangle inequality then gives adaptive bounds for these variations.

Example: Simple Network with a Semiconductor Switch

This way of determining adaptive thresholds for parameter variations of ARR residuals is illustrated by means of the network displayed in Fig. 5.10 which is a simplification of the switched circuit in Fig. 4.1.

Figure 5.11 shows a diagnostic bond graph of the simple switched RC-circuit. For this circuit, ARRs (4.6)–(4.7) simplify to

$$0_2: \quad r_1 = 0 = f - C_1(\dot{V}_i - R_1\dot{f}) - \frac{b}{R_{\text{on}} + R_2}(V_i - R_1 f - e_1) \tag{5.34a}$$

$$0_4: \quad r_2 = 0 = \frac{b}{R_{\text{on}} + R_2}(V_i - R_1 f - e_1) - C_2\dot{e}_1 \tag{5.34b}$$

Figure 5.12 displays the corresponding incremental bond graph.

It is assumed that the small ON-resistance of the switch and resistance R_2 are not uncertain. Therefore, the switch model Sw : b and resistor R : R_2 haven't been

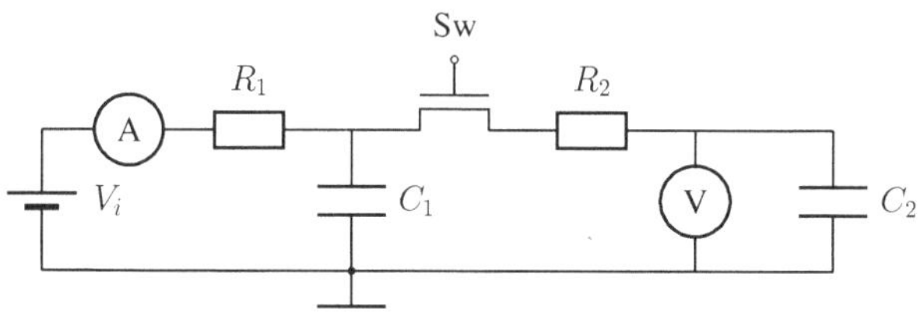

Fig. 5.10 Simple RC-circuit with a semiconductor switch

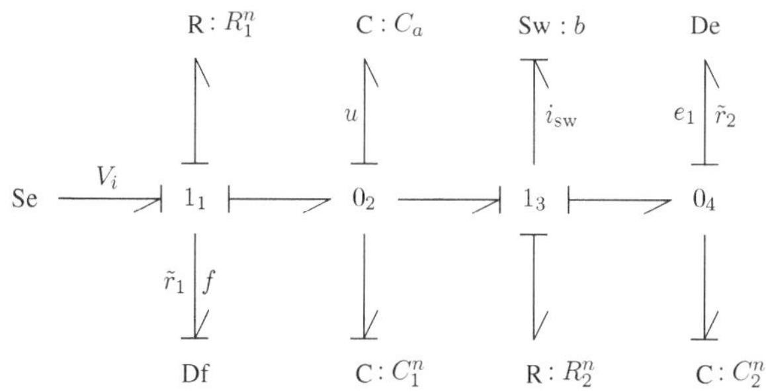

Fig. 5.11 Diagnostic bond graph of the simple switched RC-circuit

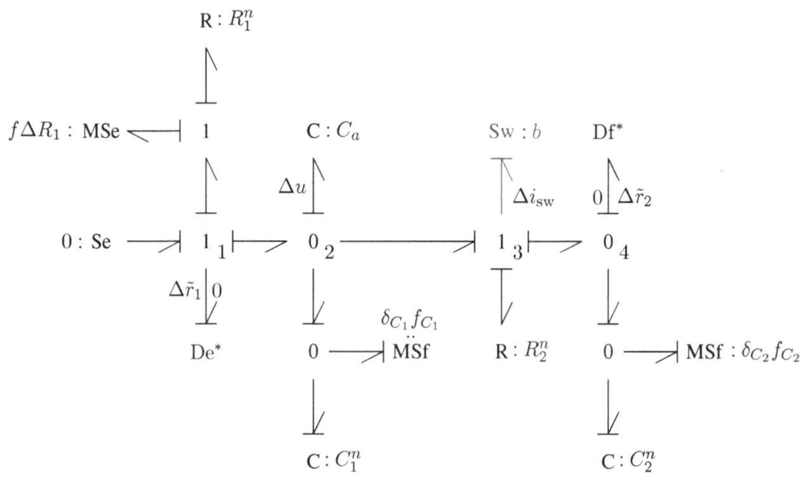

Fig. 5.12 Incremental bond graph of the simple circuit in Fig. 5.10

replaced by their incremental model. Summation of variations of power variables at the junctions yields

$$1_1: \quad \Delta\tilde{r}_1 = 0 - f\Delta R_1 - \Delta u \tag{5.35a}$$

$$0_2: \quad 0 = C_1^n \Delta\dot{u} + \delta_{C_1} f_{C_1} + \underbrace{C_a}_{\approx 0} \Delta\dot{u} + \Delta i_{sw} \tag{5.35b}$$

$$1_3: \quad \Delta i_{sw} = \frac{b}{R_{on}^n}(\Delta u - 0 - R_2^n \, \Delta i_{sw}) \tag{5.35c}$$

$$0_4: \quad \Delta\tilde{r}_2 = \Delta i_{sw} - \delta_{C_2} f_{C_2} \tag{5.35d}$$

Combination of (5.35a)–(5.35d) gives for variations of the residuals Δr_1, Δr_2:

$$\begin{aligned} \Delta r_1 := -C_1^n \Delta\dot{\tilde{r}}_1 - \frac{b}{R_{on}^n + R_2^n}\Delta\tilde{r}_1 \; &= \; C_1^n \dot{f} \Delta R_1 - \delta_{C_1} f_{C_1} + \frac{b}{R_{on}^n + R_2^n} f \Delta R_1 \\ &= \delta_{R_1} R_1^n C_1^n \dot{f} - \delta_{C_1} \underbrace{C_1(\dot{V}_i - R_1 \dot{f})}_{f_{C_1}} + \frac{b}{R_{on}^n + R_2^n} \delta_{R_1} R_1^n f \end{aligned} \tag{5.36a}$$

$$\begin{aligned} \Delta r_2 := \Delta\tilde{r}_2 + \frac{b}{R_{on}^n + R_2^n}\Delta\tilde{r}_1 \; &= \; -\frac{b}{R_{on}^n + R_2^n} f \Delta R_1 - \delta_{C_2} f_{C_2} \\ &= -\frac{b}{R_{on}^n + R_2^n}\delta_{R_1} R_1 f - \delta_{C_2} C_2 \dot{e}_1 \end{aligned} \tag{5.36b}$$

Adaptive mode dependent thresholds thr_1, thr_2 for parameter variations of ARR residual thus can be chosen as

$$|\Delta r_1| \leq |\delta_{R_1} R_1^n C_1^n \dot{f}| + |(\Delta C_1)(\dot{V}_i - R_1 \dot{f})| \\ + |\frac{b}{R_{\mathrm{on}}^n + R_2^n} \delta_{R_1} R_1 f| =: thr_1(t) \tag{5.37a}$$

$$|\Delta r_2| \leq |\frac{b}{R_{\mathrm{on}}^n + R_2^n} \delta_{R_1} R_1^n f| + |(\Delta C_2)\dot{e}_1| =: thr_2(t) \tag{5.37b}$$

Remark 5.1 The same results for Δr_1 and Δr_2 are obtained by taking the total differential of ARRs (5.34a)–(5.34b). □

Fault Scenario: The Value of Capacitance C_1 Is Changed for Some Time Interval

The simulated fault scenario assumes that capacitance $C_1(t)$ is reduced to 20 % of its initial nominal value C_1^n for $2\,\mathrm{s} < t < 4\,\mathrm{s}$ and is restored to C_1^n for $t > 4\,\mathrm{s}$.

$$C_1(t) = C_1^n - dC_1 \cdot \texttt{pulse(t,2,4)} \tag{5.38}$$

where $dC_1 = 80/100 \cdot C_1^n$ and `pulse(t,2,4)` denotes a unit pulse lasting from 2 to 4 s.

According to (5.34a)–(5.34b), residual r_1 is sensitive to a change in capacitance C_1 independent of the two system modes determined by the switch state b, while residual r_2 is insensitive of C_1 in both modes. To show this mode independence, the closed switch is opened at $t = 3.5\,\mathrm{s}$. That is, a mode change happens while parameter C_1 has a significantly reduced value. For the determination of adaptive threshold bounds a constant relative parameter variation of 2 % has been adopted for parameters R_1, R_2, C_1 and C_2. Table 5.1 lists the parameter used by the simulation of the fault scenario.

First, Fig. 5.13a, b indicate the time evolutions of the capacitor voltages u_{C_1}, $\tilde{u}_{C_1}$, u_{C_2}, $\tilde{u}_{C_2}$ and of the currents $\tilde{i}_{R_1}$ and $\tilde{i}_{C_1}$ respectively as would be expected. The tilde and the prefixing letter ‘t’ in the figures denote variables of the perturbed

Table 5.1 Parameters used by the simulation of the fault scenario

Parameter	Value	Units
E	5	V
R_1^n	100	kΩ
C_1^n	10	μF
dC_1	8 (2s < t < 4s)	μF
C_2^n	10	μF
δ_{R_1}	2 %	–
δ_{R_2}	2 %	–
δ_{C_1}	2 %	–
δ_{C_2}	2 %	–

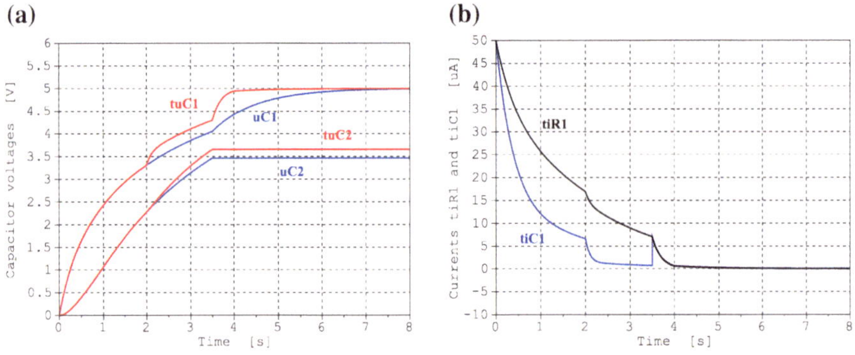

Fig. 5.13 Time evolution of the capacitor voltages and of currents $\tilde{i}_{R_1}$ and $\tilde{i}_{C_1}$. **a** Time history of capacitor voltages. **b** Time history of currents $\tilde{i}_{R_1}$ and $\tilde{i}_{C_1}$

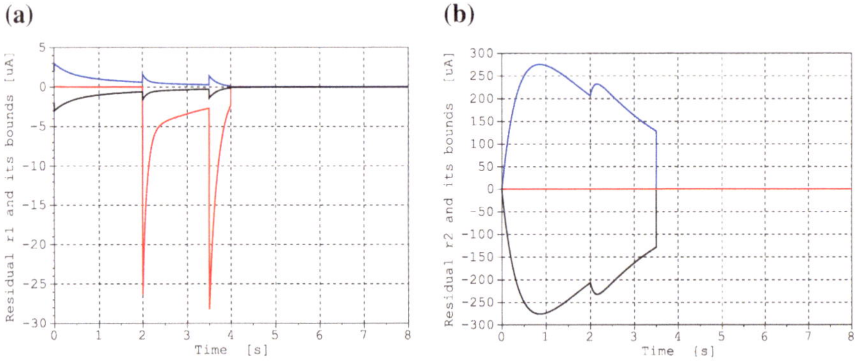

Fig. 5.14 Residuals r_1, r_2 and their adaptive thresholds. **a** Time evolution of residual r_1 and its adaptive thresholds $\pm\, thr_1$ **b** Time evolution of residual $r_1$and its thresholds $\pm\, thr_2$

system. Both figures clearly show the effect of the reduction of capacitance C_1 for the time interval [2 s, 4 s] and the effect of the opening the switch at $t = 3.5$ s.

Figure 5.14a, b display the time evolution of residual r_1, r_2 and their adaptive thresholds $\pm\, thr_1$, $\pm\, thr_2$.

Figure 5.14b confirms that residual r_2 is insensitive to a change in capacitance C_1 independent of the system mode. The time history of residual r_1 in Fig. 5.14a shows a spike at $t = 2$ s due to the abrupt decrease of capacitance C_1 and another spike at $t = 3.5$ s due to the opening of the switch. Moreover, for the time interval [2 s, 4 s] in which C_2 is significantly reduced, values of r_1 are clearly below the adaptive lower bound $-thr_1$ indicating a fault. Outside of this time interval, the values of r_1 are well inside the narrow adaptive bounds. That is, r_1 is sensitive in both system modes to the temporary parametric fault in C_1 but insensitive to a 2 % relative parameter variation of parameters R_1, R_2, C_1 and C_2.

Moreover, Fig. 5.14a shows that for $0 < t < 3.5$ s (mode 1: closed switch, b = 1) the bounds are much wider than for $t > 3.5$ s (mode 0: open switch, b = 0). This would be expected because in (5.37a)–(5.37b) $b = 0$ and f_{C_1} as well as $\dot{f}$ are very small for $t > 3.5$ s (cf. Fig. 5.13b for the time history of the currents $f = i_{R_1}$ and $f_{C1} = i_{C_1}$).

5.3.3 Measurement Uncertainties

Inputs into an incremental bond graph are relative parameter variations $\Delta\Theta_i/\Theta_i$ multiplied by a power variable from the diagnostic bond graph with nominal parameters. If inputs into the diagnostic bond graph obtained either by measurements from the real system or from a behavioural model replacing the real system carry measurement uncertainties then this affects power variables in the diagnostic bond graph that control modulated sinks of the incremental bond graph. As a result, measurement uncertainties have an impact on the variations of ARR residuals and thus on their thresholds.

Figure 5.15 illustrates this situation assuming that measurement uncertainties are additive. A flow $\tilde{f} = \tilde{f}' + \Delta\tilde{f}'$ with a predicted part $\tilde{f}'$ and an uncertain part $\Delta\tilde{f}'$ due to measurement uncertainty is the output of a non-ideal sensor and an input into the diagnostic bond graph. The input $\tilde{f}$ into the diagnostic bond graph results in an effort $e = e' + \Delta e'$ that controls a modulated sink MSe where e' denotes the predicted part and $\Delta e'$ the uncertain part. The output $w_\Theta := \delta(e' + \Delta e')\Delta\Theta$ of the modulated sink is an input into the incremental bond graph that is needed to compute the variation Δr of an ARR residual r.

If measurement uncertainties can be assumed to be bounded, then application of the triangle inequality may yield thresholds for parameter variations of ARR residuals that are independent of measurement uncertainties. For instance, let z' be the predicted part of an output variable z of the diagnostic bond graph that controls a modulated sink of the incremental bond and let $|\Delta z'| \leq b_z$ be the bounded measurement uncertainty. Furthermore, let Δr be the variation of an ARR residual r that depends on z and its derivative. Then

$$
\begin{aligned}
|\Delta r| &\leq |n_1\,(z' + \Delta z')| + |n_2\,\frac{d}{dt}(z' + \Delta z')| \\
&\leq \underbrace{|n_1\,z'| + |n_2\frac{d}{dt}z'|}_{\leq\, thr(t)} + |n_1\,b_z| + |n_2\frac{d}{dt}\Delta z'|
\end{aligned}
\tag{5.39}
$$

where n_1 and n_2 are constants.

Fig. 5.15 Accounting for measurement uncertainties

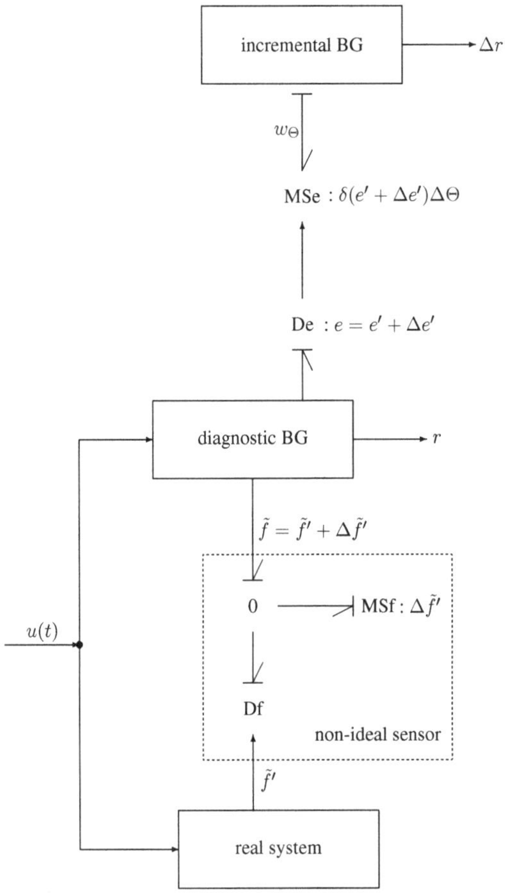

If the derivative $\frac{d}{dt}\Delta z'$ is approximated by the difference quotient then

$$|n_2 \frac{d}{dt} \Delta z'| \leq |n_2 \frac{2b_z}{\Delta t}| \tag{5.40}$$

where Δt denotes the sampling time of the measurement.

The result is a threshold $thr'(t) \geq thr(t)$ that is independent of measurement uncertainties but that depends on the sampling rate of the measurement. In [14], Touati et al. have accounted for measurement uncertainties by adding modulated sources to an uncertain bond graph.

The simple model of a non-ideal sensor in Fig. 5.15 accounts for measurement uncertainties by means of a modulated sink. However, a sensor may deliver wrong readings because it operates in a faulty mode due to external disturbances that are caused by changes in the ambient or by internal disturbances such as parametric faults. That is, the computation of ARR residuals and of adaptive thresholds for their parameter variations may give misleading results. If the build-up and the parameters of a sensor are known, parametric faults as well as a sensor's sensitivity with regard to known external disturbances can be taken into account by a more elaborate bond graph model replacing a simple model such as the one in Fig. 5.15. If, however, details of its internal build-up are not known, then the sensor's dynamic behaviour may be approximately captured for small deviations from its operating point by a transfer function with parameters that at least account for the sensor's time delay and its gain.

5.3.4 Uncertain Excitations

Constant excitations to a system are represented by an effort or a flow source that provides an output of constant value. In the incremental bond graph these sources are replaced by sources of value zero. If a constant excitation, however, is to be considered uncertain, its source may be replaced in the incremental bond graph by a source modulated by the nominal value. For instance, let Se : E_n represent a constant voltage or constant hydraulic pressure supply. If there is a relative uncertainty $\delta_E = \Delta E / E_n$, then the constant effort source may be replaced in the incremental bond graph by an effort source MSe : $\delta_E E_n$ modulated by the nominal effort E_n obtained from the bond graph with nominal parameters. If the internal structure and the parameters of the device are known that provides the excitation and if possible disturbances acting on the device can be modelled, then an incremental bond graph model can be constructed that accounts for the uncertainty of the excitation.

5.4 Summary

ARR residuals as fault indicators should be distinctly sensible to true faults and robust with regard to parameter uncertainties. That is, if parameters varies, the time evolution of ARR residuals should be within prescribed bounds. For real systems described by a hybrid model bounds should be adapted to system modes as the dynamic behaviour can be quite different in different system modes.

This chapter briefly recalls the basic idea of incremental bond graphs and extends their application to switched LTI systems. Incremental bond graphs can be systematically constructed from an original bond graph and retains its structure. They differ from an original bond graph of a hybrid model only by additional sinks. These sinks are introduced by the replacement of bond graph elements with varying parameters by their incremental element model. The additional sinks represent parameter variations and are modulated by a power variable from the original bond graph.

In contrast to an original bond graph, the bonds of an incremental bond graph carry variations of power variables. Outputs of interest with regard to FDI are variations of ARR residuals. For a switched LTI system, these variations are a weighted sum of the inputs and their derivatives into the incremental bond graph. The weighting factors are the parameter sensitivities of an ARR residual. Model equations and ARRs can be derived in the same manner from an incremental bond graph as from an original bond graph. That is, variations of ARR residuals due to parameter variations and parameter sensitivities of ARR residuals can be derived from an incremental bond graph, while the nominal part of an ARR is obtained from a diagnostic bond graph with nominal parameters. Once the time evolution of parameter sensitivities of ARRs is known, their assessment may give rise to simplify a structural FSM by setting some entries to zero. For switched LTI systems, the principle of superposition may be applied. That is, if only some parameter sensitivities are of interest, only their corresponding bond graph elements may be replaced by their incremental model and increments of ARR residuals be derived from the incremental bond graph.

Furthermore, the expression of variations of ARR residuals as a weighted sum of inputs and their derivatives suggests in a natural way to apply the triangle inequality to obtain adaptive mode-dependent thresholds for variations of ARRs due to parameter variations. That is, parameters may vary. As long as ARR residuals are within these bounds, no fault is reported.

References

1. Borutzky, W. (2009). Bond graph model-based fault detection using residual sinks. *Proc of the Institution of Mechanical Engineers Part I Journal of Systems and Control Engineering.*, *223*(3), 337–352.
2. Samantaray, A. K., & Ghoshal, S. K. (2007). Sensitivity bond graph approach to multiple fault isolation through parameter estimation. *Proceedings of the Institution of Mechanical Engineers Part I: Journal of Systems and Control Engineering*, *221*(4), 577–587.
3. Gawthrop, P. J. (2000). Sensitivity bond graphs. *Journal of the Franklin Institute*, *337*, 907–922.
4. Cabanellas, J. M., Félez J., & Vera C. (1995) A formulation of the sensitivity analysis for dynamic systems optimization based on pseudo bond graphs. In: Cellier F. E., & Granda J. J., (Eds.), *ICBGM'95, International Conference on Bond Graph Modeling and Simulation*. Vol. 27(1) of Simulation Series. Las Vegas, Nevada: SCS Publishing. (pp. 135–144).
5. Borutzky, W., & Granda, J. J. (2002). Bond graph based frequency domain sensitivity analysis of multidisciplinary systems. *Proc Instn Mech Engrs, Part I, Journal of Systems and Control Engineering*, *216*(1), 85–99.
6. Borutzky W., & Granda J. J. (2001) Determining sensitivities from an incremental true bond graph. In: J. J. Granda & G. Dauphin-Tanguy (Eds.), *2001 International Conference on Bond Graph Modeling, and Simulation (ICBGM 2001)*. Vol. 33(1) of Simulation Series. Phoenix, Arizona: SCS Publishing (pp. 3–8).
7. Borutzky, W. (Ed.). (2011). *Bond Graph modelling of engineering systems—theory, applications and software support*. New York: Springer.
8. Merzouki, R., Samantaray, A., Pathak, P., & Ould Bouamama, B. (2013). *Intelligent mechatronic systems*. Berlin: Springer.
9. Djeziri, M. A., Merzouki, R., Ould Bouamama, B., & Dauphin-Tanguy, G. (2007). Robust fault diagnosis by using bond graph approach. *IEEE/ASME Transactions on Mechatronics*, *12*(6), 599–611.
10. Djeziri, M. A. (2007) Diagnostic des sysèmes incretains par l'Approche bond graph [PhD thesis]. Ecolde Centrale de Lille, France. Available from: http://hal.archives-ouvertes.fr/docs/00/20/00/30/PDF/These-Djeziri-07-12-2007.pdf.

11. CadSim Engineering. CAMPG, Available from: http://www.bondgraph.com.
12. The Mathworks: MATLAB—the language of technical computing. Available from: http://www.mathworks.com/products/matlab/.
13. Scilab Enterprises: Scilab. 78000 Versailles, France. Available from: http://www.scilab.org/.
14. Touati, Y., Merzouki, R., & Ould Bouamama, B. (2011) Fault detection and isolation in presence of input and output uncertainties using bond graph approach. In: A. Bruzzone & G. Dauphin-Tanguy, S. Junco, M. A. Piera (Eds.) *Proceedings of the 5th International Conference on Integrated Modeling and Analysis in Applied Control and Automation (IMAACA 2011)*. DIPTEM University of Genoa (pp. 221–227).

Chapter 6
Isolation of Multiple Parametric Faults from a Hybrid Model

Let m denote the number of component parameters. If the number $m' < m$ of parametric fault candidates exceeds the number n of sensors then a set of structured ARRs in which each ARR is sensitive to only one parameter cannot be achieved. That is, the FSM is not diagonal. Some rows in the FSM will have the same component fault signature so that some faults cannot be isolated. This result cannot be improved with regard to a further isolation of faults if the real system does not permit to add more sensors. That is, faults cannot be structurally isolated.

Provided that faults can be mapped onto deviations of parameters from their nominal values, i.e. component faults can be represented by parameter values, and assuming that for a given input and known output vector a unique set of parameter values exists, then multiple faults may be isolated by parameter estimation techniques. If a model of a system is available, then the problem of determining system parameters is known as partial system identification [1]. If faults in a system cause a change in some parameter values without changing the system structure, then the evaluation of ARRs using nominal parameters will give residuals that are distinguishably beyond parameter uncertainty thresholds for some time intervals. This will not be the case if ARRs are evaluated by using parameter values obtained as the result of a parameter estimation algorithm. That is, the estimated parameter values can be considered the new nominal parameter values of the system after some faults have occurred. By comparing estimated parameter values with known nominal parameter values multiple faults may be isolated. Moreover, the magnitude of a fault can be estimated so that its severity can be assessed and appropriate fault accommodation can be triggered. Of course, only those parametric fault candidates that have an effect on the unstructured residuals will have to be estimated.

The next sections first recall standard least square optimisation. Subsequently, this technique is applied to ARR residuals assuming that they can be obtained in closed symbolic form so that their expressions can be used in an objective function to be minimised. If unknowns cannot be eliminated from the sum of power variables at junctions to which a detector has been attached, then the time evolution of ARR residuals can be obtained by either computing ARRs as outputs of a

W. Borutzky, *Bond Graph Model-based Fault Diagnosis of Hybrid Systems*,
DOI 10.1007/978-3-319-11860-4_6

diagnostic bond graph or by computing a behavioural model of the system coupled to a model of the non-faulty system my means of residual sinks as has been shown in Chap. 4.

The use of a behavioural BG model of the faulty system and a DBG as a residual generator in conjunction with fault isolation by means of parameter estimation is illustrated in Sect. 6.6.

6.1 Parameter Estimation by Least Squares Output Error Minimisation

Let $\tilde{\boldsymbol{y}}(t, \tilde{\boldsymbol{\Theta}})$ be the vector of either n measured outputs $\tilde{y}_1(t, \tilde{\boldsymbol{\Theta}}), \ldots, \tilde{y}_n(t, \tilde{\boldsymbol{\Theta}})$ of the real system provided by n sensors or n outputs of a behavioural model of the real system and $\tilde{\boldsymbol{\Theta}}$ a vector of $m > n$ system parameters $\tilde{\Theta}_1, \ldots, \tilde{\Theta}_m$ of which some components have changed due to some faults. It is assumed that fault effects do not cancel each other. Furthermore, let $\boldsymbol{\Theta}$ be an estimate of the unknown changed parameter vector $\tilde{\boldsymbol{\Theta}}$. Then the 'measured' output vector $\tilde{\boldsymbol{y}}(t, \tilde{\boldsymbol{\Theta}})$ differs from the computed output $\boldsymbol{y}(t, \boldsymbol{\Theta})$ of a model of the system. An iterative parameter estimation over some time window of length T, however, can identify the set of changed parameter values if abrupt faults are assumed to remain constant over the considered time window. Slowly progressive faults reflected by slowly varying parameters can also be identified if the time window is sufficiently small in comparison to the rate of fault progression. Recursive least squares optimisation then aims at estimating parameter values that give a minimum error $\boldsymbol{e}(t, \boldsymbol{\Theta}) := \tilde{\boldsymbol{y}}(t, \tilde{\boldsymbol{\Theta}}) - \boldsymbol{y}(t, \boldsymbol{\Theta})$ by minimising an objective or cost function $f : \mathbb{R}^m \rightarrow \mathbb{R}$ over some time interval of length T.

$$f(\boldsymbol{\Theta}) = \frac{1}{2} \sum_{j=1}^{q+1} \boldsymbol{e}^T (t_{k-q-1+j}, \boldsymbol{\Theta}) \mathbf{W} \boldsymbol{e}(t_{k-q-1+j}, \boldsymbol{\Theta}) \tag{6.1}$$

where $\boldsymbol{\Theta} \in \mathbb{R}^m$, $\boldsymbol{e}(t_{k-q-1+j}) \in \mathbb{R}^n$ and $\boldsymbol{W} \in \mathbb{R}^{n \times n}$ a positive semi-definite weighting matrix. The index j denotes the jth sample, k the present sample and $q \geq 0$ the number of past samples.

In general, the error $\boldsymbol{e}(t_{k-q-1+j}, \boldsymbol{\Theta})$ is a non-linear function of the parameter vector $\boldsymbol{\Theta}$. Therefore, the above problem is a well-known nonlinear least squares problem (NLSP) that may be solved by various optimisation algorithms such as the Levenberg-Marquardt algorithm [2], the quasi-Newton method or the Gauss-Newton (GN) algorithm [3].

6.2 Gauss-Newton Method

Let $\boldsymbol{e}_j := \boldsymbol{e}(t_{k-q-1+j}, \boldsymbol{\Theta})\ j = 1 \dots q+1$ and $\boldsymbol{e} := [\boldsymbol{e}_1\ \boldsymbol{e}_2 \dots \boldsymbol{e}_{q+1}]^T$. If $\mathbf{W}$ is assumed to be an identity matrix then the objective function $f : \mathbb{R}^m \to \mathbb{R}$ in (6.1) may be written in the short form

$$f(\boldsymbol{\Theta}) = \frac{1}{2}\boldsymbol{e}^T\boldsymbol{e} = \frac{1}{2}||\boldsymbol{e}||_2^2 \tag{6.2}$$

A *local* minimum of f requires

$$\frac{\partial f(\boldsymbol{\Theta})}{\partial \boldsymbol{\Theta}} = \nabla f(\boldsymbol{\Theta}) = 0 \tag{6.3}$$

In (6.3), ∇f denotes the gradient of f.

$$\nabla f(\boldsymbol{\Theta}) = \left[\frac{\partial \boldsymbol{e}(\boldsymbol{\Theta})}{\partial \boldsymbol{\Theta}}\right]^T \boldsymbol{e}(\boldsymbol{\Theta}) \tag{6.4}$$

where

$$\mathbf{J}_e(\boldsymbol{\Theta}) = \frac{\partial \boldsymbol{e}(\boldsymbol{\Theta})}{\partial \boldsymbol{\Theta}} = \left[\frac{\partial \boldsymbol{e}_1(\boldsymbol{\Theta})}{\partial \boldsymbol{\Theta}} \dots \frac{\partial \boldsymbol{e}_j(\boldsymbol{\Theta})}{\partial \boldsymbol{\Theta}} \dots \frac{\partial \boldsymbol{e}_{q+1}(\boldsymbol{\Theta})}{\partial \boldsymbol{\Theta}}\right]^T \tag{6.5}$$

is the Jacobian of $\boldsymbol{e}(\boldsymbol{\Theta})$ and $\partial \boldsymbol{e}_j(\boldsymbol{\Theta})/\partial \boldsymbol{\Theta}$ is the Jacobian of $\boldsymbol{e}_j$

$$\begin{bmatrix} \frac{\partial e_{j1}(\boldsymbol{\Theta})}{\partial \Theta_1} & \cdots & \frac{\partial e_{j1}(\boldsymbol{\Theta})}{\partial \Theta_m} \\ \vdots & \ddots & \vdots \\ \frac{\partial e_{jn}(\boldsymbol{\Theta})}{\partial \Theta_1} & \cdots & \frac{\partial e_{jn}(\boldsymbol{\Theta})}{\partial \Theta_m} \end{bmatrix} \tag{6.6}$$

Newton iteration then gives

$$\boldsymbol{\Theta}^{(\kappa+1)} = \boldsymbol{\Theta}^{(\kappa)} - \mathbf{H}^{-1}(\boldsymbol{\Theta}^{(\kappa)})\nabla f(\boldsymbol{\Theta}^{(\kappa)}) \tag{6.7}$$

where κ denotes the iteration index and $\mathbf{H} := \nabla^2 f(\boldsymbol{\Theta})$ the Hessian matrix. The latter can be split into a matrix containing first order derivatives and another matrix $\mathbf{Q}(\boldsymbol{\Theta})$ containing second order derivatives.

$$\mathbf{H} = \mathbf{J}_e^T(\boldsymbol{\Theta})\mathbf{J}_e(\boldsymbol{\Theta}) + \mathbf{Q}(\boldsymbol{\Theta}) \tag{6.8}$$

where

$$\mathbf{Q}(\boldsymbol{\Theta}) := \sum_{j=1}^{n(q+1)} e_j(\boldsymbol{\Theta})\nabla^2 e_j(\boldsymbol{\Theta}) \tag{6.9}$$

The Gauss-Newton method ignores the matrix with second order derivatives. As a result,

$$\mathbf{H} \approx \mathbf{J}_e^T(\boldsymbol{\Theta})\mathbf{J}_e(\boldsymbol{\Theta}) = \left[\frac{\partial \boldsymbol{e}(\boldsymbol{\Theta})}{\partial \boldsymbol{\Theta}}\right]^T \frac{\partial \boldsymbol{e}(\boldsymbol{\Theta})}{\partial \boldsymbol{\Theta}} \tag{6.10}$$

The iteration terminates when either the step $\Delta\boldsymbol{\Theta} := \boldsymbol{\Theta}^{(\kappa+1)} - \boldsymbol{\Theta}^{(\kappa)}$ or the value of the cost function $f(\boldsymbol{\Theta})$ falls below a predefined threshold.

Remark 6.1 If the Jacobian $\mathbf{J}_e(\boldsymbol{\Theta})$ has full rank, then the approximation $\mathbf{J}_e^T(\boldsymbol{\Theta})\mathbf{J}_e(\boldsymbol{\Theta})$ of the Hessian $\mathbf{H}$ is positive definite and the Gauss-Newton search direction $\Delta\boldsymbol{\Theta}$ is a downhill direction.

For numerical reasons the inverse of the Hessian $\mathbf{H}$ is not actually computed, instead $\Delta\boldsymbol{\Theta}$ is computed as the solution of

$$\mathbf{H}\,\Delta\boldsymbol{\Theta} = -\nabla f(\boldsymbol{\Theta}^{(\kappa)}) \tag{6.11}$$

The use of the full Hessian in a Newton iteration requires the computation of $(n \times m)^2$ second order derivatives. Instead, for the Gauss-Newton approximation of the Hessian only $n \times m$ first order derivatives are to be computed in each iteration step.

Convergence of the Gauss-Newton method is not guaranteed (cf. Theorem C.1 in Appendix C). If the algorithm convergences, it may converge to a local minimum and give wrong results. □

6.3 The Weighted Nonlinear Least Square Problem

Measurements $\tilde{\boldsymbol{y}}(t, \tilde{\boldsymbol{\Theta}})$ obtained by n sensors from the real system carry some uncertainty. Therefore, the residuals $\boldsymbol{e}(t, \boldsymbol{\Theta}) := \tilde{\boldsymbol{y}}(t, \tilde{\boldsymbol{\Theta}}) - \boldsymbol{y}(t, \boldsymbol{\Theta})$ between measured and computed vectors, the cost function and the estimated parameters $\boldsymbol{\Theta}^*$ are uncertain. Each residual $\boldsymbol{e}_j(\boldsymbol{\Theta}) = \boldsymbol{e}(t_{k-q-1+j}, \boldsymbol{\Theta})$ has an error $\boldsymbol{\varepsilon}_j$. It is assumed that all these errors $\boldsymbol{\varepsilon}_j$, $j = 1, \ldots, q+1$, are independent and that each of its n components is normally distributed with mean zero and a known variance $(\sigma_i^j)^2$, $i = 1, \ldots, n$. If the known variances are quite different, a weighted least squares problem may be considered.

$$\min_{\boldsymbol{\Theta}} \frac{1}{2} \sum_{j=1}^{q+1} \boldsymbol{e}_j(\boldsymbol{\Theta})^T \mathbf{W}_j \boldsymbol{e}_j(\boldsymbol{\Theta}) \tag{6.12}$$

The weighting matrix $\mathbf{W}_j$ may be chosen as a diagonal matrix with $w_{ii}^j = 1/(\sigma_i^j)^2$, $i = 1, \ldots, n$. That is, less uncertain residuals will have a stronger influence on the cost function and the solution of the parameter estimation problem will be more accurate.

For a weighted least squares problem, the equation

$$\mathbf{J}(\boldsymbol{\Theta})^T \mathbf{W} \mathbf{J}(\boldsymbol{\Theta}) \Delta \boldsymbol{\Theta} = -\mathbf{J}(\boldsymbol{\Theta})^T \mathbf{W} \boldsymbol{e}(\boldsymbol{\Theta}) \tag{6.13}$$

then is to be solved where $\mathbf{W}$ is a diagonal matrix with diagonal matrices $\mathbf{W}_{jj} = (1/\sigma_j^2)\mathbf{I}_{n \times n}$.

6.4 Multiple Fault Isolation by Least Squares ARR Residuals Minimisation

The parameter estimation presented in the previous section is based on a least squares minimisation of the errors between 'measured' system outputs and outputs of a system model evaluated by using estimated parameter values. If the real system is replaced by a model in preferred integral causality, 'measured outputs' can be obtained by solving the model equations for given initial conditions and can be used for offline parameter estimation in order to isolate multiple faults deliberately introduced into the system model. In real-time FDI, initial conditions are either not known or difficult to obtain. Therefore, in online parameter estimation, they have to be considered as additional unknowns that are to be estimated.

Alternatively, multiple faults may be isolated by least squares minimisation of ARR residuals. The latter are indicators for the errors between measurements from a faulty system and outputs of a model computed by using estimated parameters. ARRs obtained from a DBG do not depend on initial conditions but use derivatives of measurements with respect to time which entails the drawback that differentiation carried out in discrete time amplifies noise if not properly filtered.

ARRs may be obtained as outputs of a diagnostic bond graph. If unknowns in ARR candidates can be eliminated, they are functions in closed form of known inputs $\boldsymbol{u}(t)$, known measurements $\boldsymbol{y}(t)$, known system parameters $\boldsymbol{\Theta}$ and the system mode denoted by all discrete switch states $\boldsymbol{m}(t)$. Let r_i denote the ith residual, then

$$r_i(t) = g_i(\boldsymbol{u}(t), \boldsymbol{y}(t), \boldsymbol{\Theta}, \boldsymbol{m}(t)) \tag{6.14}$$

If the vector of the estimated parameters $\boldsymbol{\Theta}$ equals the vector of the actual parameters in the faulty system then residuals are within their parameter uncertainty thresholds or vanish.

For least squares minimisation of ARR residuals only the m' parameters contributing to the unstructured part of a FSM need to be considered. Residuals in that part structurally depend on more than one parametric fault and the unstructured part of a FSM can be further subdivided into subspaces (rows of the FSM) with common component fault signatures. Which parameters in which subspaces can be identified as possible fault candidates depends on which of the residuals in the unstructured part of a FSM are abnormal at a sampling point and result in an entry equal to one in the coherence vector. As abnormal residuals in the unstructured part of a structural FSM may have more than one entry in a fault subspace, all parameters in a subspace identified as fault candidate need to be estimated in order to identify those parameters that are actually faulty [4].

Let $m' \leq m$ and let $\boldsymbol{\Theta}$ now denote the subset of parameters $\{\Theta_1, \dots \Theta_{m'}\} \subseteq \{\Theta_1, \dots \Theta_m\}$ that need to be estimated. Then the cost function f is of the same form as the one in (6.1) except that the output error $\boldsymbol{e}$ is to be replaced by the vector $\boldsymbol{r}$ of $n' \leq n$ ARR residuals.

Clearly, if the parameters $\boldsymbol{\Theta}_0$ of the healthy system are used in the ARRs of the faulty system, then their evaluation does not provide residuals close to zero.

$$\exists\, j \quad 0 \neq g_j(\boldsymbol{u}(t), \boldsymbol{y}(t), \boldsymbol{\Theta}_0, \boldsymbol{m}(t)) \tag{6.15}$$

However, if the parameters are varied so that the value of the cost function

$$f(\boldsymbol{\Theta}) = \frac{1}{2} \sum_{j=k-q}^{k} \boldsymbol{r}^T(t_j, \boldsymbol{\Theta}) \boldsymbol{r}(t_j, \boldsymbol{\Theta}) \tag{6.16}$$

becomes minimal, then all residuals $\boldsymbol{r}(t_j, \boldsymbol{\Theta}) = [r_1(t_j, \boldsymbol{\Theta}) \dots r_n(t_j, \boldsymbol{\Theta})]^T$ are close to $\mathbf{0}$ for $j = k - q, \dots, k$. The parameters $\boldsymbol{\Theta}$ that make the residuals $\boldsymbol{r}(t_j, \boldsymbol{\Theta})$ close to $\mathbf{0}$ are then the ones of the faulty system. For $m' > n'$ an observation window of $q > 0$ additional consecutive samples of measurements is needed. In order to over constrain the estimation solution a larger time window may be chosen as long the parameters can be assumed to be constant during the chosen window.

Let the weighting matrix in (6.1) be the identity matrix, i.e. $\mathbf{W} = \mathbf{I}$. The derivative of the cost function f with respect to Θ_μ may then be written in the form

$$\frac{\partial f(\boldsymbol{\Theta})}{\partial \Theta_\mu} = \sum_{j=k-q}^{k} \sum_{\nu=1}^{n} \frac{\partial r_\nu(t_j)}{\partial \Theta_\mu} r_\nu(t_j) \tag{6.17}$$

The second order derivative of $\partial f / \partial \Theta_\mu$ with respect to Θ_σ reads

$$\frac{\partial^2 f(\boldsymbol{\Theta})}{\partial \Theta_\sigma \partial \Theta_\mu} = \sum_{j=k-q}^{k} \sum_{\nu=1}^{n} \left(\frac{\partial^2 r_\nu}{\partial \Theta_\sigma \partial \Theta_\mu} r_\nu + \frac{\partial r_\nu}{\partial \Theta_\mu} \frac{\partial r_\nu}{\partial \Theta_\sigma} \right) \tag{6.18}$$

The term with second order derivatives is ignored by the Gauss-Newton method. Let $\boldsymbol{\Theta}^*$ denote the set of parameters that makes the value of the objective function a minimum. If any $r_\nu(\boldsymbol{\Theta}^*)$ $(1 \leq \nu \leq n)$ is not small then the approximation of the Hessian matrix $\mathbf{H}$ (cf. (6.10)) is poor and a line search may be needed for the method to be convergent.

As (6.17) shows, the Gauss-Newton method needs residual sensitivities with respect to parameters. If ARRs can be derived in closed symbolic form, they can be partially differentiated with respect to a parameter Θ_μ and evaluated at a time point $t_j > 0$

$$\frac{\partial r_i(t_j)}{\partial \Theta_\mu} = \frac{\partial}{\partial \Theta_\mu} g_i(\boldsymbol{u}(t), \boldsymbol{y}(t), \boldsymbol{\Theta}, \boldsymbol{m}(t))\Big|_{t=t_j} \tag{6.19}$$

The derivatives $\partial r_i / \partial \Theta_\mu$ are termed residual sensitivity functions.

Sensitivities of the outputs of a model with respect to a parameter can be derived from a sensitivity bond graph [5–7]. Sensitivities of ARR residuals with respect to a parameter can be obtained from incremental bond graphs (Chap. 5), from sensitivity pseudo bond graphs [8] and from diagnostic sensitivity bond graphs [9].

The least squares problem may be solved numerically by computing a quasi-Newton direction.

$$\Delta\boldsymbol{\Theta} = -\mathbf{H}_{k-1}^{-1} \nabla f(\boldsymbol{\Theta}^{(k-1)}) \tag{6.20}$$

The inverse of the Hessian matrix is approximated by a symmetric positive definite matrix that is constructed iteratively. To that end, the Scilab function `optim()` [10] uses the Broyden, Fletcher, Goldfarb and Shanno (BFGS) update [11]. Alternatively, the Levenberg-Marquardt algorithm implemented in the Scilab function `lsqrsolve()` may be used.

If the numerical computation of the gradient of an objective function shall be avoided, and if accuracy requirements are not too high, a direct method such as the Nelder-Mead simplex algorithm [12] implemented in the Scilab function `fminsearch()` may be used that allows for noise in the cost function.

6.5 A Fault Parameter Estimation Procedure Based on User Defined Scilab Functions

For illustration of parametric fault isolation by means of least squares ARR residual minimisation, the simple hybrid network in Fig. 4.1 shall be considered once again. The all-mode FSM (Table 4.1) indicates that resistors R_1 and R_2 have the same component signature when the switch is on so that a parametric fault in one of the two resistors cannot be isolated by inspection of the coherence vector.

As a fault scenario it is assumed that the value of resistance R_1 abruptly doubles at $t = 2$ s and remains at that value for $t > 2$ s (Fig. 6.1). As soon as this parametric fault is detected, the values of the two parameters R_1 and R_2 are estimated over a time window from 2 to 8 s. That is, the changed parameter value remains constant over the observation window.

If mode changes are not caused by externally controlled signals as in the case of the simple network but happen autonomously then their time instant is not known in

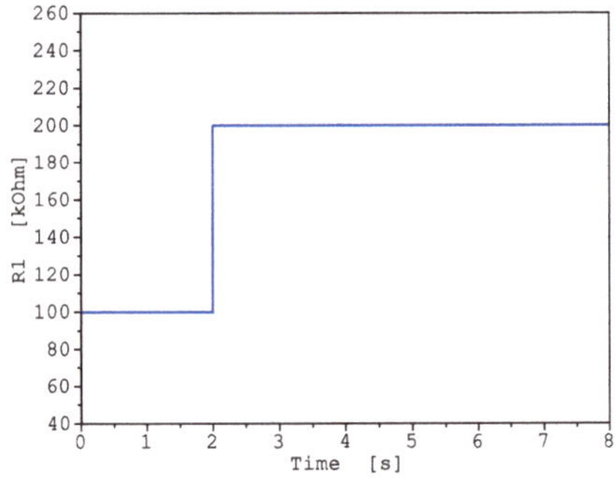

Fig. 6.1 Abrupt fault in resistance R_1

advance and the observation window used for the parameter estimation may extend over the mode change. To account for a mode change in the observation window, it is assumed that the switch Sw : $b(t)$ is on at $t = 0$ and is off for $t \geq 5$ s. ARR residuals obtained from hybrid system models are mode dependent. However, even if a mode change happens in the window used for parameter estimation, not all residuals will drop to zero due to a change of discrete switch state variables. As long as the parametric fault persists, the absolute value of at least one residual will be beyond a given threshold so that parameter estimation can continue to minimise the value of a cost function. Let t_f be the time point of an abrupt parameter fault that has been detected but cannot be isolated and let $t_s > t_f$ be the time instant of a mode change. Another strategy would be to collect sampled data for $t \geq t_f$ until a mode change is detected at $t = t_s$ and to use the window $t_f \leq t \leq t_s$ for fault parameter estimation. That is, during parameter estimation the system mode does not change. Accordingly, the structure of ARRs does not change. This means that not only parameter faults but also system mode changes are to be detected. System mode identification is addressed in Chap. 7.

First, 'measurements' of the voltages across the capacitors are generated by an off-line simulation run with the model of the faulty system and a sampling time dt $= 0.01$ s. This generation of measured data may be achieved by calling a user defined Scilab function `process()` (Fig. 6.2).
The user defined Scilab function `network()`

```
function dy = network(t,y,p) ... endfunction
```

Fig. 6.2 User defined Scilab function `process()` for generating measurements

```
function [ycalc] = process()
// Generates measurements
   y0 = [0 0 0];
   ptrue = [R1 R2];
   p = ptrue;
   ycalc = ode(y0',t0,tt,list(network,p));
endfunction
```

provides the right hand side of the ODE system of the network model. The target fault parameters $R1$, $R2$ are collected in a vector `p` that is passed as an argument of the function `network()`. Further model parameters beyond $R1$, $R2$ are made available as global variables.

Measurement data is then generated by the Scilab script lines

```
// generate measured data
t0 = 0; dt=0.01;tf=8; tt=t0:dt:tf;
ycalc = process()
```

Clearly, in real-time fault parameter estimation, the function `process()` is not needed. Sampled measurements may be stored in the array `ycal[]`.

ARR residuals r_1, r_2 (cf. (4.6) and (4.7) are expressed by means of the state variables $u_1 = u_{C_1}$, $e_1 = u_{C_2}$, $e_2 = u_{C_3}$ of the network. The reformulated ARRs then read:

$$r_1 = (V_i - u_1) \cdot \frac{1}{R_1} - C_1 \dot{u}_1 - \frac{b}{R_{\text{on}} + R_2}(u_1 - e_1) \tag{6.21}$$

$$r_2 = \frac{b}{R_{\text{on}} + R_2}(u_1 - e_1) - C_2 \dot{e}_1 - \frac{1}{R_3}(e_1 - e_2) \tag{6.22}$$

and have been coded in a user defined Scilab function `residuals()` (Fig. 6.3).

Derivatives of the state variables with respect to time are computed in discrete time by means of the Scilab function `diff()`. Variable `b` denotes the switch state. At $t = 0$, the switch is closed ($b = 1$). The switch is then turned off ($b = 0$) at $t = 5$ s during parameter estimation. That is, a mode change takes place in the observation window used for fault parameter estimation.

```
function [res] = residuals (p)
// Computes ARR residuals
   ⋮
   R1 = p(1); R2 = p(2);
   ⋮
   u11  = uC1(1:$-1);
   e1   = uC2(1:$-1);
   e2   = uC3(1:$-1);
   du1  = diff(uC1)/dt;
   de1  = diff(uC2)/dt;
   res1 = (Vi - u11) ./ R1 - C1 .* du1 - ...
   b(1:$-1)' .* (u11 - e1) ./ (Ron + R2)
   res2 = b(1:$-1)' .* (u11 - e1) ./ (Ron + R2) - ...
   C2 .* de1 - (e1 - e2) ./ R3
   res = [res1;res2]
endfunction
```

Fig. 6.3 User defined Scilab function `residuals()` computing ARR residuals

```
function val = LSquares (p)
// Computes the sum of the squares of the residuals
   res = residuals(p)
   val = 0.5*sum(res.^2)
endfunction
```

Fig. 6.4 User defined Scilab function `LSquares()` computes the least squares cost function

The Scilab function `LSquares()` (Fig. 6.4) then computes the sum of the squares of the residuals. It is used to define an objective function `Costf-1()` that evaluates the sum of squares of residuals and provides the gradient of `LSquares` with respect to the parameter vector `p`.

In function `Costf-1()` (Fig. 6.5), the gradient is numerically computed by means of the Scilab function `derivative()`. The parameter values and the values of the objective function obtained by the iterative estimation are stored for later plotting in the global vectors `Theta` and `fval` respectively.

Least squares ARR residual minimisation then may be started by a call of the Scilab function `optim()`.

```
p0 = [0.7*R1;0.2*R2]
[fopt,popt] = optim(Costf-1,p0,"ar",1000,1000,imp=2)
```

The starting vector `p0` is given to the cost function. As a result of the iterative optimisation, `optim()` returns the optimal parameter values in the vector `popt` and the final value of the cost function in the variable `fopt`. The arguments of function `optim()` following the keyword `"ar"` specify the maximum number of calls to the cost function `Costf()` and the maximum number of iterations allowed. The flag `imp=2` causes the output of information for every iteration step. If the algorithm to be used is not specified, `optim()` uses the Quasi-Newton algorithm in conjunction with the BFGS method [11].

```
function [f,g,ind] = Costf-1(p, ind)
// Evaluates the least squares cost function
// and provides its gradient with regard to p
   f = LSquares(p)
   g = derivative(LSquares, p)
   global Theta;
   Theta=[Theta;p'];
   global fval;
   fval = [fval;f]
endfunction
```

Fig. 6.5 User defined Scilab function `costf-1()` evaluates the least squares cost function and its gradient

The output

```
popt  =

     200154.69
     100176.1
  fopt  =

     7.057D-14
```

identifies resistance R_1 as the one that has changed. The initial values at $t = 0$ are $R_1 = R_2 = 100\,\text{k}\Omega$. Moreover, fault parameter estimation not only identifies faults among several potential fault candidates but also provides information about the magnitude of faults that can serve to obtain a refined form of the FSM used by a decision support system.

Figure 6.6 depicts the estimated parameter values at each iteration step, the path from the initial parameter guess $\boldsymbol{p}_0 = [0.7 * R_1; 0.2 * R_2]$ to the final estimate $\boldsymbol{p}_0 = [2 * R_1; R_2]$, and the decrease of the cost function values.

Figure 6.7 shows how the time evolution of the residuals converges towards zero during the iteration of the parameter estimation.

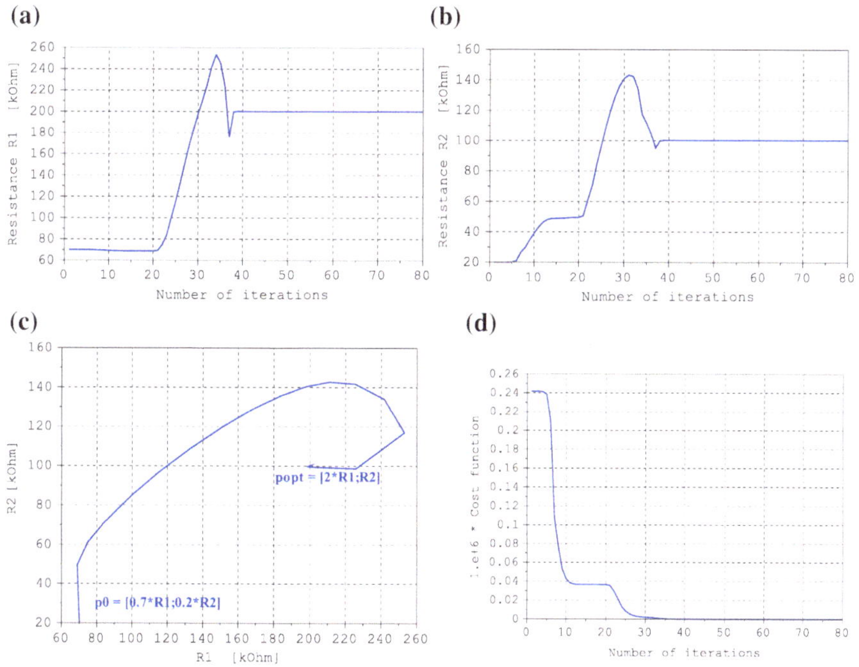

Fig. 6.6 Estimated parameters R_1 and R_2 and the decrease of the cost function values. **a** Resistance R_1. **b** Resistance R_2. **c** Path from the initial parameter guess to the final estimate. **d** Decrease of the cost function values

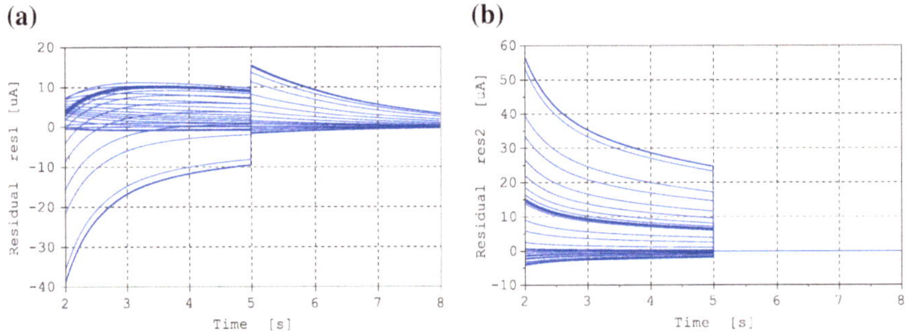

Fig. 6.7 Convergence of the residuals $r_1(t)$ and $r_2(t)$ towards zero during the parameter estimation. **a** Residual r_1. **b** Residual r_2

In the example, the gradient of the least squares cost function with respect to the target fault parameters has been computed numerically. For the simple network, ARRs in closed symbolic form can be derived from the diagnostic bond graph so that partial derivatives of the cost function with respect to parameters can be provided in symbolic form.

$$\frac{\partial r_1}{\partial R_1} = C_1 \dot{f} + \frac{b}{R_{\text{on}} + R_2} f \tag{6.23a}$$

$$\frac{\partial r_1}{\partial R_2} = -\frac{b}{(R_{\text{on}} + R_2)^2}(V_i - R_1 f - e_1) \tag{6.23b}$$

$$\frac{\partial r_2}{\partial R_1} = -\frac{b}{(R_{\text{on}} + R_2)} f \tag{6.23c}$$

$$\frac{\partial r_2}{\partial R_2} = -\frac{\partial r_1}{\partial R_2} \tag{6.23d}$$

Accordingly, the computation of the gradient of the least squares cost function with respect to targeted fault parameters R_1, R_2 can be implemented straightforward in a user defined Scilab function `gradient()` (Fig. 6.8).

Numerical computation of the gradient of the least squares cost function may then be replaced by the evaluation of analytical expressions for the partial derivatives (Fig. 6.9).

A call to the Scilab function `optim()` then gives the result

```
popt  =

    204514.26
    101265.58
 fopt  =

    4.637D-12
```

```
function z = gradient(p)
// Computes the gradient of LSquares with respect to p
   R1 = p(1); R2 = p(2);
   ⋮
   u11 = u1(1:$-1);
   e1  = u2(1:$-1);
   de1 = diff(u2)/dt;
   e2  = u3(1:$-1);
   f   = (Vi - u11) ./R10
   df  = - diff(u1)/dt ./R10

   dres1dR1 = C1 .* df + b(1:$-1)' ./ (Ron + R2) .* f
   dres1dR2 = b(1:$-1)' ./ (Ron + R2)^2 .* (Vi - R10 .* f - e1)
   dres2dR1 = - b(1:$-1)' ./ (Ron + R2) .* f
   dres2dR2 = - dres1dR2

   z1 = res1' * dres1dR1 + res2' * dres2dR1
   z2 = res1' * dres1dR2 + res2' * dres2dR2
   z  = [z2 z1]
endfunction
```

Fig. 6.8 User defined Scilab function `gradient()` providing the gradient of the least squares cost function in analytical form

Fig. 6.9 User defined Scilab function `Costf-2()` that evaluates the least squares cost function and its gradient in analytical form

```
function [f,g,ind] = Costf-2(p, ind)
// Evaluates the least squares cost function
// and provides its gradient with regard to p
   f = LSquares(p)
   g = gradient(p)
   global Theta;
   Theta=[Theta;p'];
   global fval;
   fval = [fval;f]
endfunction
```

In this example, the evaluation of analytical expressions for the partial derivatives does not give more accurate values for the estimated parameters. It has to be taken into account that differentiation of 'measured' variables is still performed in discrete time in both cases and that the step size is fairly coarse. However, the number of iterations is considerably lower which is imported for real-time parameter optimisation.

Figures 6.10 and 6.11 depict the convergence of the parameter values of resistors R_1 and R_2 towards their targeted values, the decrease of the cost function value, and the convergence of the ARR residuals r_1 and r_2 towards their minimal values.

Figures 6.7 and 6.11 indicate that the time evolutions of residual $r_2(t)$ are almost identical zero for $t > 5$ s, i.e. as soon as the switch Sw is turned off. This can be

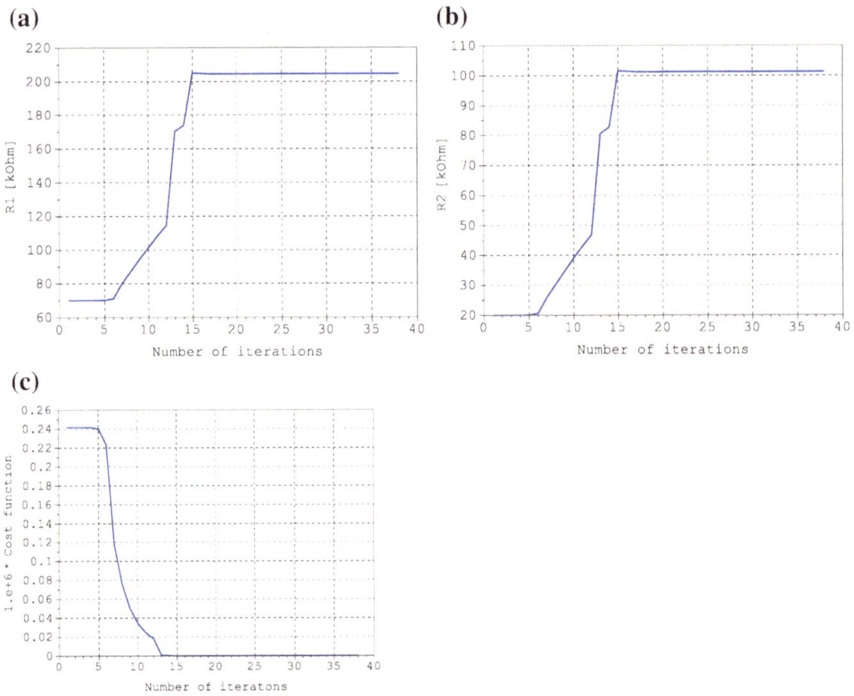

Fig. 6.10 Convergence of the estimated parameter R_1 and R_2 towards their targeted values in the case of given analytical gradients of the ARR residuals. **a** Resistance R_1. **b** Resistance R_2. **c** Decrease of the cost function values

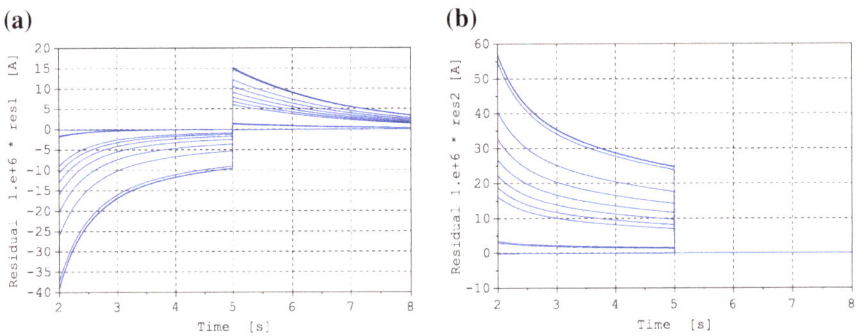

Fig. 6.11 Convergence of the ARR residuals r_1 and r_2 towards their minimal values. **a** Residual r_1. **b** Residual r_2

explained by considering the time evolution of the voltages across the capacitors (Fig. 6.12).

When the switch Sw : b is turned off ($b = 0$), capacitors C : C_2 and C : C_3 are no longer loaded. The difference $e_1 - e_2 := u_{C_2} - u_{C_3}$ is small and is divided

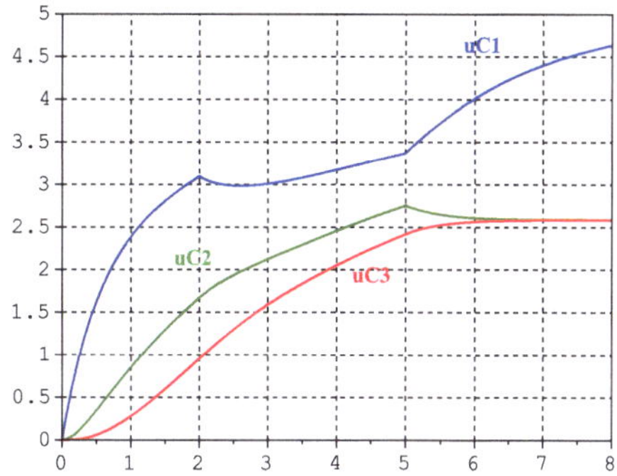

Fig. 6.12 Time evolution of the voltages across the capacitors

by resistance R_3. Moreover, for $t > 6\,\text{s}$, the two voltages are constant. Accordingly, $\dot{e}_1 = 0$. Hence,

$$r_2(t) = \frac{b}{R_{\text{on}} + R_2}(V_i - R_1 f(t) - e_1(t)) - C_2\dot{e}_1(t) - \frac{1}{R_3}(e_1(t) - e_2(t)) \approx 0 \quad \forall t > 6\,\text{s} \tag{6.24}$$

In this fault scenario, the value of resistance R_1 abruptly doubles at $t = 2\,\text{s}$ and remains at this elevated value for $t > 2\,\text{s}$, while the switch is turned off at $t = 5\,\text{s}$. After the parameter fault has been detected, parameter estimation also provides the correct values for $R_1 = 200\,\text{k}\Omega$ and $R_2 = 100\,\text{k}\Omega$ if the switch is turned off before $t = 5\,\text{s}$, e.g. at $t_s = 3\,\text{s}$.

6.6 Computing Implicitly Given ARR Residuals

As to the previously considered simple hybrid network (Fig. 4.1), ARRs in closed symbolic form could be derived from its diagnostic bond graph (Fig. 4.2). As a result, analytical expressions for the ARR residuals to be used in the formulation of the least squares cost function are available and symbolic partial derivatives of ARR residuals with respect to the targeted fault parameters can be exploited.

If the sum of power variables at junctions in the DBG with a detector attached to it, i.e. ARR candidates cannot be turned into ARRs by eliminating unknowns, ARR residuals are given implicitly. In an off-line simulation, they can be obtained by either feeding outputs of a behavioural bond graph model of the faulty system into a DBG (Sect. 4.8.1) that serves as a residual generator or by coupling the bond graph of the faulty system to a bond graph model of the non-faulty system by means of residual sinks (Sect. 4.8.2).

Fault parameter estimation based on Scilab functions as previously illustrated by means of a simple hybrid network may be easily adapted to this case. Instead of evaluating analytical expressions for the ARR residuals, the equations of the coupled models are solved in the function `residual()` in order to provide values of the ARR residuals used by the least squares cost function. Note that `residual()` is called by `LSquares()` which in turn is called by `Costf-1()`. The latter function is called in each iteration step of the parameter optimisation. That is, the two coupled models are evaluated in each iteration step. If the two models of the faulty and the non-faulty system are coupled by residual sinks, a DAE system has to be solved in each iteration step of the parameter optimisation and the nominal parameters of the non-faulty system model are varied such that the cost function converges to a minimum. If a DBG is used as an ARR residual generator, the behavioural model is to be evaluated and its sensor outputs $\tilde{\boldsymbol{y}}$ and model inputs $\boldsymbol{u}$ are fed into the DBG. Due to the preferred derivative causality of a DBG inputs from the behavioural model into storage elements are to be differentiated in discrete time. With these inputs, a set of nonlinear algebraic equations is then to be solved for the ARR residuals as outputs of the DBG model in each iteration step as indicated in Fig. 6.13.

The computation of the behavioural model and the DBG takes place in a modified function `residuals-2()` (Fig. 6.14).

The computation of the ARR residuals is actually performed by the function `DBG()` displayed in Fig. 6.15.

As a result, the computational cost of a parameter optimisation may be much higher in comparison to the case in which residuals are computed from available measured or computed data points either by using numerical differentiation or by evaluating analytical expressions. Clearly, if two coupled models are to be evaluated in order to get numerical values of ARR residuals, the gradient of the least squares cost function with respect to the vector of the targeted fault parameters is to be determined numerically. All other parts of the parameter estimation procedure remain unchanged.

If parameters are to be estimated in real-time, only present measured values and past values sampled over a time window of finite length stored in an array are available. Nevertheless, they are inputs into the diagnostic model and solving its equations yields values for the ARR residuals. As a diagnostic bond graph model used for

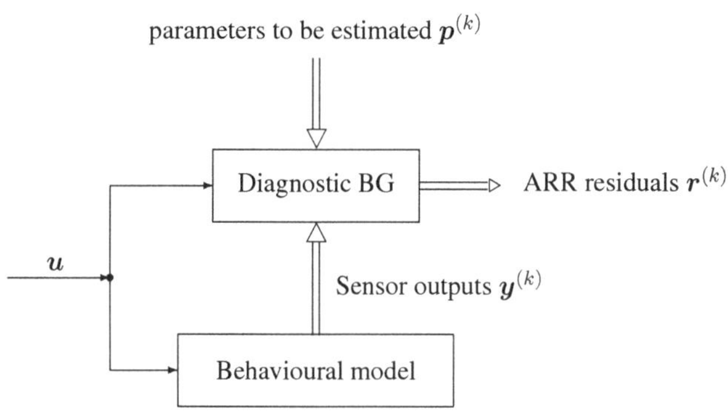

Fig. 6.13 Numerical computation of implicitly given ARR residuals in the kth parameter estimation step

```
function [res] = residuals-2 (p)
//
// Compute outputs of the behavioural model
//
ycalc = process()
//
// Provide sensor outputs f, e1=uC2, and e2=uC3
// as inputs into the DBG
//
u1  = ycalc(1,N0:N)';
u2  = ycalc(2,N0:N)';
u3  = ycalc(3,N0:N)';
f   = (Vi - u1) ./ R1
e1  = u2(1:$);
e2  = u3(1:$);
//
// Solve the equations of the DBG model for ARR residuals
//
res = DBG(f,e1,e2,p)
endfunction
```

Fig. 6.14 Modified user defined Scilab function `residuals-2()` that computes implicitly given ARR residuals

```
function [res] = DBG(f,e1,e2,p)
//
// Get inputs f,e1,e2 from the behavioural BG
// and solve the equations of the DBG for ARR residuals
//
   R1   = p(1)
   R2   = p(2)

   df   = diff(f)/dt
   de1  = diff(e1)/dt
   de2  = diff(e2)/dt
//
// In this case, the equations of the DBG can be solved
// symbolically for the ARR residuals
//
   res1 = f(1:$-1) + C1 * R1 .* df - b(1:$-1)' ./ (Ron + R2) .* ...
          (Vi - R1 .* f(1:$-1) - e1(1:$-1))
   res2 = b(1:$-1)' .* (Vi - R1 .* f(1:$-1) - e1(1:$-1)) ./ (Ron + R2) - ...
          C2 .* de1 - (e1(1:$-1) - e2(1:$-1)) ./ R3
   res  = [res1;res2]
endfunction
```

Fig. 6.15 User defined Scilab function `DBG()` that solves the equations of the DBG for the ARR residuals

on-line FDI relies on preferred derivative causality in order to get rid of initial conditions, differentiation of measured variables has to be performed in discrete time which is not very accurate. This suggests to compute ARR residuals by using only present values of measured variables and those values one sampling time step back [4]. As a result, a set of algebraic equations is to be solved for the ARR residuals delayed by one step. That is, measured values may be constantly written into a buffer and are taken for on-line computation of ARR residuals which are then used for fault parameter estimation over some time window of past sampling steps.

6.7 Sensitivity Pseudo Bond Graphs

Multiple parameter fault isolation by means of minimisation of least squares of ARR residuals needs residual parameter sensitivity functions if a gradient search based method is used. If ARRs can be derived in closed symbolic form from a bond graph, their analytical expressions can be used in the formulation of the least squares cost function and can be differentiated with respect to the vector of targeted parameters either numerically or residuals as functions of the targeted parameters can be differentiated symbolically. If ARRs are not available in symbolic form, they can be numerically computed by solving the equations of a DBG.

Section 5.3.1 has shown that parameter sensitivities of ARR residuals may be obtained from an incremental bond graph. The latter bond graph can be systematically developed from an initial bond graph with nominal parameters by replacing elements with parameters to be estimated by their incremental component model. Inputs into the incBG are variations of the parameters to be estimated multiplied by a power variable of the initial BG. Outputs may be parameter variations of ARR residuals. They are a weighted sum of the parameter variations and the weighting factors are just the residual sensitivity functions.

Another option is to use *sensitivity* pseudo bond graphs (SPBGs) introduced by Cabanellas et al. in 1995 [7]. In sensitivity pseudo bond graphs, bonds carry partial derivatives of power variables with respect to a parameter θ instead of variations of power variables as in an incremental bond graph. Accordingly, there are as many such sensitivity pseudo bond graphs as there are parameters in a model. However, all these sensitivity bond graphs have the same structure. At first, the focus of sensitivity bond graphs has been on the parameter sensitivity of outputs of continuous time models. In [8], they have been used to obtain parameter sensitivities of ARRs. Sensitivity pseudo bond graphs can be systematically developed from a behavioural bond graph model and provide sensitivity functions that are needed in gradient based parameter estimation as rapid convergence is essential in real-time parameter estimation. In [5], Gawthrop consideres the problem of minimising a cost function $J(\boldsymbol{\Theta})$ of the form:

$$J(\boldsymbol{\Theta}) = \frac{1}{2}\int_0^T (y(t,\boldsymbol{\Theta}) - \hat{y}(t,\boldsymbol{\Theta}))^T \mathrm{Q}(t)(y(t,\boldsymbol{\Theta}) - \hat{y}(t,\boldsymbol{\Theta}))dt \quad (6.25)$$

where $\boldsymbol{y}(t, \boldsymbol{\Theta})$ is the simulated system output and $\hat{\boldsymbol{y}}(t)$ the target output. For a solution of this problem by means of the Quasi-Newton method only first order sensitivity functions $\partial \boldsymbol{y}(t, \boldsymbol{\Theta})/\partial \boldsymbol{\Theta}$ are needed and this gradient information can be provided by a sensitivity pseudo bond graph.

In the following, first, sensitivity pseudo bond graphs are briefly reviewed and are then used to obtain residual sensitivity functions needed for the previously presented least squares ARR residuals minimisation. The simple hybrid network in Fig. 4.1 is used again for illustration of the approach.

6.7.1 Parameter Sensitivity Models of Bond Graph Elements

Consider a 1-port resistor that may be given by a nonlinear constitutive relation

$$f_R(t) = \Phi_R^{-1}(e_R(t), \theta_R) \tag{6.26}$$

It is assumed that all parameters are uncorrelated. Partial differentiation with respect to a parameter θ then yields

$$\underbrace{\frac{\partial f_R}{\partial \theta}}_{=:\, f_{R_\theta}} = \frac{\partial \Phi_R^{-1}}{\partial e_R} \underbrace{\frac{\partial e_R}{\partial \theta}}_{=:\, e_{R_\theta}} + \frac{\partial \Phi_R^{-1}}{\partial \theta_R} \frac{\partial \theta_R}{\partial \theta} \tag{6.27}$$

and can be represented by the parameter sensitivity component bond graph model MR^s depicted in Fig. 6.16.

The sensitivity component bond graph model in Fig. 6.16 reduces to a 1-port resistor if $\theta \neq \theta_R$.

Similarly, a sensitivity component bond graph model is obtained for a nonlinear 1-port C storage element in derivative causality given by the constitutive relation

$$q(t) = \Phi_C(e_C(t), \theta_C) \tag{6.28}$$

Differentiation with respect to time and subsequent partial differentiation with respect to a parameter θ gives

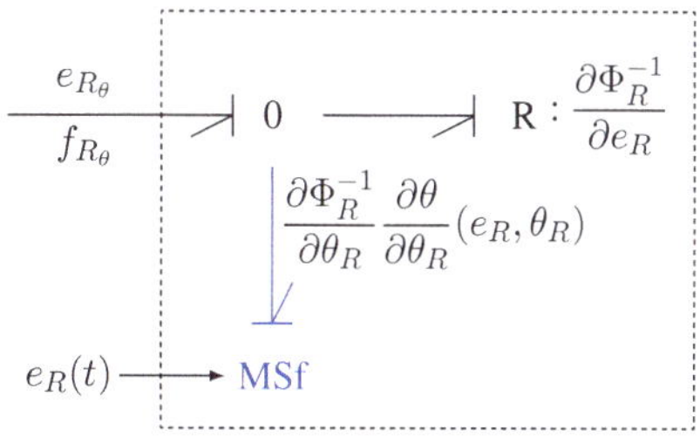

Fig. 6.16 Parameter sensitivity component bond graph model MR^s of a nonlinear 1-port R element

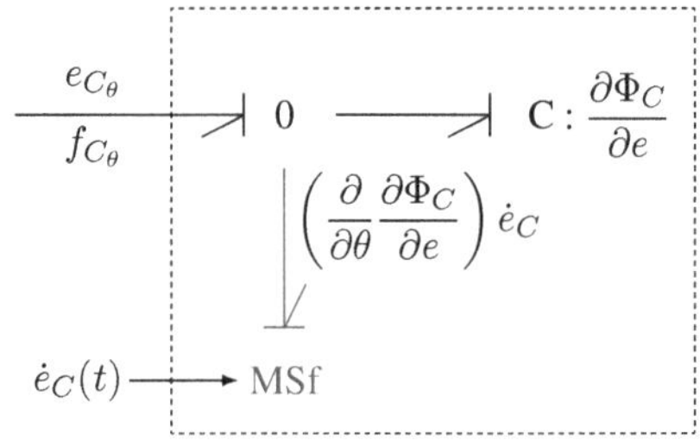

Fig. 6.17 Parameter sensitivity component bond graph model MC^s of a nonlinear 1-port C storage element

$$\underbrace{\frac{\partial f_C}{\partial \theta}}_{=:\, f_{C_\theta}} = \left(\frac{\partial}{\partial \theta}\frac{\partial \Phi_C}{\partial e_C}\right) \dot{e}_C + \frac{\partial \Phi_C}{\partial e_C} \underbrace{\frac{\partial \dot{e}_C}{\partial \theta}}_{=:\, \dot{e}_{C_\theta}} \tag{6.29}$$

Figure 6.17 displays a sensitivity component bond graph model MC^s according to (6.29).

Like in the previously considered case of a resistor, the modulated flow sink disappears in the sensitivity bond graph model of a C storage element if θ does not equal the element parameter

In the case of a linear 1-port C element (6.29) reads

$$f_{C_\theta} = C \cdot \dot{e}_{C_\theta} + \frac{\partial C}{\partial \theta} \dot{e}_C = C \cdot \dot{e}_{C_\theta} + \frac{\partial C}{\partial \theta} \cdot \frac{1}{C} f_C \tag{6.30}$$

In the same manner, sensitivity component models can be obtained for the other bond graph elements. As junctions do not depend on parameters they remain junctions in a sensitivity pseudo bond graph. Sources that provide a constant become sources of value zero. Sensitivity component models of other elements differ from their element only by additional sinks. As a result, a sensitivity pseudo bond graph is of the same structure as the behavioural system bond graph. Moreover, causalities of the latter one are retained.

6.7.2 Deducing Residual Sensitivity Functions from a Sensitivity Pseudo Bond Graph

A sensitivity pseudo bond graph from which residual sensitivity functions for parameter estimation can be deduced is constructed by simply replacing those elements in a DBG by their sensitivity component model whose parameters are to be estimated. Equations for parameter sensitivities of ARR residuals can then be deduced from the SPBG in the same way as the equations of a state space model are deduced from a behavioural BG or equations for ARR variations from an incBG.

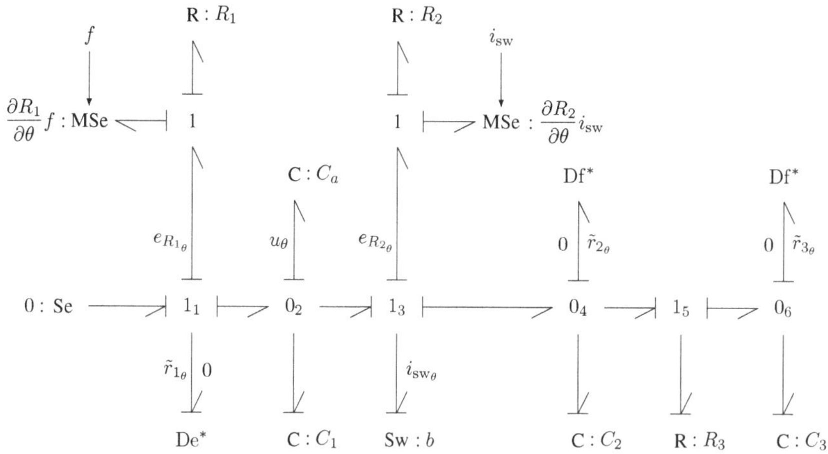

Fig. 6.18 Sensitivity pseudo bond graph of the network in Fig. 4.1

Example

This is illustrated for the simple hybrid network considered in the previous section. The parameters to be estimated are the resistances R_1 and R_2. Accordingly, resistors R : R_1 and R : R_2 are replaced in the DBG (Fig. 4.2). The resulting SPBG is depicted in Fig 6.18. Again, the auxiliary storage element C : C_a is only added to resolve the causal conflict at junction 0_2. During equation formulation its capacitance C_a is set to zero.

From the SPBG in Fig. 6.18 the following equations are deduced.

$$1_1: \quad \tilde{r}_{1_\theta} = 0 - f \frac{\partial R_1}{\partial \theta} - u_\theta \tag{6.31a}$$

$$0_2: \quad 0 = C_1 \dot{u}_\theta + \underbrace{C_a}_{=0} \dot{u}_\theta + i_{sw_\theta} \tag{6.31b}$$

$$\begin{aligned} 1_3: \quad i_{sw_\theta} &= \frac{b}{R_{on}}(u_\theta - e_{R_{2\theta}} - 0) \\ &= \frac{b}{R_{on}}(u_\theta - R_2 i_{sw_\theta} - \frac{\partial R_2}{\partial \theta} i_{sw}) \end{aligned} \tag{6.31c}$$

$$0_4: \quad \tilde{r}_{2_\theta} = i_{sw_\theta} \tag{6.31d}$$

Differentiating (6.31a) with respect to time and substituting the subsequent two equations gives

$$\begin{aligned} r_{1_\theta} &:= -C_1 \dot{\tilde{r}}_\theta - \frac{b}{R_{on} + R_2} \tilde{r}_{1_\theta} \\ &= C_1 \frac{\partial R_1}{\partial \theta} \dot{f} + \frac{b}{R_{on} + R_2} [f \frac{\partial R_1}{\partial \theta} + \frac{\partial R_2}{\partial \theta} i_{sw}] \end{aligned} \tag{6.32}$$

Substituting (6.31c) into (6.31d) yields

$$\begin{aligned} r_{2_\theta} &:= \tilde{r}_{2_\theta} + \frac{b}{R_{\text{on}} + R_2}\tilde{r}_{1_\theta} \\ &= -\frac{b}{R_{\text{on}} + R_2}\left[f\frac{\partial R_1}{\partial \theta} + \frac{\partial R_2}{\partial \theta} i_{\text{sw}}\right] \end{aligned} \tag{6.33}$$

Let $\theta = R_1$. Then

$$\frac{\partial r_1}{\partial R_1} = C_1 \dot{f} + \frac{b}{R_{\text{on}} + R_2} f \tag{6.34}$$

$$\frac{\partial r_2}{\partial R_1} = -\frac{b}{R_{\text{on}} + R_2} f \tag{6.35}$$

Likewise, partial derivatives with respect to R_2 are obtained by letting $\theta = R_2$.

$$\begin{aligned} \frac{\partial r_1}{\partial R_2} &= \frac{b}{R_{\text{on}} + R_2} i_{\text{sw}} \\ &= \frac{b}{(R_{\text{on}} + R_2)^2}[V_i - R_1 f - e_1] \end{aligned} \tag{6.36}$$

$$\frac{\partial r_2}{\partial R_2} = -\frac{\partial r_1}{\partial R_2} \tag{6.37}$$

These results obtained from the SPBG can be confirmed by symbolic partial differentiation of ARRs (4.6) and (4.7) with respect to R_1 and R_2 respectively.

Figures 6.19 and 6.20 display the time evolution of the residual functions as they have been used by the fault parameter estimation in Sect. 6.5.

Residual functions $\partial r1/\partial R_2$, $\partial r2/\partial R_1$, and $\partial r2/\partial R_2$ vanish for $t > 5$ s because for $t > 5$ s the switch Sw : b is off ($b = 0$).

In the illustrating example, residual sensitivity functions have been deduced from a single SPBG as partial derivatives of ARR residuals with respect to a non-specific parameter θ. The parameter sensitivities of ARR residuals needed for the parameter estimation procedure are obtained by letting successively θ be one of the parameters

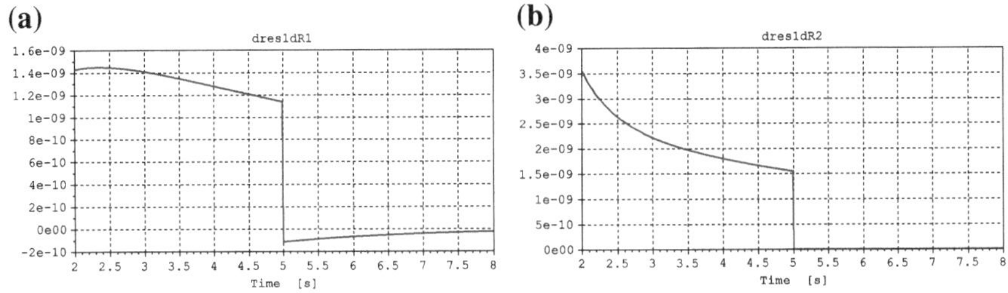

Fig. 6.19 Time evolution of the residual functions $\partial r_1/\partial R_1$, $\partial r_1/\partial R_2$ as used by the fault parameter estimation. **a** Residual functions $\partial r_1/\partial R_1$. **b** Residual functions $\partial r_1/\partial R_2$

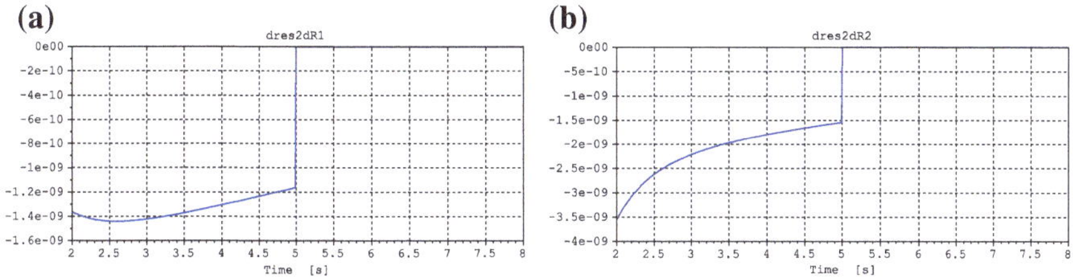

Fig. 6.20 Time evolution of the residual functions $\partial r_2/\partial R_1$, $\partial r_2/\partial R_2$ as used by the fault parameter estimation. **a** Residual functions $\partial r_2/\partial R_1$. **b** Residual functions $\partial r_2/\partial R_2$

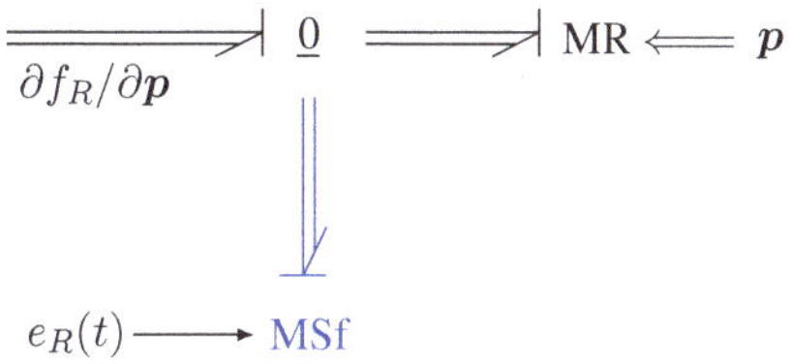

Fig. 6.21 Multibond graph sensitivity model of an R element with a one-dimensional port

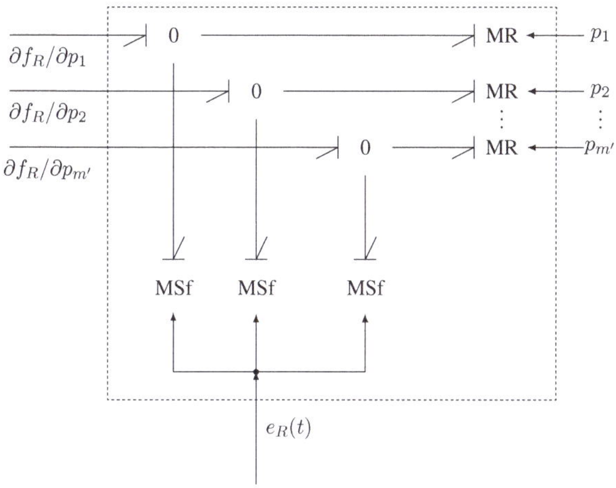

Fig. 6.22 Multiple sensitivity models of an R element with a one-dimensional port

to be estimated. Alternatively, each of the latter parameters may be an input into one SPBG. That is, there are as many SPBGs as there are parameters to be estimated. They all have the same structure, provide different outputs and can be evaluated in parallel. This suggests to condense all SPBGs into one single multibond sensitivity graph. The dimension of its bonds is equal to the number of parameters to be estimated. Figure 6.21 displays all sensitivity models of a 1-port R element condensed into one single multibond graph sensitivity model.

The multibond graph sensitivity model in Fig. 6.21 is a concise representation of the multiple sensitivity models of an R element in Fig. 6.22.

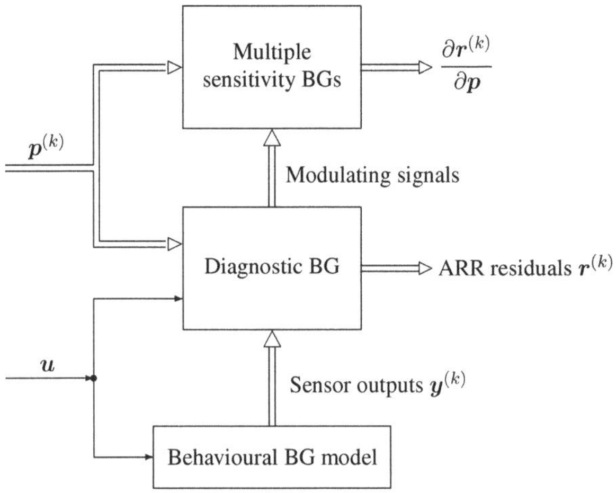

Fig. 6.23 Simultaneous numerical computation of ARR residuals and their parameter sensitivities in the kth iteration of the parameter estimation

Each of the multiple sensitivity BGs constructed from a DBG needs the same modulating signals as inputs which can be provided by the DBG. By coupling the multiple sensitivity BGs to the DBG, ARR residuals and their partial derivatives with respect to the parameters to be estimated can be numerically computed at the same time for each step of the parameter estimation procedure. This is indicated in Fig. 6.23.

Once the ARR residuals $r_\nu(t_j, p_1, p_2, \ldots, p_{m'})$ $(\nu = 1, \ldots, n')$ and their partial derivatives $\frac{\partial r_\nu}{\partial p_\mu}(t_j, p_1, p_2, \ldots, p_{m'})$ $(\mu = 1, \ldots, m'$ and $n' \leq m')$ are available for each point t_j of the time window used for the parameter estimation, the gradient of the cost function can be computed according to (6.17).

6.8 Summary

If there are rows in a FSM with the same component signature and if the real system does not permit to add more sensors then detected faults cannot be isolated by inspection of the FSM. If faults in a system cause a change in some parameters without changing the system structure and if it can be assumed that multiple simultaneous faults do not cancel each other then least squares parameter estimation may be used to identify parameters whose values deviate distinguishably from their nominal values. These deviations indicate simultaneous multiple faults. Moreover, parameter estimation allows to assess the magnitude of faults. If some parameter changes are small in comparison to others, they may be neglected and the FSM may be simplified accordingly.

A cost function to be minimised in an iterative parameter estimation procedure may be formulated by using either differences between outputs from a real system and computed outputs from a model or by means of ARR residuals. As output errors, as well as ARR residuals are generally nonlinear functions of the component parameters, multiple fault parameter isolation becomes a well-known nonlinear least squares problem. For real-time FDI, ARR residuals obtained from a DBG have the advantage that they make the parameter estimation independent of any initial conditions of the process that are hardly known and will have to be estimated along with component parameters. In off-line simulation, the real system may be replaced by a behavioural model. 'Measured' data is then generated by assuming realistic consistent initial conditions and by solving the equations of the behavioural model.

If ARRs can be established in symbolic form, they may be differentiated analytically with respect to the parameters to be estimated. That is, the Jacobian of the cost function to be minimised can be provided and a gradient based method such as the Gauss-Newton, or the Levenberg-Marquardt algorithm can be used. As a result, the number of iteration is much lower in comparison to a gradient free algorithm such as the one of Nelder-Mead which is important in real-time FDI.

ARR residuals may be obtained from a DBG. If ARRs cannot be deduced from a DBG in closed symbolic form because of nonlinear implicit equations indicated by causal paths, ARR residuals are given implicitly. Their numerical values can be obtained by solving the entire DBG model in each step of the parameter estimation iteration. Derivatives of 'measured' variables with respect to time that are needed in the evaluation of the DBG model are to be performed in discrete time. Once residuals are available for the time points of an observation window, the cost function can be built. If a gradient based parameter estimation method is used, the gradient of the least squares cost function can be obtained by using discrete derivatives.

Whether ARRs are available in closed symbolic form or not, parameter estimation can be well performed by running a Scilab script that uses appropriate user defined functions and functions provided by Scilab for the iterative parameter estimation such as `optim()`. A DBG can be obtained from a BG of the behavioural model by just changing causalities. The equations of both bond graphs can be established automatically by a bond graph processor such as CAMPG [13]. Alternatively, bond graph software such as 20-sim [14] can export equations to Matlab®/Simulink® [15]. The latter program offers similar support for parameter estimation as Scilab.

Beyond computing ARRs residuals either analytically or numerically from a DBG, bond graph modelling can also support the computation of residual sensitivity functions. The outputs of incremental bond graphs are parameter variations of residuals. They are a weighted sum of parameters variations and the weighting factors are just the residual sensitivity functions to be determined. Similarly, bonds in sensitivity pseudo bond graphs carry sensitivities of power variables with respect to a parameter. Residual sensitivity functions are outputs of SPBGs. Both types of bond graphs can be systematically constructed from the DBG. If power variables of the DBG that are needed for modulation of elements in the SPBG are input into the SPBG then ARR residuals as outputs of the DBG and their parameter sensitivities as outputs of a set of SPBGs can be computed simultaneously for a given set of values for the parameters to be evaluated as displayed in Fig. 6.23. This information for each time point of an observation window allows to compute the gradient of the cost function even if closed symbolic expressions for ARRs are not possible.

References

1. Gawthrop, P., Jones, R., & Mackenzie, S. (1992). Identification of partially-known systems. *Automatica*, *28*(4), 831–836.
2. Ranganathan, A. (2004). The Levenberg-Marquardt algorithm, from http://ananth.in/Notes_les/lmtut.pdf.
3. Björck, A. (1996). *Numerical methods for least squares problems*. Philadelphia: SIAM.
4. Samantaray, A. K. (2008). *Model-based process supervision—A bond graph approach*. Advances in Industrial Control. London: Springer.
5. Gawthrop, P. J. (2000). Sensitivity bond graphs. *Journal of the Franklin Institute*, *337*, 907–922.
6. Gawthrop, P.J., & Ronco, E. (1999). A sensitivity bond graph approach to estimation and control of mechatronic systems. Centre for systems and control. University of Glasgow, Faculty of Engineering, CSC-99018.
7. Cabanellas, J.M., Félez, J., & Vera, C. (1995). A formulation of the sensitivity analysis for dynamic systems optimization based on pseudo bond graphs. In F.E. Cellier & J.J. Granda (Eds.), *ICBGM'95, International Conference on Bond Graph Modeling and Simulation. Volume 27(1) of Simulation Series* (pp. 135–144). Las Vegas, Nevada: SCS Publishing
8. Borutzky, W. (2009). Bond graph model-based fault detection using residual sinks. *Proceedings of the Institution of Mechanical Engineers Part I Journal of Systems and Control Engineering*, *223*(3), 337–352.
9. Samantaray, A. K., & Ghoshal, S. K. (2007). Sensitivity bond graph approach to multiple fault isolation through parameter estimation. *Proceedings of the Institution of Mechanical Engineers Part I: Journal of Systems and Control Engineering*, *221*(4), 577–587.
10. Scilab Enterprises. Scilab, 78000 Versailles, France, from http://www.scilab.org/.
11. Dennis, J., & Schnabel, R. (1996). *Numerical methods for unconstrained optimization and nonlinear equations*. Philadelphia: SIAM.
12. Nelder, J., & Mead, R. (1965). A simplex method for function minimization. *Computer Journal*, *7*(4), 308–313.
13. CadSim Engineering. CAMPG, from http://www.bondgraph.com.
14. Controllab Products. 20-sim The Power in Modeling, from: http://www.20sim.com.
15. The Mathworks. MATLAB®/Simulink®, from http://www.mathworks.com/products/simulink/.

Chapter 7
ARR Based System Mode Identification

In the previous chapters, the focus has been on ARR-based FDI of systems represented by a hybrid model. ARRs derived from a DBG of a hybrid system model depend on the discrete states of the switches used in the hybrid model. The values of all switch states at a time instant $t > 0$ define the system mode that lasts until at least one switch changes its state giving rise to a new system mode. If ARRs valid for the current system mode are used beyond a mode change, their evaluation may give residuals that exceed current thresholds although no parametric fault has happened. In order to avoid reporting non-existent faults to a supervisor system, it is essential to detect discrete mode changes. The system mode must be known to make sure that the correct ARRs valid for the current system mode are evaluated. This chapter shows that ARRs obtained from a DBG with non-ideal switches and thus with fixed, mode independent causalities can not only be used for FDI in systems represented by a hybrid model but also for system mode identification. In [1], the method has been formulated for hybrid LTI systems and has been applied to a DC–DC buck converter. The concept, however, is not limited to linear systems. Another recent bond graph approach makes use of controlled junctions and a sequential causality assignment procedure (SCAP) adapted for FDI [2, 3].

7.1 Bond Graph Based System Mode Identification Using ARRs

The time instances of discrete mode changes may be either known or may be given by contraints. For instance, the pass transistor in a buck power converter is switched on and off by an external control signal, while the diode switches autonomously conversely to the pass transistor. Once conditions for controlled and autonomous switching have been formulated, e.g. for the bouncing ball problem, a Petri net (PN) or a finite state machine can be set up that determines the new mode after a mode change. For each mode, a DBG model can be developed. Its evaluation provides the ARR residuals valid for that mode and the PN governs the transition from the current DBG to the one valid for the next mode. This corresponds to the presentation

W. Borutzky, *Bond Graph Model-based Fault Diagnosis of Hybrid Systems*,
DOI 10.1007/978-3-319-11860-4_7

of hybrid models by a combination of a PN and a set of bond graphs as has been outlined in Sect. 2.2.

7.1.1 Identifying a Set of ARR Residuals Close to Zero

Alternatively, a DBG with non-ideal switches and mode invariant causalities can be developed that holds for all system modes. It is assumed that the system under consideration is healthy and that no faults occur during system mode identification. For simplicity it is assumed that mode dependent ARRs in closed symbolic form can be deduced from the DBG. Let s be the number of switches in the model and $n_f \leq 2^s$ the number of physically feasible switch state combinations, i.e. n_f denotes the number of system modes. Furthermore, let $\boldsymbol{u} = (u_1\, u_2 \cdots u_N)^T$ be the vector of N known system inputs and $\boldsymbol{y} = (y_1\, y_2 \cdots y_M)^T$ the vector of M inputs into a DBG either obtained by measurements from the real system or by evaluating a behavioural model of the real system. Then each ARR residual $r_i(t)$, $i = 1, \ldots s$, is the weighted sum of known system inputs, known measurements and derivatives of measurements.

$$
\begin{aligned}
r_i^{(\sigma)}(t) = & \sum_{\nu=1}^{N} c_{i\nu}^{(1)}(\boldsymbol{\Theta}, \boldsymbol{m}^{(\sigma)}(t))u_\nu(t) + \sum_{\mu=1}^{M} c_{i\mu}^{(2)}(\boldsymbol{\Theta}, \boldsymbol{m}^{(\sigma)}(t))y_\mu(t) \\
& + \sum_{\mu=1}^{M} c_{i\mu}^{(3)}(\boldsymbol{\Theta}, \boldsymbol{m}^{(\sigma)}(t))\dot{y}_\mu(t)
\end{aligned}
\tag{7.1}
$$

where $\boldsymbol{m}^{(\sigma)}(t) = (m_1^{(\sigma)}(t)\, m_2^{(\sigma)}(t) \ldots m_s^{(\sigma)}(t))^T$ with $m_j^{(\sigma)}(t) \in 0, 1$, $j = 1, \ldots, s$ and $1 \leq \sigma \leq n_f$ represents the system mode at time instant t and $\boldsymbol{\Theta}$ denotes the vector of all system parameters.

For each of the n_f technically feasible system modes, there is a set of $n_s^{(\sigma)}$ ARRs. Some of them may be system mode independent and can be discarded with regard to system mode identification. Furthermore, for each sampled time instant t, the weighting factors $c_{i\nu}^{(1)}(\boldsymbol{\Theta}, \boldsymbol{m}^{(\sigma)}(t))$, $c_{i\mu}^{(2)}(\boldsymbol{\Theta}, \boldsymbol{m}^{(\sigma)}(t))$, and $c_{i\mu}^{(3)}(\boldsymbol{\Theta}, \boldsymbol{m}^{(\sigma)}(t))$ are constants. That is, the ARRs for system mode $\boldsymbol{m}^{(\sigma)}(t)$ are a weighted sum of $\boldsymbol{u}(t)$, $\boldsymbol{y}(t)$, and $\dot{\boldsymbol{y}}(t)$. Their number is $n_r^{(\sigma)} \leq n_s^{(\sigma)}$.

Now, for each of the n_f feasible system modes, the set of $n_r^{(\sigma)} \leq n_s^{(\sigma)}$ ARRs is evaluated for discrete time instances t yielding ARR residuals $\boldsymbol{r}^{(\sigma)}(t), \sigma = 1, \ldots, n_f$. The resulting ARR residuals may be arranged into a matrix $(\boldsymbol{r}^{(\sigma)}(t))$ with s_f columns and n_s rows. For $n_r^{(\sigma)} < n_s^{(\sigma)}$ rows with an index $n_r^{(\sigma)} \leq i \leq n_s^{(\sigma)}$ are filled up with zeros so that all residuals build a $n_s \times n_f$ matrix. The current system mode is identified by searching for a column of which all entries have values close to zero. As there are no faults, only one out of the n_f sets of ARRs will give residuals close to zero within given bounds because the discrete state switch variables used in this set of

ARRs are the ones that reflect the current system mode. Once some discrete switch states change, the system mode changes. Accordingly, output variables, obtained by measurement of a real system or by a numerical evaluation of a behavioural model that may encompass models of the sensors, change significantly. Accordingly, all entries of another column in the matrix become close to zero indicating the new system mode.

In conclusion, there is a unique set of ARRs for each system mode. A system mode is characterised by a set of discrete switch states that are variables in the ARRs. When switches change their state, the system mode of operation changes. As a result, residuals of the current set of ARRs may take significantly different values invalidating the set of ARRs that characterises the current system mode. Sets of ARRs can be used to identify system modes and changes of system modes.

This system mode identification may require considerable computational costs especially if ARRs cannot be deduced in closed symbolic form so that the entire DBG model is to be evaluated to obtain numerically the time history of ARR residuals. However, as it is one and same the problem with different sets of discrete switch state variables, this computation can be easily and efficiently performed in parallel on multicore processors or multiprocessor computers.

Measurements from the real system are overloaded with noise. In order to be able to clearly identify the set of ARRs with residuals close to zero, measured data should undergo appropriate filtering before it is used in the evaluation of ARRs. Moreover, some of the sensors providing measured values may operate in a faulty mode due to external disturbances caused by changes in the ambient or by internal parametric faults. If details of the internal build-up of a sensor are not fully known so that a bond graph model cannot be developed, then for small changes, its dynamic behaviour may be approximately captured by a transfer function that, at least, accounts for the sensor's delay and its gain.

After all, it should be kept in mind that ARRs are deduced from a bond graph that is the outcome of a modelling process based on physical first principles. In this process, modelling assumptions have been made, effects have been conceptually captured in an idealised manner or have been neglected. Accordingly, because of these modelling simplifications, because of a possible linearisation of model equations or a reduction of the model order and because of model parameter uncertainties some ARR residuals may deviate from zero even if no fault has happened.

7.1.2 Identification of System Mode Changes in a Healthy System

Once an initial system mode is known, this knowledge can be used to identify candidate sets of ARRs for the next system mode and only these sets of equations need to evaluated to identify the next system mode [3]. To that end, similar to a FSM, a so-called Mode-change Signature Matrix (MCSM) is introduced in [3] with the assumption that any mode change is caused by the change of one single discrete state variable. In this matrix, there is a row for each discrete state variable q_i and a column for each ARR of index j. If q_i occurs in an ARR of index j, the matrix entry in place (i, j) is equal to one and zero otherwise. That is, a mode change represented by a change of q_i can be detected, if at least one entry is equal to one in that row. Moreover, if the row signature is unique, the mode change can be isolated. System mode changes may be viewed as a special case of switch faults . The value of a discrete switch state may indicate that an instantaneous mode change has taken place but that the system is still healthy or that an open circuit or short circuit fault has occurred. Discrete switch states in mode dependent ARRs may therefore be included in a FSM in the same way as continuous component parameters. In this book, the MCSM is part of the all-mode FSM.

If a system under consideration is faulty and if the evaluation of ARR gives residuals that are outside some given bounds then this could be due to a fault in some component, or the values of the discrete state variables used in the set of ARRs are not valid for the current mode. Changing the set of discrete state variables, i.e. accounting for mode changes affects ARR residuals differently. Some ARRs in a set of mode dependent ARRs may be even independent (cf. the example network in Sect. 4.2, ARR 4.8). If ARR based fault parameter estimation is used for fault isolation, sampled data would be collected while the system is faulty and the ARRs are no longer valid. That is, parameter estimation is likely to give wrong and misleading results.

For these reasons, in the following, it is assumed that no parametric faults happen during system mode identification and that an initial system mode is known. System mode identification in the presence of faults is more difficult. In [4], Arogeti et al. present an advanced method for this more general case that categorises ARRs into different types and provides a refined set of fault candidates to the fault parameter estimation procedure. Multiple fault detection, isolation and identification for hybrid systems with no available information on the nature of faults (abrupt or incipient) and on system mode changes has been recently addressed in [5].

Once a discrete system mode change takes places at a time instant t_1, the values of 'measured' inputs into the ARRs change and the discrete switch state variables used in the set of ARRs are no longer valid. As a result, the residuals of some ARRs will no longer be within bounds close to zero. Accordingly, some entries in the coherence vector will be equal to one. The coherence vector is compared with the rows of that part of the all-mode FSM that constitutes the MCSM to identify candidates for the switch state variable that may have changed and has caused the system mode change. Let $[m_s^0 \cdots m_2^0 \; m_1^0]$ denote the initial system mode. If the comparison of the coherence vector with the MCSM part of the all-mode FSM reveals say m_2 as candidate for the mode change then the next possible system mode is $[m_s^0 \cdots m_2 \; m_1^0]$. This is checked by evaluating the ARRs with this set of discrete switch state variables. If the residuals of the adapted ARRs are not close to zero and if there is another switch state that may have changed then the ARRs are also evaluated with a second set of switch state variables. If a new mode cannot be identified, it is assumed that the change in the coherence vector is due to a parametric fault. In that case, the all-mode FSM is inspected for fault isolation by setting the discrete switch state variables to their values of the current system mode. If the inspection indicates that faults cannot be isolated, fault parameter estimation will have to be performed.

Example

For illustration, the method for identification of mode changes is applied to a modification of the simple network in Fig. 4.1 with two more switches displayed in Fig. 7.1.

The network is easily converted into the DBG in Fig. 7.2.

The sum of flows at the two sensor junctions 0_1, 0_2 yields two independent ARRs.

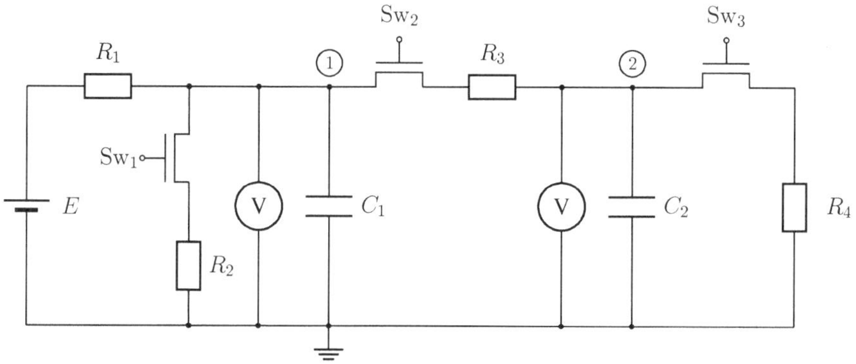

Fig. 7.1 Network with three semiconductor switches

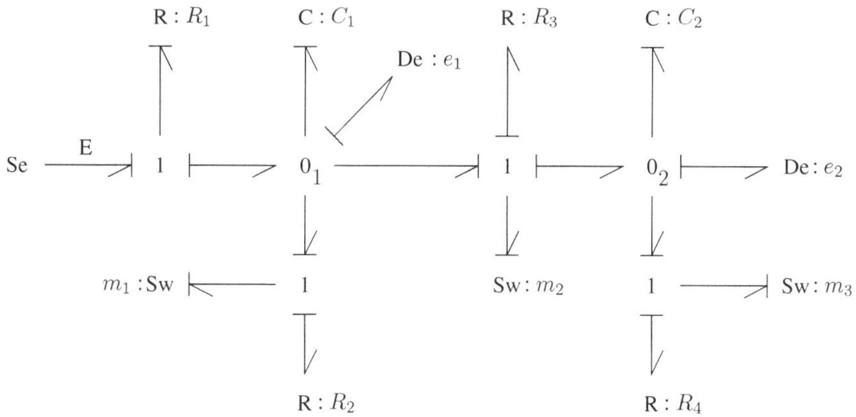

Fig. 7.2 Diagnostic bond graph of the network in Fig. 7.1

$$\text{ARR}_1\text{:}\quad 0 = \frac{1}{R_1}(E - e_1) - C_1\dot{e}_1 - \frac{m_1}{R_{\text{on}} + R_2}e_1 - \frac{m_2}{R_{\text{on}} + R_3}(e_1 - e_2) \;=\; r_1 \tag{7.2}$$

$$\text{ARR}_2\text{:}\quad 0 = \frac{m_2}{R_{\text{on}} + R_3}(e_1 - e_2) - C_2\dot{e}_2 - \frac{m_3}{R_{\text{on}} + R_4}e_2 \;=\; r_2 \tag{7.3}$$

The structural information contained in the two ARRs (7.2)–(7.3) is given by the all-mode FSM in Table 7.1.

The first three rows of the all-mode FSM constitute what has been called a Mode Change Signature Matrix (MCSM) in [3]. These rows indicate that ARR_1 is sensitive to system mode changes caused by a change of the discrete state of switches Sw_1 or Sw_2, while ARR residual r_2 is affected when either switch Sw_2 or Sw_3 changes its state. Moreover, it can be seen that all three system mode changes caused by one of the three switches can be detected and can be isolated. There is an entry equal to one in each of the first three rows and their mode change signatures are unique.

An inspection of the lower part of the all-mode FSM reveals that a fault in resistance R_3 can be isolated in all system modes in which Sw_2 is on ($m_2 = 1$). This is indicated by the entry m_2 in the last column. For instance, for mode $[m_3\ m_2\ m_1] = [1\ 1\ 1]$ (mode 7), the FSM is displayed in Table 7.2.

Table 7.1 All-mode FSM for the switched network in Fig. 7.1

Component	Parameter	ARR_1	ARR_2	D_b	I_b
Sw_1	R_{on}, m_1	1	0	1	1
Sw_2	R_{on}, m_2	1	1	1	1
Sw_3	R_{on}, m_3	0	1	1	1
R	R_1	1	0	1	0
C	C_1	1	0	1	0
R	R_2	m_1	0	m_1	0
R	R_3	m_2	m_2	m_2	0 (m_2)
C	C_2	0	1	1	0
R	R_4	0	m_3	m_3	0

Table 7.2 FSM in mode 7 ($[m_3\ m_2\ m_1] = [1\ 1\ 1]$) for the switched network in Fig. 7.1

Component	Parameter	ARR_1	ARR_2	D_b	I_b
R	R_1	1	0	1	0
C	C_1	1	0	1	0
R	R_2	1	0	1	0
R	R_3	1	1	1	1
C	C_2	0	1	1	0
R	R_4	0	1	1	0

Identification of a Sequence of System Mode Changes

Now, let $[m_3\ m_2\ m_1] = [1\ 1\ 1]$ (mode 7) be the initial system mode. The initial system mode is detected by evaluating the two ARRs for all eight switch state combinations. As the system is healthy, all residuals must be close to zero. Let all switches be closed for some time interval $0 \leq t < t_1$ so that the capacitors get charged to some extent. At $t = t_1$ a mode change may happen. An evaluation of the ARRs shows that residual $r_1(t_1)$ is no longer close to zero so that the coherence vector becomes $\boldsymbol{c} = [1\ 0]$. In other words, ARR_1 is no longer consistent with the system mode. Comparison of the coherence vector with the first three rows of the all-mode FSM indicates that Sw_1 must have changed its state so that the new system mode is likely to be $[1\ 1\ 0]$ (mode 6). As ARR_2 is consistent with the new system mode, only ARR_1 with the new set of discrete switch states is re-evaluated. If its residual is again close to zero, the new system mode is identified.

Let there be another mode change at $t = t_2$ and the coherence be $\boldsymbol{c} = [1\ 1]$. Comparison of its pattern with the rows of the MCSM part of the all-mode FSM indicates that m_2 has likely changed its value. As it is assumed that any system mode change is due to a single switch state change, a possible transition from mode $[1\ 1\ 0]$ (mode 6) to mode $[1\ 0\ 0]$ (mode 4) has taken place. As m_2 affects both residuals, the two ARRs must be evaluated again to validate the system mode hypothesis.

Likewise, a subsequent mode change at $t = t_3$ with the coherence vector $\boldsymbol{c} = [0\ 1]$ leads to mode $[0\ 0\ 0]$ (mode 0) and if a final mode change takes place at $t = t_4$ with the coherence vector $\boldsymbol{c} = [1\ 1]$ then mode $[0\ 1\ 0]$ (mode 2) is identified.

If $\boldsymbol{c} = [1\ 1]$ for a time instant $t_5 > t_4$, no new system mode is identified. It is then concluded that a parametric fault has happened. For this simple network, the parametric fault can be isolated by inspection of the FSM for system mode $[0\ 1\ 0]$ (mode 2) (Table 7.3) and identified as a fault in resistance R_3. No fault parameter estimation is needed. In addition, the FSM for mode 2 shows that in this mode, faults in resistances R_2 and R_4 cannot be detected.

Note that the re-evaluation of ARRs for validating a mode hypothesis is not necessary in the case of the considered simple network because the system is assumed to be healthy, the initial system mode is known, i.e. initial conditions for the discrete switch state variables are known and all system mode changes can be isolated by inspection of the MCSM part of the all-mode FSM. If the signature of a mode change

Table 7.3 FSM in mode 2 ([0 1 0]) for the switched network in Fig. 7.1

Component	Parameter	ARR_1	ARR_2	D_b	I_b
R	R_1	1	0	1	0
C	C_1	1	0	1	0
R	R_2	0	0	0	0
R	R_3	1	1	1	1
C	C_2	0	1	1	1
R	R_4	0	0	0	0

Table 7.4 System mode identification and FDI

Time instant (s)	Coherence vector	Switch states $[m_3\ m_2\ m_1]$	Detected mode	Fault
$t = t_0 = 0$	[0 0]	[1 1 1]	7	No
$t = t_1 = 10$	[1 0]	[1 1 0]	6	No
$t = t_2 = 20$	[1 1]	[1 0 0]	4	No
$t = t_3 = 30$	[0 1]	[0 0 0]	0	No
$t = t_4 = 40$	[1 1]	[0 1 0]	2	No
$t = t_5 = 45$	[1 1]	[0 1 0]	2	Yes

is not unique so that it cannot be isolated by inspection of the MCSM part of the all-mode FSM, a subset of ARRs must be re-evaluated in order to identify the new system mode among a small set of potential mode candidates. If two or more mode changes take place in rapid sequence in real-time system mode identification while mode identification based on the evaluation of a subset of ARRs is still in progress, the last known system mode as a starting point for the decision about the next possible system mode and for the evaluation of subsets of ARRs is no longer valid. In such a case, the entire set of ARRs must be evaluated for identification of the next system mode.

The identification of the above sequence of mode changes is summarised in Table 7.4.

Figure 7.3 depicts the time evolution of the signals $m_1(t)$, $m_2(t)$, and $m_3(t)$ controlling the discrete state of the switches Sw_1, Sw_2, and Sw_3. Accordingly, the time history of the system mode is displayed in Fig. 7.4. Table 7.5 lists the parameters used for the simulation of the network's dynamic behaviour. Figure 7.5 displays the time evolution of the capacitor voltages u_{C_1} and u_{C_2}.

When switch Sw_1 opens at $t = 10$ s, the current charging the two capacitors is increased and thus the rise of the voltages across the capacitors increases. This is less distinct for u_{C_1}. When switch Sw_2 opens at $t = 20$ s, the rise of the voltage u_{C_1} increases again, while capacitor $\mathrm{C} : C_2$ discharges via the resistor $\mathrm{R} : R_4$. After all, when Sw_3 opens at $t = 30$ s, capacitor $\mathrm{C} : C_2$ is neither charged nor can it discharge. It is charged again as of $t = 40$ s when switch Sw_2 is closed. Finally, a parametric

Table 7.5 Parameters of the switched network in Fig. 7.1

Parameter	Value	Units
E	5	V
R_1	100	kΩ
C_1	100	μF
R_2	150	kΩ
C_2	50	μF
R_3	400	kΩ
R_{on}	0.1	Ω

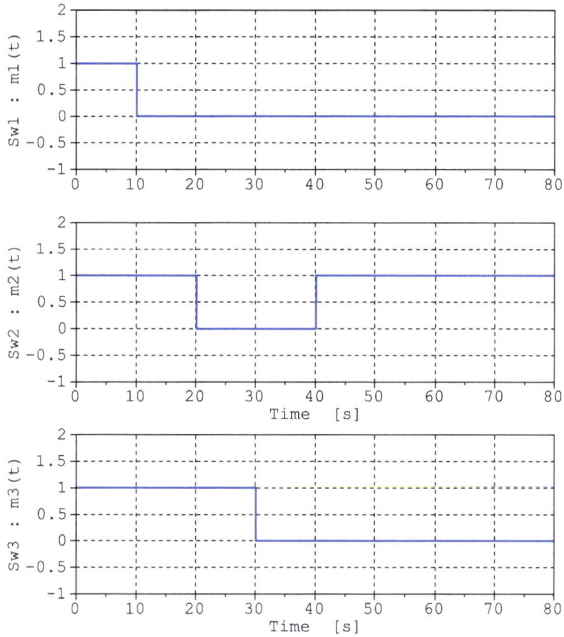

Fig. 7.3 Time evolution of the signals $m_1(t)$, $m_2(t)$, and $m_3(t)$ controlling the switches Sw_1, Sw_2, and Sw_3

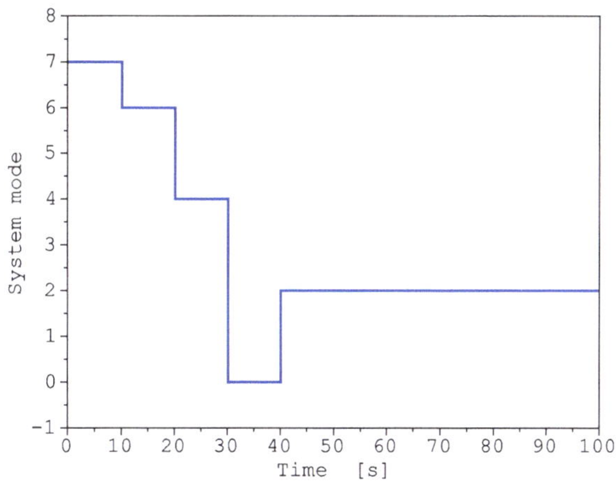

Fig. 7.4 Time history of the system mode

fault is introduced by abruptly increasing resistance R_3 at $t = 45$ s. As a result, the recharging of capacitor C: C_2 is slowed down.

The mode changes and the abrupt fault in resistance R_3 are captured by the time evolution of the residuals $r_1(t)$ and $r_2(t)$ as depicted in Fig. 7.6.

In real-time FDI and in real-time mode identification, ARRs are evaluated with sampled data. Residuals are used by a decision procedure that provides an update of the coherence vector. A system mode change is indicated by a coherence vector

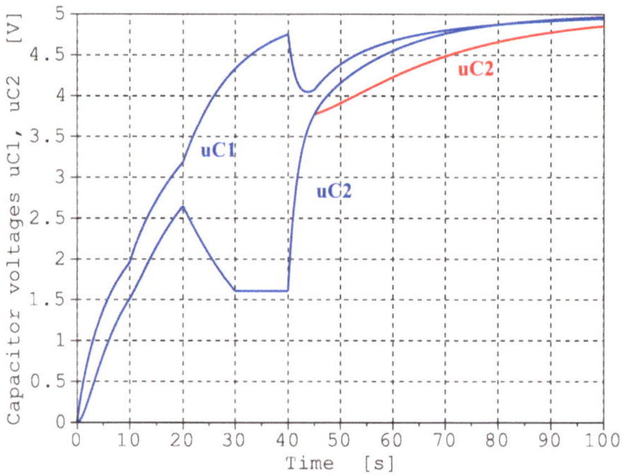

Fig. 7.5 Time evolution of the voltages across the capacitors of the network in Fig. 7.1

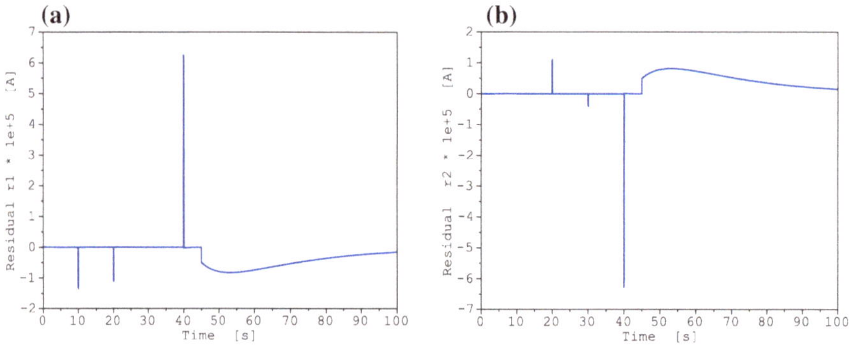

Fig. 7.6 Time evolution of residuals $r_1(t)$ and $r_2(t)$. **a** Residual $r_1(t)$ **b** Residual $r_2(t)$

($\mathbf{c} \neq \mathbf{0}$). Comparison of the coherence vector with the rows of the MCSM part of the all-mode FSM either isolates the new system mode or gives possible candidates, i.e. switch state combinations. For these mode candidates, a subset of ARRs has to be re-evaluated to isolate the new mode and that ends the mode identification.

To visualise mode changes by the time history of ARRs, their evaluation uses the last known, i.e. old and thus wrong values of the discrete switch state variables for a small time interval of length Δt once a system mode change has taken place at time $t = t_s$. This gives a peak in the affected residual at $t = t_s$ indicating the mode change. This use of wrong switch state values for a short while beyond the switching event is illustrated in Fig. 7.7 for the discrete state change of switch Sw: m_1.

The values of the signal $m_1(t)$ are used in the evaluation of the behavioural bond graph model replacing the real network, while the evaluation of the ARRs uses the signal $m_{11}(t)$.

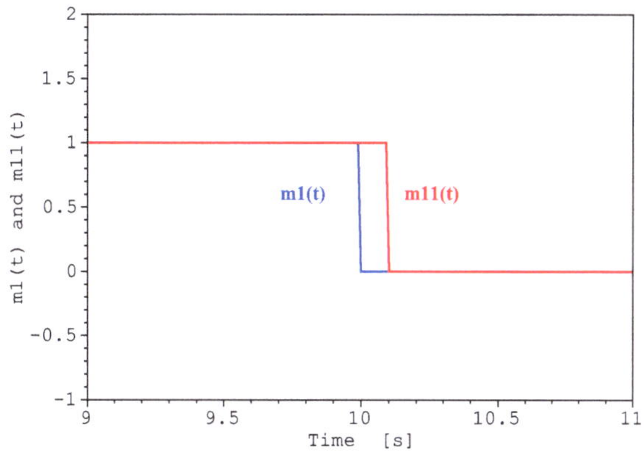

Fig. 7.7 Use of the last known state $m_1(t)$ of switch Sw: m_1 for a small time Δt beyond the switching event at $t = 10\,\mathrm{s}$

7.2 Detection and Isolation of Switch Faults

Mode identification for systems represented by a hybrid model means identification of the current discrete state of all switches in the system and can therefore be used to detect switch faults. Under normal faultless conditions, the signal controlling a semiconductor switch toggles between two discrete values with a certain frequency and the switch opens and closes accordingly. This is no longer the case when an open or a short circuit fault happens in a switch. In a case where no other parametric faults have happened, system mode identification then reveals that a discrete switch state does not toggle any more as of some time instant. Clearly, a high switching frequency leaves little time for real-time ARR-based system mode identification. However, as has already been mentioned, evaluation of ARRs for all switch state combinations can be performed in parallel on multiple processors. Once discrete switch states have been identified, rule-based reasoning can detect and isolate switch faults. For instance, the two switches in the DC-DC buck converter in Fig. 2.19 have opposite states in normal operation, i.e. if one switch is open then the other one is closed. The two of them cannot be open or closed at the same time unless an open or a short circuit fault has happened in one of them.

Identification of System Mode Changes in the Presence of Parametric Faults

In the previous section, each discrete switch state has been taken into account by a row in the FSM. System mode changes can be viewed as faults in discrete switch states. This means that there are far more fault candidates than sensors. Accordingly, discrete switch states will share the same component fault signature so that a switch fault cannot be isolated. If multiple simultaneous faults have happened only in parameters that change continuously with time then parameter estimation can be used to isolate them. However, if there are discrete switch state faults among the multiple simultaneous faults then parameter estimation may result in meaningless real values for the discrete switch states. Therefore, in [6], Alavi and her co-authors propose to chose a combination of switch states, to insert them into the functional to be minimised and to perform a parameter estimation. This task is repeated for all feasible switch state combinations. The combination of switch states that gives the minimum of all optimal values of the functional identifies the system mode. At the same time, the minimisation of the functional provides estimated values for the continuous parameters. A comparison of estimated parameter values with their nominal ones then isolates parametric faults.

7.3 Summary

ARRs derived from hybrid system models are mode-dependent. For ARR-based FDI it is therefore necessary to know the current system mode so that ARRs with the correct set of discrete switch state values are evaluated. When ARR residuals are outside admissible parameter uncertainty bounds it is not clear whether a parametric fault has occurred or a mode change has happened. If a mode change has happened then an evaluation of ARRs with the discrete switch states of the last known mode yields wrong and misleading results. This chapter considers system mode identification for healthy systems only. For the more complex case of system mode changes in the presence of faults, it is suggested to see latest publications, e.g. [4].

The current system mode of a healthy system can be identified by evaluating all ARRs for all feasible switch state combinations. This may require considerable computational effort. The computational time, however, can be reduced by distributing the task on multiple parallel processors. Moreover, this task is only necessary if an initial system mode is not known or if the last known system mode is no longer valid because rapid system modes have taken place while system mode identification is still in progress. Once the current system mode is known, the all-mode FSM can be consulted to identify a subset of ARRs that is to be evaluated to identify the new current system. This has been illustrated by application to a simple circuit with three semiconductor switches.

Finally, the knowledge of the current values of discrete switch states can be used to detect and to isolate switch faults by rule-based reasoning. If switches toggle their state at a high frequency little time is left for mode identification. However, again, parallel processing can help to cope with the time constraints.

References

1. Borutzky, W. (2014). Bond graph model-based system mode identification and mode-dependent fault thresholds for hybrid systems. *Mathematical and Computer Modelling of Dynamical Systems*, *20*(06), 585–616.
2. Low, C. B., Wang, D., Arogeti, S., & Zhang, J. B. (2010). Causality assignment and model approximation for hybrid bond graph: Fault diagnosis perspectives. *IEEE Transactions on Automation Science and Engineering*, *7*(3), 570–580.

3. Arogeti, S. A., Wang, D., & Low, C. B. (2008). Mode tracking and FDI of hybrid systems. *In: Proceedings of the 2008 19th International Conference on Control, Automation, Robotics and Vision, Hanoi* (pp. 892–897). Hanoi, Vietnam: IEEE.
4. Arogeti, S. A., Wang, D., & Low, C. B. (2010). Mode identification of hybrid systems in the presence of fault. *IEEE Transactions on Industrial Electronics*, *57*(4), 1452–1467.
5. Ming, Y., Wang, D., Luo, M., Zhang, D., & Chen, Q. (2012). Fault detection, isolation and identification for hybrid systems with unknown mode changes and fault patters. *Expert Systems with Applications*, *39*, 9955–9965.
6. Alvari, M., Luo, M., Wang, D., & Zhang, D. (2011). Fault diagnosis for power electronic inverters: A model-based approach. *In: Proceedings of 2011 IEEE International Symposium on Diagnostics for Electric Machines, Power Electronics and Drives (SDEMPED)* (pp. 221–228). Bologna, Italy: IEEE.

Chapter 8
Applications

The previous chapters address various aspects of quantitative bond graph-based FDI and system mode identification for systems represented by a hybrid model. This chapter illustrates applications of the presented methods by means of a number of small case studies. The examples chosen are widely used switched power electronic systems. Various kinds of electronic power converters, e.g. buck- or boost converters, or DC to AC converters are used in a variety of applications such as DC power supplies for electronic equipment, battery chargers, motor drives, or high voltage direct current transmission line systems [1].

Power electronic systems contain semiconductor devices, i.e. diodes, transistors such as Insulated Gate Bipolar Transistors (IGBTs), and thyristors. In circuit simulation, e.g. by means of the well known software program Spice [2, 3], sophisticated models are used for these elements. As these semiconductors exhibit fast transients in comparison to the overall dynamic system behaviour it is justified to model them as switches and to perform a hybrid system simulation especially for complex circuits with many fast switching elements. Moreover, commutation usually takes place at high frequency and can entail pulses that can cause power electronic switches to fail which in turn has an impact on the dynamic behaviour of mechatronic systems that use switched power electronic converters as a subsystem. Reference [4], for instance, reports that 38 % of the faults in variable-speed ac-drives are due to failures in power devices. Typically, short-circuit or open-circuit faults occur in power electronic switches. An open-circuit fault in a semiconductor switch means that the transistor falls into the OFF-state and remains in this state regardless of the gate voltage. Causes for an open-circuit fault in a switching transistor may be a lifting of bonding wires due to thermal cycling or a short-circuit induced rupture of a transistor [5]. In general, open-circuit switch faults do not cause the system to shutdown but the performance is degraded. However, in order to avoid malfunctions, or damages in the load of a power converter, to avoid high costs for standstill and repair in industrial applications and to improve reliability, it is important to apply FDI methods. Accordingly, FDI in power electronic systems, especially in motor drives, has been a subject of many publications [5–10]. Many publications consider open circuit faults. The majority of them, however, does not use bond graphs.

W. Borutzky, *Bond Graph Model-based Fault Diagnosis of Hybrid Systems*,
DOI 10.1007/978-3-319-11860-4_8

A bond graph model-based approach to FDI is especially suited for application to mechatronic systems with switched power electronic subsystems as bond graph modelling makes it possible to represent the components e.g. of a motor drive, i.e. the power converter, the motor, a clutch and the mechanical load in a uniform manner. Various publications have used bond graphs for modelling power electronic systems [11–16]. However, so far, few publications on a bond graph model-based approach to FDI in power electronic systems have been reported in the literature (see, for instance [17–20]).

Clearly, a bond graph approach to FDI of systems modelled as a hybrid system is not limited to switched power electronic systems but may be applied to other engineering systems as well for which a hybrid model is appropriate. In the following, the case studies consider faults in a DC to DC boost-converter, in a three-phase DC to AC inverter and in a three-phase rectifier AC to DC. In some motor drives, a rectifier and an inverter are used back-to-back [8]. Computations have been performed by means of the open source software program Scilab [21].

8.1 Switched-Mode Boost Converter

In power generation plants, e.g. in wind turbines, electronic converters and inverters are used. Faults in these components can degrade the performance of power distribution systems and lead to failures. In [19, 22], it is reported that semiconductors and electrolytic capacitors in power power electronic converter systems have a higher failure rate than other components.

For low voltage battery driven portable electronic devices, DC-DC power converters are important because they can accommodate different voltage demands from components and can boost the battery voltage as the battery charge declines. The conversion to the required DC voltage output is highly efficient. Typically DC-DC converters use power MOSFETs that switch at high frequency more efficiently than power bipolar transistors and are compact in size. Moreover, because of their ability to boost a widely varying and low voltage to a needed regulated supply voltage, DC-DC converters can be even operate with battery free power sources such as small thermoelectric generators or microbial fuel cells and are therefore important for implantable medical devices for which a battery change would require another surgery [23].

This chapter considers a simple boost converter often used in power electronic systems. Figure 8.1 depicts its circuit schematic. In this circuit, the MOSFET transistor and the diode may be considered non-ideal switches. The transistor is a controlled power switch. Boost converters are designed that they operate either in so-called *continuous conduction mode* or in *discontinuous conduction mode*. In continuous conduction mode the inductor current never falls to zero. Accordingly, the converter assumes two states per switching cycle. When the transistor is on, the diode is off and vice versa. The diode commutates autonomously and oppositely to the transistor. Hence, there are two system modes in a healthy boost converter.

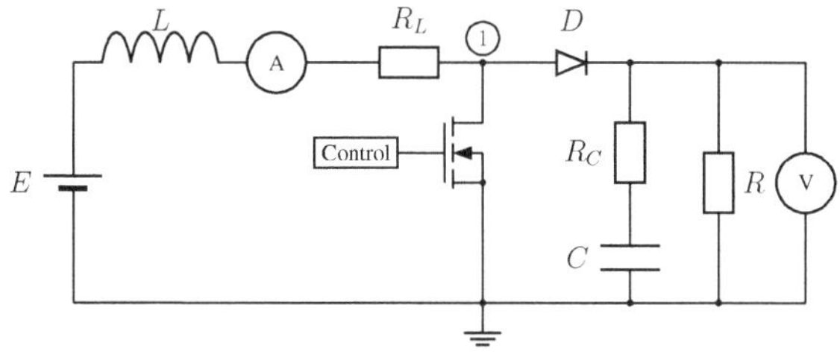

Fig. 8.1 Circuit schematic of a boost converter

Let T_s denote the switching cycle time, t_{on} the time the transistor switch is on, and t_{off} the time it is off. Then, since there are only two states per cycle, $T_s = t_{\text{on}} + t_{\text{off}}$. When the transistor is on, the inductor current increases, while it decreases when the switch is off. In continuous conduction mode, the average inductor current follows the output current. In particular, if the output current decreases then so does the average inductor current. However, if the output load current falls below a certain level then, due to the diode, the inductor current decreases until it has reached zero and then remains at zero for the rest of the switching period time until the next switching cycle begins. This operating mode of the boost converter is known as the discontinuous current or conduction mode. In this mode, there are three states during each switching cycle. The ON-time portion of the switching period is determined by the control of the transistor. It is followed by the OFF-time in which the average inductor current decreases and an idle time in which both the transistor and the diode are off. More details may be found e.g. in [24].

The conversion from one voltage level to another is achieved by storing the input energy temporarily in the inductor when the controlled switch is ON and the diode is OFF, and then releasing that energy to the output at a different voltage value when the controlled switch is turned OFF and the diode is ON. In some applications, the diode is replaced by a pFET in order to avoid the diode's forward voltage drop.

Figure 8.2 depicts a DBG of the circuit. The inverted causality of the flow sensor $\text{Df} : i_L$ and the conductance causality of the switch $\text{Sw} : m_1$ representing the diode let the causality at junction 0_1 be undetermined. In order to resolve the causal conflict an auxiliary capacitor $\text{C} : C_a$ has been added. In equations derived from the DBG, capacitance C_a is set to zero. Summing power variables at the sensor junctions 1_1, 0_2, and at junctions 0_1, 1_3 yields the equations

$$1_1 : \quad 0 = E - L\frac{di_L}{dt} - R_L i_L - u \tag{8.1}$$

$$0_2 : \quad 0 = i_D - \frac{V}{R} - i_C \tag{8.2}$$

$$0_1 : \quad 0 = C_a \dot{u} = i_L - i_{\text{sw}} - i_D \tag{8.3}$$

$$1_3 : u_C = V - R_C \cdot i_C \tag{8.4}$$

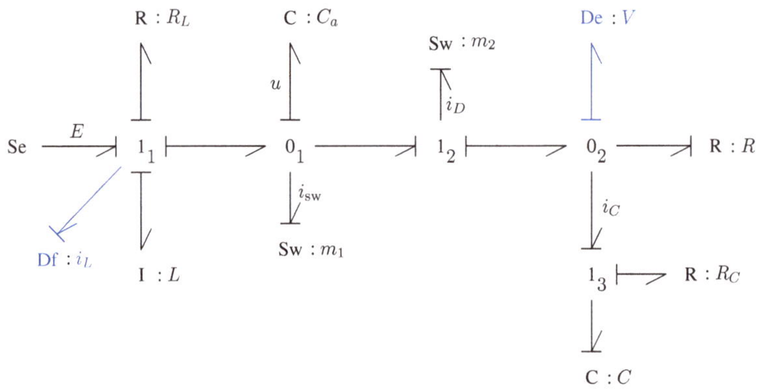

Fig. 8.2 Diagnostic bond graph of the boost converter in Fig. 8.1

Unknowns in (8.1)–(8.3) can be eliminated by means of the following constitutive element equations

$$i_C = C\dot{u}_C \tag{8.5}$$
$$i_D = \frac{m_2}{R_D}(u - V) \tag{8.6}$$
$$i_{sw} = \frac{m_1}{R_{sw}}u \tag{8.7}$$

The result are the two ARRs

$$\text{ARR}_1: \quad 0 = E - L\frac{di_L}{dt} - R_L i_L - u \tag{8.8}$$
$$\text{ARR}_2: \quad 0 = \frac{m_2}{R_D}(u - V) - \frac{V}{R} - C(1 + \frac{R_C}{R})\dot{V} + R_C \cdot C\frac{di_D}{dt} \tag{8.9}$$

where

$$\left(\frac{m_1}{R_{sw}} + \frac{m_2}{R_D}\right)u = i_L + \frac{m_2}{R_D}V \tag{8.10}$$

R_{sw} and R_D denote the ON-resistances of the power MOSFET switch and the diode respectively.

The structural parameter sensitivity of the ARRs is reflected by the FSM in Table 8.1. All possible faults are detectable but none can be isolated with the two detectors.

In a healthy system operating in continuous conduction mode, the switch and the diode open and close oppositely ($m_1 + m_2 = 1$). Let $R_{sw} = R_D = R_{on}$. Then the ARRs simplify and the dynamic behaviour of a correctly operating boost converter is given by the state equations

Table 8.1 Structural fault signature matrix of the boost converter with sensors Df : i_L and De : u_C

Component	Parameter/output	ARR_1	ARR_2	D_b	I_b
Control	m_1	1	1	1	0
	m_2	1	1	1	0
Supply voltage	E	1	0	1	0
Switch	R_{sw}	m_1	m_1	m_1	0
Diode	R_D	m_2	m_2	m_2	0
Inductor	L	1	0	1	0
	R_L	1	0	1	0
Capacitor	C	0	1	1	0
	R_C	0	1	1	1
Load resistor	R	0	1	1	0
Sensor of i_L	i_L	1	1	1	0
Sensor of u_C	u_C	1	1	1	0

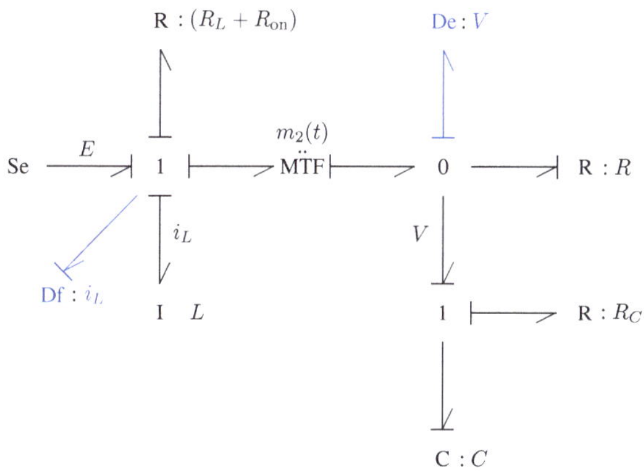

Fig. 8.3 Simplified diagnostic bond graph of the boost converter in the case $m_1 + m_2 = 1 \wedge R_{sw} = R_D = R_{on}$

$$\begin{bmatrix} L & 0 \\ -m_2 R_C C & C(1+\frac{R_C}{R}) \end{bmatrix} \frac{d}{dt} \begin{bmatrix} i_L \\ V \end{bmatrix} = \begin{bmatrix} -(R_{on}+R_L) & -m_2 \\ m_2 + R_C C \dot{m}_2 & -\frac{1}{R} \end{bmatrix} \begin{bmatrix} i_L \\ V \end{bmatrix} + \begin{bmatrix} 1 \\ 0 \end{bmatrix} [E] \tag{8.11}$$

For a correctly operating boost converter with $m_1 + m_2 = 1 \wedge R_{sw} = R_D = R_{on}$ a simpler DBG depicted in Fig. 8.3 can be found.

In subsequent simulation runs, resistor R : R_C representing the equivalent series resistance of the capacitor is set to zero, the inductor resistance R : R_L, however, is kept as its value affects the output voltage of the boost converter as well as the power efficiency of the converter. Taking the average of the expressions on both sides of the state equations (8.11) over the switching time period T_s yields

$$\begin{bmatrix} L & 0 \\ 0 & C \end{bmatrix} \frac{d}{dt} \begin{bmatrix} i_{La} \\ u_{Ca} \end{bmatrix} = \begin{bmatrix} -(R_L + R_{\text{on}}) & -(1-d) \\ (1-d) & -\frac{1}{R} \end{bmatrix} \begin{bmatrix} i_{La} \\ u_{Ca} \end{bmatrix} + \begin{bmatrix} 1 \\ 0 \end{bmatrix} [E] \quad (8.12)$$

where an index 'a' denotes averaged values, t_{on} denotes the duration the semiconductor switch Sw : $m_1(t)$ is on, T_s the switching time period, and $d := t_{\text{on}}/T_s$ the duty ratio. Let V_o denote the steady state output voltage. Equation (8.12) then gives for the ratio V_o/E the expression

$$\frac{V_o}{E} = \frac{R(1-d)}{(R_{\text{on}} + R_L) + R(1-d)^2} \quad (8.13)$$

Equation (8.13) holds under the assumption that the boost converter operates in continuous conduction mode and reduces for $R_L = 0$ into the well known formula for the voltage conversion

$$\frac{V_o}{E} = \frac{1}{1-d} \quad (8.14)$$

which indicates that the averaged output voltage V_o is always greater than the input voltage E as $0 \leq d < 1$. Therefore, boost converters are sometimes called step-up converters. By choosing the duty ratio d, i.e. by determining the length of the transistor ON-time t_{on} the averaged output voltage can be adjusted to the needs of the application.

Figure 8.4 indicates the dependency of the normalised output voltage V_o/E from the inductor resistance R_L for $R = 10\,\Omega$, $R_L = R_{\text{on}} = 0.1\Omega$.

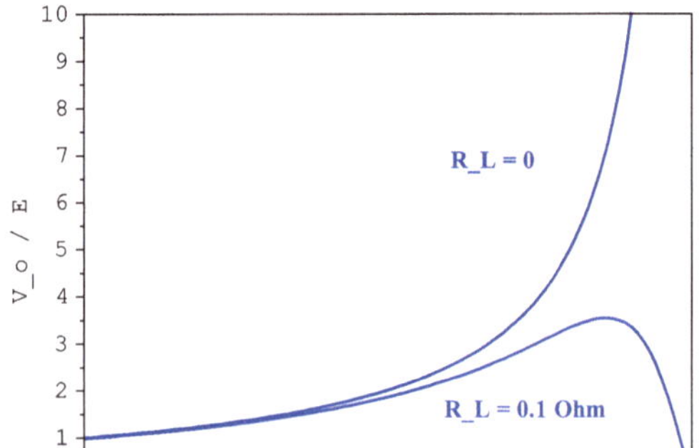

Fig. 8.4 Normalised output voltage V_o/E versus the duty ratio d for different values of the inductor resistance R_L

The expression for the voltage conversion in (8.14) is to be replaced by a more complicated one in case the boost converter is operated in discontinuous conduction mode.

$$\frac{V_o}{E} = \frac{1}{2}\left(1 + \sqrt{1 + \frac{4d^2}{K}}\right) \tag{8.15}$$

where $K := 2L/(RT_s)$ [24, 25].

Let P_o denote the output power of the converter in steady state and I_L the steady state value of the inductor current. In steady state, (8.11) yield

$$0 = (1-d)I_L - \frac{1}{R}U_C \; = \; (1-d)I_L - \frac{1}{R}V_o \tag{8.16}$$

and the power efficiency η then reads:

$$\begin{aligned}\eta = \frac{P_o}{P_o + P_{\text{loss}}} &= \frac{V_o^2/R}{V_o^2/R + R_L I_L^2} \\ &= \frac{R(1-d)^2}{R_L + R(1-d)^2}\end{aligned} \tag{8.17}$$

Figure 8.5 depicts how the function power efficiency η versus the duty ratio d depends on the inductor resistance R_L.

Figures 8.4 and 8.5 show that the inductor resistance should be small in order to reduce the power loss $P_{\text{loss}} = R_L I_L^2$ and to maximise the power efficiency and the gain V_o/E.

In the sequel, some fault scenarios are considered. Simulation runs use the parameters given in Table 8.2.

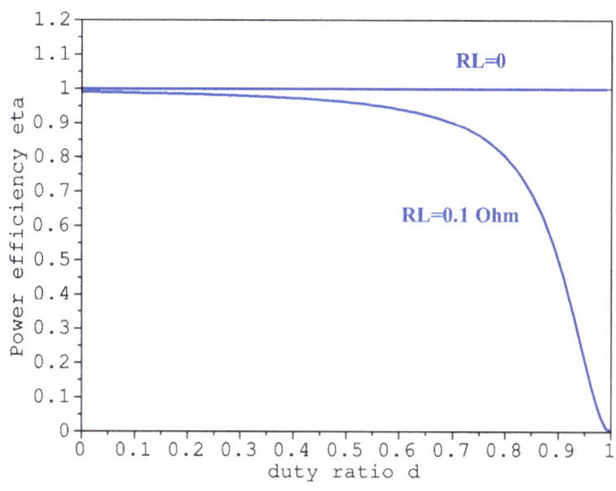

Fig. 8.5 Power efficiency η versus the duty ratio d for different values of the inductor resistance R_L

Table 8.2 Parameters of the boost converter circuit

Parameter	Value	Units	Meaning
E	12.0	V	Voltage supply
V_o	20.0	V	Required output voltage
L	1.0	mH	Inductance
R_L	0.1	Ω	Inductor resistance
R_{on}	10.0	mΩ	ON resistance of the switch
C	500	μF	Capacitance
R	5.0	Ω	Load resistance
f_s	1.0	kHz	Switching frequency

First, it is assumed that the input voltage varies. In normal operation, the control of the MOSFET switch would adjust the duty ratio to keep the output voltage constant. In case the switch is faulty, this adjustment does not happen properly.

8.1.1 Fault Scenario 1: The Input Voltage Changes Temporarily

Let the input voltage drop from its initial value 12 V to a value of 8 V for $t \in [0.04\,\text{s}, 0.06\,\text{s}]$. According to the FSM (Table 8.1), ARR residual r_1 is sensitive to a change of the input voltage, while r_2 is not. This is confirmed by the time evolution of the ARR residuals depicted in Fig. 8.6.

Figure 8.7 shows the effect of the temporary drop of the input voltage on the inductor current $i_L(t)$ and the output voltage $V(t)$. The time evolutions of their mean values i_{L_a} and u_{C_a} respectively are also plotted.

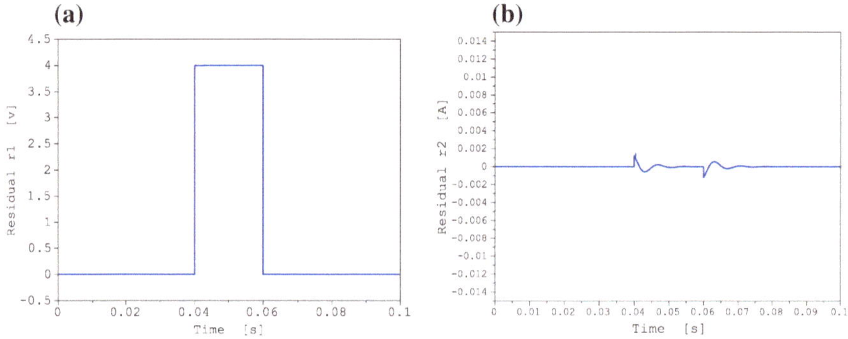

Fig. 8.6 Time evolution of the averaged ARR residuals $r_{1_a}(t)$ and $r_{2_a}(t)$ in the case of a temporary drop of the input voltage $t \in [0.04\,\text{s}, 0.06\,\text{s}]$. **a** Averaged ARR residual $r_{1_a}(t)$. **b** Averaged ARR residual $r_{2_a}(t)$

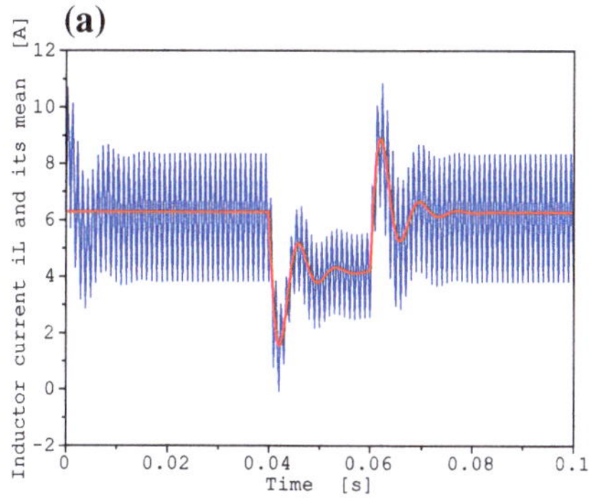

Fig. 8.7 Time evolution of inductor current $i_L(t)$ and the output voltage $V(t)$ in the case of a temporary drop of the input voltage. **a** Inductor current $i_L(t)$. **b** Output voltage $V(t)$

The steady-state values in the simulation results agree with those that can be obtained from (8.12). The analytical values are $i_{L_a}(t \to \infty) = 6.2827\,\text{A}$ and $u_{C_a}(t \to \infty) = 18.848\,\text{V}$ given the parameters in Table 8.2.

In practice, a control circuit adjusts the duty ratio so that the output voltage remains constant. This is achieved by increasing the ON-time of the control signal $m_1(t)$. Let V_o be the steady state output voltage. Neglecting the inductor resistance yields

$$d = \frac{t_{\text{on}}}{T_s} = 1 - \frac{E}{V_o} \tag{8.18}$$

Given the parameters listed in Table 8.2, the ON-time of the control signal $m_1(t)$ would have to be reduced by a factor of 1.5.

8.1.2 Fault Scenario 2: Temporary Drop of the Duty Ratio

In this fault scenario, the duty ratio $d = 0.4$ drops to the value 0.34 for $t \in [0.04\,\text{s}, 0.06\,\text{s}]$. According to the FSM, both ARR residuals are sensitive to this fault. This is displayed in Fig. 8.8.

Figure 8.9 shows the impact of a temporary drop of the duty ratio on the inductor current $i_L(t)$ and the output voltage $V(t)$.

8.1.3 Fault Scenario 3: Non-blocking Diode

This fault scenario considers the case that the diode does not block anymore as of $t \geq t_3 = 0.05\,\text{s}$. That is, the diode acts as a resistor with a low resistance which allows for a current in both directions. Let $\tilde{m}_1(t)$, $\tilde{m}_2(t)$ denote the control signals of

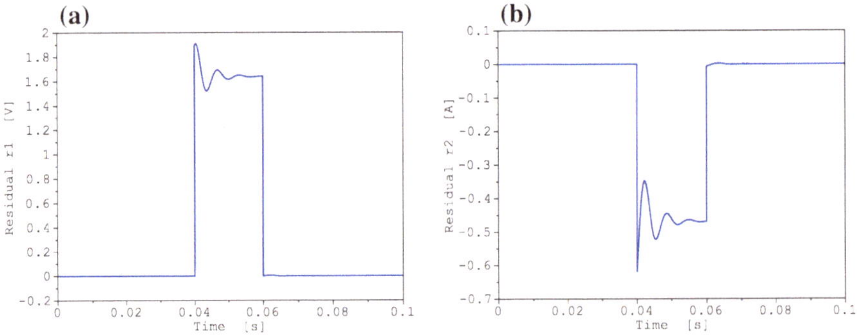

Fig. 8.8 Time evolution of the averaged ARR residuals $r_{1_a}(t)$ and $r_{2_a}(t)$ in the case of a temporary drop of the duty ratio for $t \in [0.04\,\text{s}, 0.06\,\text{s}]$. **a** Averaged ARR residual $r_{1_a}(t)$. **b** Averaged ARR residual $r_{2_a}(t)$

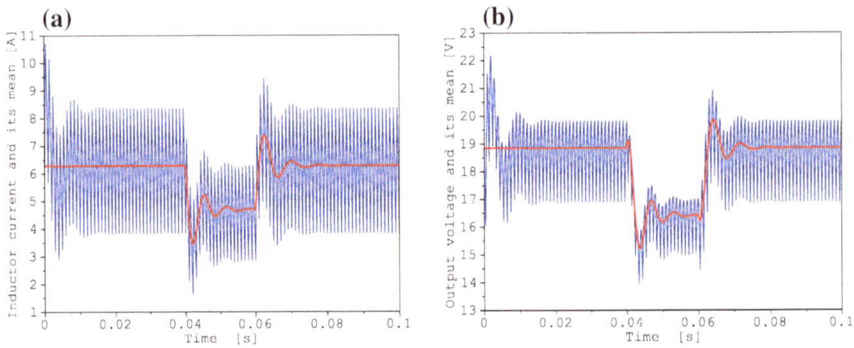

Fig. 8.9 Time evolution of inductor current $i_L(t)$ and the output voltage $V(t)$ in the case of temporary drop of the duty ratio. **a** Inductor current $i_L(t)$. **b** Output voltage $V(t)$

the switches in the model of the faulty system. The fault in the diode can be taken into account by setting $\tilde{m}_2(t) = 1\ \forall\, t > t_3$, while in the faultless system, both switches commutate conversely so that $m_1(t) + m_2(t) = 1\ \forall\, t \in \mathbb{R}^+$. The output signals from the faulty system fed into the DBG will then lead to ARR residuals no longer close to zero as of $t > t_3$. Again, $R_{sw} = R_D = R_{on}$ and $\tilde{m}_1 = m_1$ and $\tilde{m}_1 + \tilde{m}_2 = 1\ \forall\, t < t_3$. The potential u of switch node ① in Fig. 8.1 then reads for the faulty system

$$u = \frac{1}{m_1 + \tilde{m}_2} R_{on} i_L + \frac{\tilde{m}_2}{m_1 + \tilde{m}_2} V \tag{8.19}$$

The time history of $\tilde{m}_2(t)$ can be expressed by means of a pulse() and a unit step() function.

$$\tilde{m}_2(t) = m_2(t) \cdot \texttt{pulse}(t, 0, t_3) + \texttt{step}(t, t_3) \tag{8.20}$$

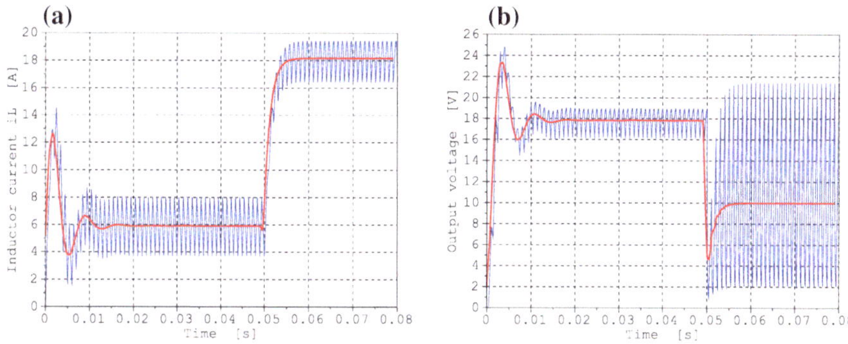

Fig. 8.10 Time evolution of inductor current $iL(t)$ and the output voltage $V(t)$ in case of a faulty non-blocking diode. **a** Inductor current $i_L(t)$. **b** Output voltage $V(t)$

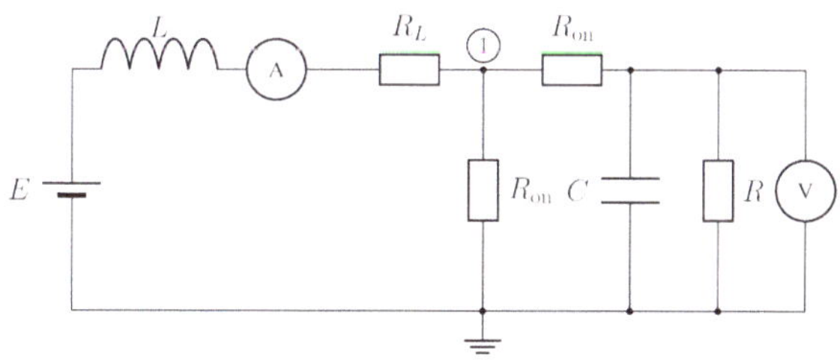

Fig. 8.11 Boost converter with a non-blocking diode in the case the MOSFET switch is closed

The pulse starts at $t = 0$ s and stops and $t = t_3$, the unit step function jumps to the value one at $t = t_3$.

Figure 8.10 depicts the time evolution of the inductor current $i_L(t)$ and the output voltage $V(t)$. In this case, the ON-resistance of the switch and of the diode has been assumed to be 100 mΩ.

As the diode does not block anymore for $t > t_3$, the capacitor can discharge via a low ON-resistance whenever the MOSFET switch is closed. Accordingly, the voltage across the capacitor sharply drops. On the contrary, the inductor current rises. This can be checked by considering the circuit for the time interval $t_3 < t < t_3 + d \cdot T_s$ in which the MOSFET switch is closed. Figure 8.11 depicts the faulty circuit for that time interval. Figure 8.12 shows the bond graph constructed from the circuit schematic in Fig. 8.11.

The state equations deduced from the bond graph in Fig. 8.12 read

$$u = \frac{1}{2}(R_{on} i_L + u_C) \tag{8.21a}$$

$$L\frac{di_L}{dt} = E - R_L i_L - u \tag{8.21b}$$

$$C\frac{du_C}{dt} = \frac{1}{R_{on}}(u - u_C) - \frac{1}{R}u_C \tag{8.21c}$$

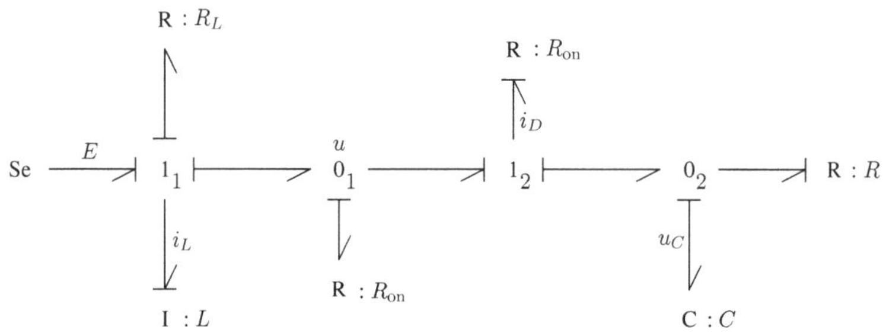

Fig. 8.12 Bond graph of the faulty boost converter circuit in the case the MOSFET switch is closed

The steady state averaged values of the inductor current and of capacitor for $t < t_3$ are used as initial conditions for these state equations. Solving (8.12) for the steady state values gives $i_{L_a}(t \to \infty) = 6\,\text{A}$, $u_{C_a}(t \to \infty) = 18\,\text{V}$. The numerical solution of (8.21a–8.21c) yields $i_L(t = t_3{+}0.4\,\text{ms}) \approx 9.4\,\text{A}$ and $u_C(t = t_3{+}0.4\,\text{ms}) \approx 1.1\,\text{V}$ in accordance with the waveforms in Fig. 8.10.

A rough estimate of the increase of the inductor current at t_3 may be obtained by assuming that the capacitor voltage drops to zero. The analytical solution of the state equation for the inductor current then is

$$i_L(t) = \frac{E}{R_L + R_{\text{on}}/2}(1 - e^{-at}) \tag{8.22}$$

where $a := (R_L + R_{\text{on}}/2)/L$. The slope of the tangent at $t = t_3$ equals E/L. For $\Delta t = d{\cdot}T_s = 0.4\,\text{ms}$, the increase of the inductor current is approximately 12 V/1 mH $\cdot 0.4\,ms = 4.8\,\text{A}$. That is, $i_L(t_3 + d \cdot T_s) \approx 6 + 4.8 = 10.8\,\text{A}$ in accordance with the waveform of the inductor current in Fig. 8.10.

According to (8.8)–(8.10), and the FSM in Table 8.1, both ARR residuals r_1, r_2 should deviate from values close to zero in the case of a faulty diode. This is displayed by their averaged time evolution in Fig. 8.13. Notice the difference in magnitude of the two residuals.

Another fault scenario may consider the degradation of the capacitor. Reference [19] lists various causes for a failure of an electrolyte capacitor and considers the current ripple which causes internal heating, i.e. an increase of the core temperature which results in a gradual aging of the capacitor. Another possible cause for a failure of the capacitor is a leakage current that may lead to a short circuit. Such a leakage can be accounted for by adding a resistor in parallel to the capacitor.

8.1.4 Analytical Computation of ARR Residuals

Section 4.9 has addressed the analytical determination of ARR residuals for switched LTI systems. In the following, this is illustrated for the example of the boost converter. It is assumed that the diode is no longer blocking in reverse direction as of $t > t_3$.

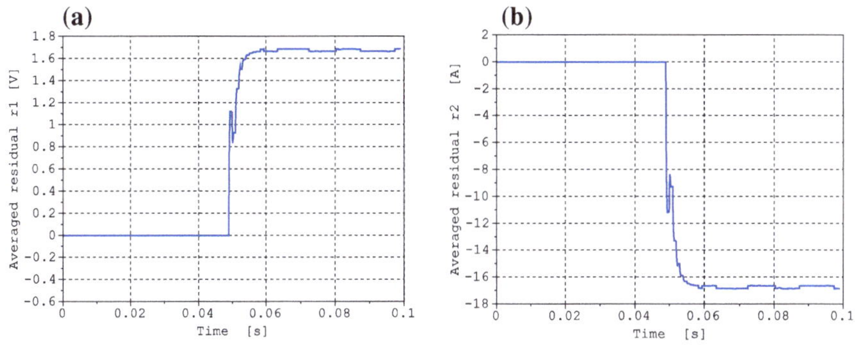

Fig. 8.13 Time evolution of the averaged ARR residuals in the case of a faulty non-blocking diode. **a** Averaged ARR residual $r_1(t)$. **b** Averaged ARR residual $r_2(t)$

For the faultless system with nominal parameters, an evaluation of the ARRs yields zero.

$$0 = E - (R_L + R_{\text{on}})i_L - L\frac{di_L}{dt} - m_2 u_C \tag{8.23}$$

$$0 = m_2 i_L - \frac{u_C}{R} - C\frac{du_C}{dt} \tag{8.24}$$

If a behavioural model of the faulty system is available it can be coupled to the model of the faultless system by means of residual sinks which force the faultless system to adapt its behaviour to that of the faulty system. Let parameters and variables of the faulty system be denoted by a tilde. In the faulty system model, an evaluation of the ARRs results in a residual equal to zero.

$$0 = E - (R_L + \frac{1}{m_1 + \tilde{m}_2} R_{\text{on}})\tilde{i}_L - L\frac{d\tilde{i}_L}{dt} - \frac{\tilde{m}_2}{m_1 + \tilde{m}_2}\tilde{u}_C \tag{8.25}$$

$$0 = \frac{\tilde{m}_2}{m_1 + \tilde{m}_2}\tilde{i}_L - \tilde{m}_2 \frac{m_1 + \tilde{m}_2 - 1}{m_1 + \tilde{m}_2}\tilde{u}_C \tag{8.26}$$

If the outputs $\tilde{i}_L$, $\tilde{u}_C$ are used in the ARRs of the faultless system with nominal parameters then their evaluation gives zero only if a residual is added.

$$0 = E - (R_L + R_{\text{on}})\tilde{i}_L - L\frac{d\tilde{i}_L}{dt} - m_2\tilde{u}_C + r_1 \tag{8.27}$$

$$0 = m_2\tilde{i}_L - \frac{\tilde{u}_C}{R} - C\frac{d\tilde{u}_C}{dt} + r_2 \tag{8.28}$$

Analytical expressions for the nominal ARR residuals are obtained by subtracting corresponding equations.

$$r_1 = \frac{m_1 + \tilde{m}_2 - 1}{m_1 + \tilde{m}_2} R_{\text{on}} \tilde{i}_L - \frac{m_1 \tilde{m}_2}{m_1 + \tilde{m}_2} \tilde{u}_C \tag{8.29}$$

$$r_2 = -\frac{m_1 \tilde{m}_2}{m_1 + \tilde{m}_2} \tilde{i}_L - \tilde{m}_2 \frac{m_1 + \tilde{m}_2 - 1}{m_1 + \tilde{m}_2} \frac{\tilde{u}_C}{R_{\text{on}}} \tag{8.30}$$

For $t < t_3$, the system is healthy and the two residuals in fact vanish as $\tilde{m}_2 = m_2$, $m_1 + m_2 = 1$, and $m_1 \cdot m_2 = 0$.

8.1.5 Parameter Uncertainty Thresholds for ARR Residuals

As explained in Sect. 5.3.2, variations of ARR residuals due to parameter variations are obtained by summing increments of power variables at junctions of an incremental bond graph to which a virtual detector has been attached. Replacement of the switch model as well as the other bond graph elements in the DBG of the boost converter in Fig. 8.2 by their incremental model gives the incremental DBG depicted in Fig. 8.14.

In this incBG, the capacitor C : C_a has been added to resolve the causal conflict at junction 0_1. During formulation of equations for variations of ARR residuals capacitance C_a is set to zero. Again, virtual detectors have been distinguished from detectors representing real sensors by an asterisk.

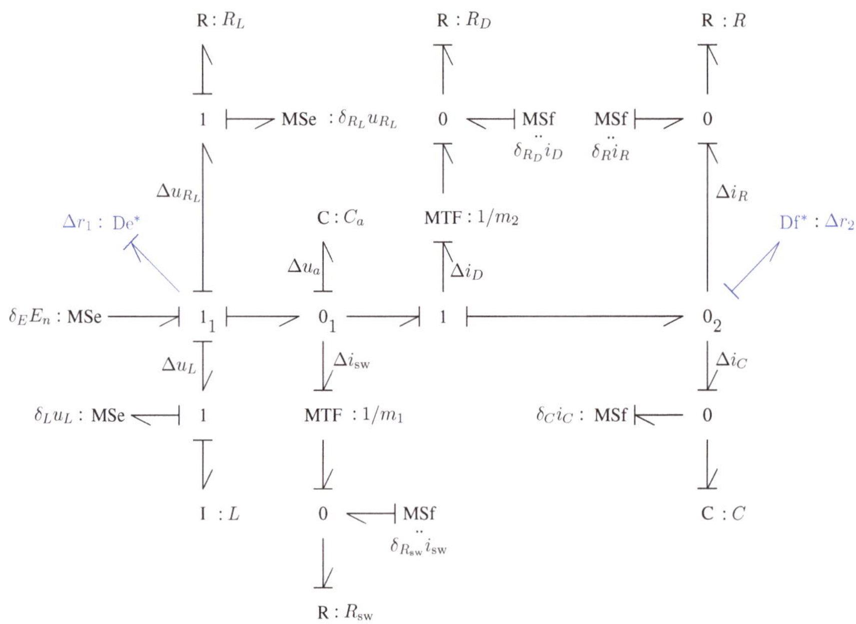

Fig. 8.14 Incremental diagnostic bond graph of the boost converter in Fig. 8.1

Summation of incremental power variables at junctions 1_1 and 0_2 yields the part of the two ARR residuals that is due to parameter uncertainties.

$$
\begin{aligned}
1_1: \quad \Delta r_1 &= \Delta E - \Delta u_{R_L} - \Delta u_L - \Delta u_a \\
&= \delta_E E_n - \delta_{R_L} R_L i_L - \delta_L L \frac{di_L}{dt} - \Delta u_a
\end{aligned}
\tag{8.31}
$$

$$
\begin{aligned}
0_2: \quad \Delta r_2 &= \Delta i_D - \Delta i_C - \Delta i_R \\
&= \Delta i_D - \delta_C \cdot C \frac{du_C}{dt} - \delta_R \frac{u_C}{R}
\end{aligned}
\tag{8.32}
$$

where

$$
\begin{aligned}
\left(\frac{m_1}{R_{\text{sw}}} + \frac{m_2}{R_D} \right) \Delta u_a &= m_1 \delta_{R_{\text{sw}}} i_{\text{sw}} + m_2 \delta_{R_D} i_D \\
&= \delta_{R_{\text{sw}}} i_{\text{sw}} + \delta_{R_D} i_D
\end{aligned}
\tag{8.33}
$$

$$
\begin{aligned}
\Delta i_D &= m_2 \left(\frac{m_2}{R_D} \Delta u_a - \delta_D i_D \right) \\
&= \frac{m_2}{m_1 + m_2} (m_2 i_{\text{sw}} - m_1 i_D)
\end{aligned}
\tag{8.34}
$$

Equations (8.31) and (8.32) simplify if $R_{\text{sw}} = R_D = R_{\text{on}}$ and if the MOSFET switch and the diode commutate conversely, i.e. $m_1 + m_2 = 1 \ \wedge \ m_1 \cdot m_2 = 0$.

$$
\Delta r_1 = \delta_E E_n - \delta_{R_L} R_L i_L - \delta_L L \frac{di_L}{dt} - \delta_{R_{\text{on}}} R_{\text{on}} i_L \tag{8.35}
$$

$$
\Delta r_2 = -\delta_C \cdot C \frac{du_C}{dt} + \delta_R \frac{u_C}{R} \tag{8.36}
$$

Equations (8.35)–(8.36) can be verified by either deducing them from the simplified incBG in Fig. 8.15 or by taking the total differential of the simplified state Eq. (8.11).

Upper bounds for the absolute values of the variations Δr_1 and Δr_2 may be found by application of the triangle inequality.

$$
|\Delta r_1| \leq \delta_E |E_n| + \delta_{R_L} R_L |i_L| + \delta_L L |\frac{di_L}{dt}| + \delta_{R_{\text{on}}} R_{\text{on}} |i_L| \ =: thr_1(t) \tag{8.37}
$$

$$
|\Delta r_2| \leq \delta_C \cdot C |\frac{du_C}{dt}| + \delta_R |\frac{u_C}{R}| \ =: thr_2(t) \tag{8.38}
$$

The thresholds defined by (8.37), (8.38) shall be computed for fault scenario 1 in which the input voltage temporarily drops from 12 to 8 V. As depicted in Fig. 8.6, residual r_1 is sensitive to this fault while residual r_2 is not. For the sake of simplicity, all relative parameter uncertainties are assumed to be 2 %. Figure 8.16 displays the waveforms of the averaged residuals along with their thresholds.

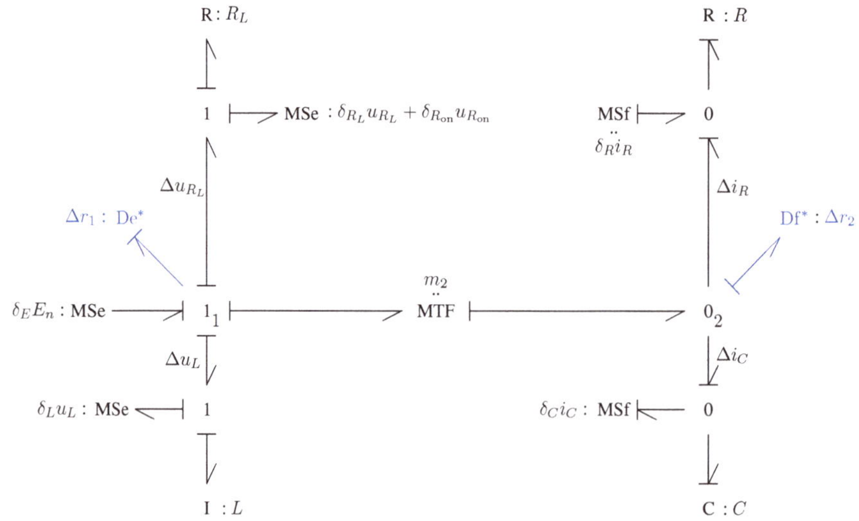

Fig. 8.15 Simplified incremental bond graph of the boost converter in the case the switches commutate conversely

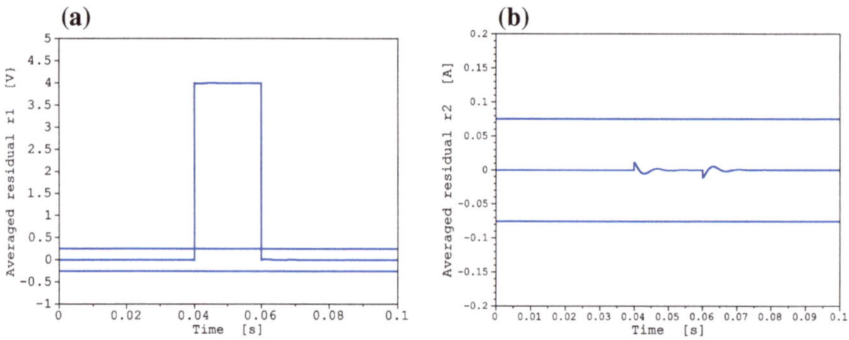

Fig. 8.16 Time evolution of the averaged ARR residuals and their parameter uncertainty bounds in the case of a temporary input voltage drop for $t \in [0.04\,\text{s}, 0.06\,\text{s}]$. **a** Averaged ARR residual $r_{1_a}(t)$ and its thresholds. **b** Averaged ARR residual $r_{2_a}(t)$ and its thresholds

As the input voltage temporarily drops by more than 2 %, averaged residual $r_{1_a}(t)$ temporarily exceeds the parameter uncertainty threshold $thr_1(t)$ indicating a fault.

The magnitude of the thresholds obtained by using simulation results can be analytically checked. Assuming an uncertainty δ for all parameters the variation of the averaged residual r_{1_a} reads

$$\Delta r_{1_a} = \delta \left[E_n - (R_L + R_{\text{on}}] i_{L_a} - L \frac{d i_{L_a}}{dt} \right] \tag{8.39}$$

For the faultless system with nominal parameters residual r_{1_a} equals zero.

$$0 = E_n - (R_L + R_{\text{on}})i_{L_a} - L\frac{di_{L_a}}{dt} - (1-d)u_{C_a} \tag{8.40}$$

Hence, $\Delta r_{1_a} = \delta(1-d)u_{C_a} \approx 0.1508\,V$ for which $thr_1 \approx 0.25\,\text{V}$ is a reasonable bound.

A similar computation of Δr_{2_a}, the variation of the averaged residual r_{2_a}, yields $\Delta r_{2_a} = \delta[2u_{C_a}/R - (1-d)i_{L_a}] \approx 0.073$ which is in good agreement with the threshold thr_2 in Fig. 8.16b.

8.1.6 System Mode Identification

The boost converter circuit has got two switches. In a healthy system, they commutate oppositely so that there are only two system modes. As two sensors have been attached to the circuit, each healthy mode is identified by two ARRs.

Mode 1: MOSFET switch closed, diode open ($m_1 = 1 \wedge m_2 = 0$):

$$0 = r_{11} = E - (R_L + R_{\text{on}})i_L - L\frac{di_L}{dt} \tag{8.41a}$$

$$0 = r_{12} = -\frac{u_C}{R} - C\frac{du_C}{dt} \tag{8.41b}$$

Mode 2: MOSFET switch open, diode closed ($m_1 = 0 \wedge m_2 = 1$):

$$0 = r_{21} = E - (R_L + R_{\text{on}})i_L - L\frac{di_L}{dt} - u_C \tag{8.42a}$$

$$0 = r_{22} = i_L - \frac{u_C}{R} - C\frac{du_C}{dt} \tag{8.42b}$$

The first of the two subscripts denotes the system mode, the second one denotes the index of the ARR. If the system is healthy, evaluation of the two sets of ARRs must give residuals close to zero for one of the two sets. The set of ARRs with residuals close to zero identifies the system mode. If both sets of ARRs have got residuals quite different from zero then the system is in a third faulty mode.

8.1.6.1 System Mode Identification in the Case of a Healthy System

Figure 8.17 displays the time evolution of the residuals of the four ARRs. It can be seen that both residuals r_{11}, r_{12} are equal to zero whenever both residuals r_{21}, r_{22} differ significantly from zero and vice versa.

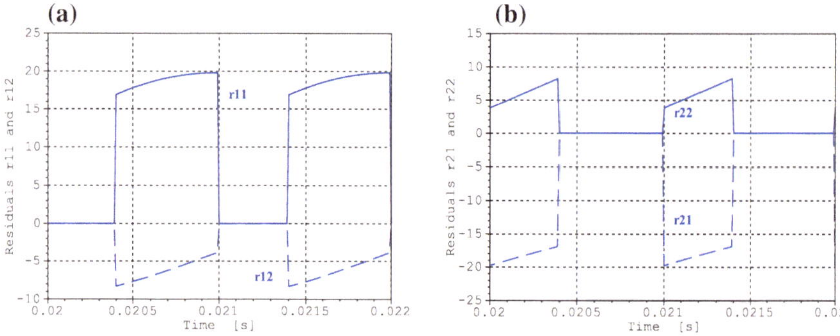

Fig. 8.17 Time evolution of pairs of ARR residuals indicating system modes. **a** Residuals r_{11} and r_{12}. **b** Residuals r_{21} and r_{22}

For instance, $r_{11} = r_{12} = 0$ for $t \in [0.02\,\text{s}, 0.0204\,\text{s}]$, i.e. for $t_{on} = d \cdot T_s = 0.4\,\text{ms}$. This is the portion of the switching time interval T_s the MOSFET switch is on ($m_1 = 1$) while the diode is off ($m_2 = 0$). That is, $r_{11} = r_{12} = 0$ means that the system is in mode 1. On the contrary, $r_{21} = r_{22} = 0$ for $t \in [0.0204\,\text{s}, 0.021\,\text{s}]$, i.e. for $(1 - d) \cdot T_s = 0.6\,\text{ms}$. In this time interval, the MOSFET switch is off and the diode is on. That is, $r_{21} = r_{22} = 0$ identifies system mode 2.

8.1.6.2 System Mode Identification in the Case of a Faulty System

In the case of a temporary drop of the input voltage (Fault scenario 1) the waveforms displayed in Figs. 8.18 and 8.19 are obtained for the four ARR residuals.

In the interval $[0.04\,\text{s}, 0.06\,\text{s}]$, residuals r_{11} and r_{21} are different from zero. There are no time intervals in $[0.04\,\text{s}, 0.06\,\text{s}]$ where both residuals r_{11} and r_{21} or r_{21} and r_{22} are zero. That is, for $t \in [0.04\,\text{s}, 0.06\,\text{s}]$ the system is neither in mode 1 nor in mode 2 but in a third faulty mode.

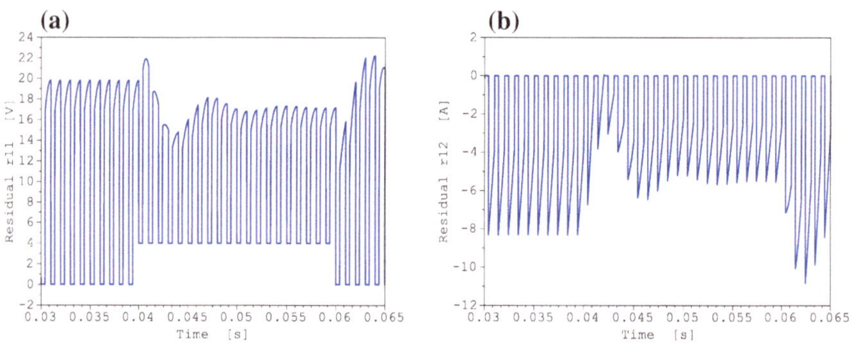

Fig. 8.18 Time evolution of residuals r_{11} and r_{12} in the case of a faulty input voltage. **a** Residuals r_{11}. **b** Residuals r_{21}

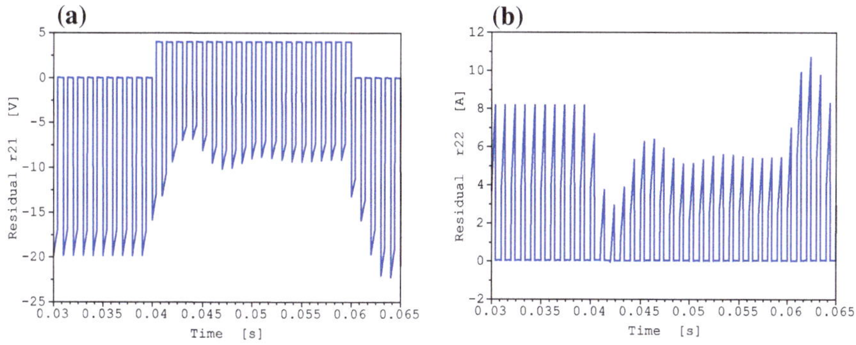

Fig. 8.19 Time evolution of residuals r_{21} and r_{22} in the case of a faulty input voltage. **a** Residuals r_{21}. **b** Residuals r_{22}

8.2 Switched Three-Phase Power Inverter

The previous section considers a simple DC to DC boost converter with two switches controlled by two complementary signals. The healthy boost converter thus may be in one of two feasible modes. Reference [18] studies switch faults in a simple single phase half-bridge inverter. In [17], bond graph-based FDI is applied to a single phase H-bridge inverter. Both works represent switches by means of controlled junctions, i.e. use hybrid bond graphs.

Three-phase DC to AC inverters have got three inverter half bridges with two power switches in each one of them that are controlled by two complementary pulse width modulated (PWM) signals. Thus, the healthy three-phase inverter has six active modes in which one switch in a phase leg is on while the other one is off. Moreover, there may be two so-called zero modes in which all upper three or all three lower switches are turned on at the same time so that the output terminals are shorted. In traditional voltage source inverters (VSIs), the upper and the lower power switch of any phase leg are not gated at the same time as a shoot-through short circuit in a phase leg may damage the inverter. The shoot-through mode is usually a forbidden switching mode. An inverter that allows for the shoot-through mode and intentionally uses this mode to boost the DC voltage is the Z-source inverter (ZSI) introduced by Peng [26] (cf. Sect. 8.3.4).

The power switches suggest to capture the dynamic behaviour of a three-phase inverter by a hybrid model and to study the effect of switch failures on the behaviour of this type of power converter. Three-phase inverters are a common component in many power electronic systems. They are used, for instance,

- in three-phase induction motor drives to produce three-phase variable frequency voltages,
- to convert the DC power produced by the solar panels in a solar power farm into three-phase ac-power, or
- in electric and in fuel cell powered vehicles.

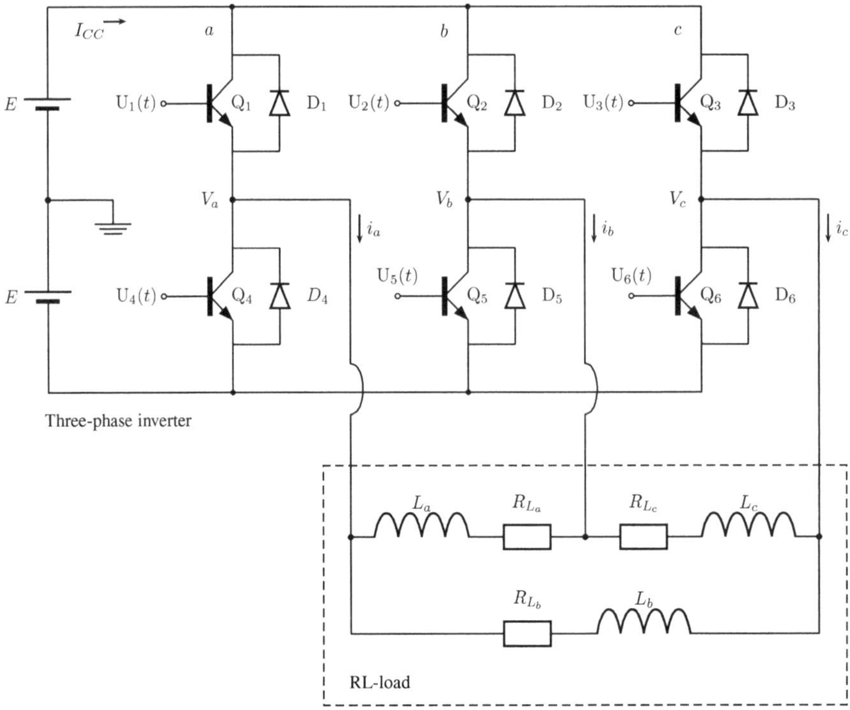

Fig. 8.20 Circuit schematic of a switched three-phase DC to AC power inverter with an RL-load in delta configuration [27]

Figure 8.20 depicts the typical structure of a three-phase DC to AC voltage source inverter (VSI) with a three-phase RL-load in delta configuration.

Control signals $U_1(t)$, $U_2(t)$, and $U_3(t)$ are sinusoidal waveforms with negative values set to zero and positive values set to one and with a phase displacement of 120° between them. The remaining control signals are $U_4(t) = 1 - U_1(t)$, $U_5(t) = 1 - U_2(t)$, and $U_6(t) = 1 - U_3(t)$ respectively. Figure 8.21 displays the waveforms of the signals $U_1(t)$, $U_2(t)$, $U_3(t)$.

Junco et al. use this power inverter as an example for the application of their implementation of the non-standard switched power junctions (SPJs) (Chap. 2) in the modelling language of the 20sim®[1] modelling and simulation software [28]. In [29], and [30, Chap. 8], they model the transistor-diode pair in each half-bridge by means of a SPJ and address the dynamic behaviour of the healthy system.

The transformation of the circuit schematic (Fig. 8.20) into a bond graph is straightforward. Figure 8.22 shows the result. The transistor-diode pair is modelled as a non-ideal switch Sw. For all switches the same ON-resistance R_{on} is assumed.

[1] 20sim is a registered trademark of Controllab Products, Hengelosestraat 500, 7521 AN Enschede, the Netherlands.

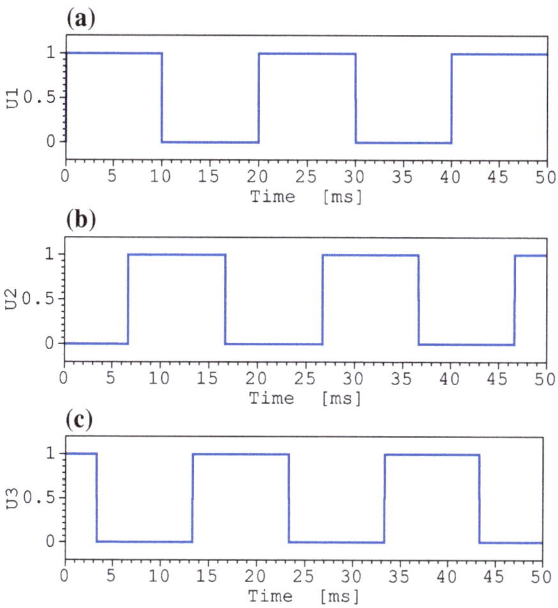

Fig. 8.21 Waveforms of the control voltages $U_1(t)$, $U_2(t)$, $U_3(t)$. **a** Signal $U_1(t)$. **b** Signal $U_2(t)$. **c** Signal $U_3(t)$

The switch state independent conductance causality of the non-ideal switches results in an undetermined causality at 0-junctions 0_a, 0_b, 0_c representing the potentials V_a, V_b, V_c of the nodes between upper and lower switches in the half-bridges. The causal conflicts have been removed by attaching an auxiliary C-storage element to each of these 0-junctions. Equations derived from the bond graph are formulated such that the auxiliary capacitances can be set to zero. Consequently, the small ON-resistance of the switches and the node capacitances do not lead to small time constants.

The three half-bridges are denoted by an index a, b, c respectively. Moreover, an index 'p' is used for the upper switch of a half-bridge. An index 'n' indicates the lower switches. Accordingly, $m_{a_p}(t)$ denotes the discrete state of the upper switch in the half-bridge with the index 'a' at time instant t.

The network of the RL-load is also easily transformed into a bond graph displayed in Fig. 8.23.

From the two bond graphs, the following equations can be deduced. Summation of flows at the junction 0_a in Fig. 8.22 gives

$$\begin{aligned} C_a \dot{V}_a &= i_{a_p} + i_{a_n} - i_a \\ &= \frac{m_{a_p}}{R_{a_p}}(E - V_a) + \frac{m_{a_n}}{R_{a_n}}(-E - V_a) - i_a \end{aligned} \tag{8.43}$$

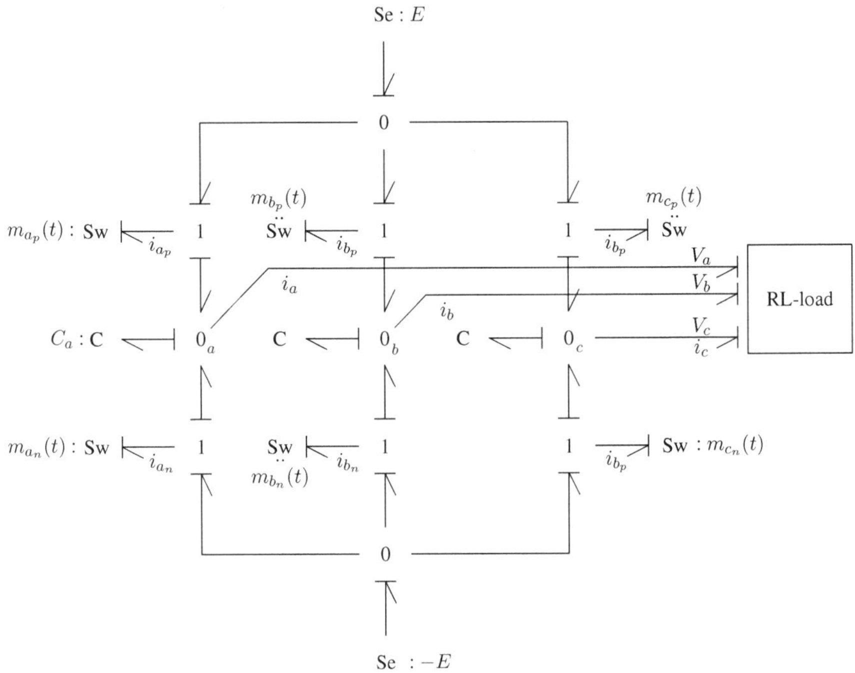

Fig. 8.22 Bond graph of the three-phase power inverter in Fig. 8.20

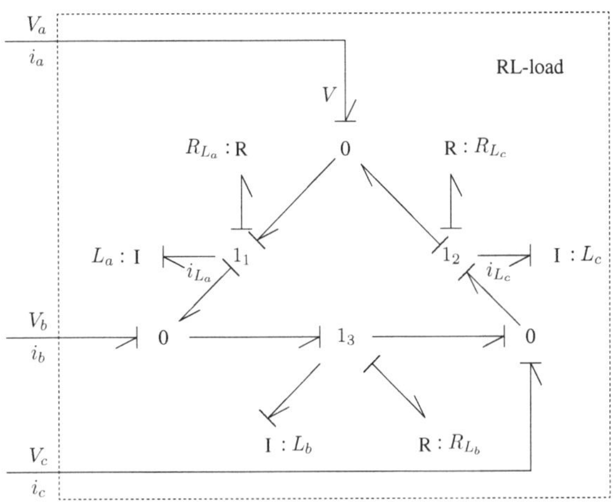

Fig. 8.23 Bond graph of the RL-load in Fig. 8.20

Set $C_a = 0$. Then (8.43) becomes an algebraic equation that can be solved for V_a. Assuming that $R_{a_n} = R_{a_p} = R_a$, the result is

$$V_a = \frac{m_{a_p} - m_{a_n}}{m_{a_p} + m_{a_n}} E - \frac{R_a}{m_{a_p} + m_{a_n}} i_a \tag{8.44}$$

In a healthy inverter, both switches commutate in a complementary fashion so that $m_{a_n} = 1 - m_{a_p}$. Equation (8.44) then simplifies.

$$V_a = (2m_{a_p} - 1)E - R_a i_a \tag{8.45}$$

As the ON-resistance R_a of the switches in leg 'a' is rather small, the second term in (8.45) may become negligible which means that V_a basically switches between $+E$ and $-E$ as would be expected.

Likewise, equations are obtained for the other two legs of the inverter. Since any energy storage in the inverter has been neglected, the model is a purely resistive network with switched resistors.

The equations deduced from the bond graph of the RL-load in Fig. 8.23 read

$$\frac{d}{dt} i_{L_a} = \frac{1}{L_a}[V_a - R_{L_a} i_{L_a} - V_b] \tag{8.46}$$

$$\frac{d}{dt} i_{L_b} = \frac{1}{L_b}[V_b - R_{L_b} i_{L_b} - V_c] \tag{8.47}$$

$$\frac{d}{dt} i_{L_c} = \frac{1}{L_c}[V_c - R_{L_c} i_{L_c} - V_a] \tag{8.48}$$

$$i_a = i_{L_a} - i_{L_c} \tag{8.49}$$

$$i_b = i_{L_b} - i_{L_a} \tag{8.50}$$

$$i_c = i_{L_c} - i_{L_b} \tag{8.51}$$

Summation of the line currents i_a, i_b, i_c yields the nodicity property

$$0 = i_a + i_b + i_c \tag{8.52}$$

not directly displayed by the BG of the RL-load.

The voltages imposed on the RL-load are a switched function of the voltage supply E and of the respective line current. The equations of the inverter and the RL-load indicate that if line currents i_a, i_b and load current i_{L_a} are measured, all other currents and thus potentials V_a, V_b, and V_b can be determined. That is, the following ARRs are obtained.

$$ARR_1: \quad 0 = \underbrace{\frac{m_{a_p} - m_{a_n}}{m_{a_p} + m_{a_n}} E - \frac{R_a}{m_{a_p} + m_{a_n}} i_a}_{V_a} - \underbrace{\frac{m_{b_p} - m_{b_n}}{m_{b_p} + m_{b_n}} E - \frac{R_b}{m_{b_p} + m_{b_n}} i_b}_{V_b}$$
$$- L_a \frac{di_{L_a}}{dt} - R_{L_a} i_{L_a} \tag{8.53}$$

$$ARR_2: \quad 0 = \underbrace{\frac{m_{b_p} - m_{b_n}}{m_{b_p} + m_{b_n}} E - \frac{R_b}{m_{b_p} + m_{b_n}} i_b}_{V_b} - \underbrace{\frac{m_{c_p} - m_{c_n}}{m_{c_p} + m_{c_n}} E - \frac{R_c}{m_{c_p} + m_{c_n}} (-i_a - i_b)}_{V_c}$$
$$- L_a \frac{d}{dt}(i_b - i_{L_a}) - R_{L_b} \underbrace{(i_b - i_{L_a})}_{i_{L_b}} \tag{8.54}$$

$$ARR_3: \quad 0 = \underbrace{\frac{m_{c_p} - m_{c_n}}{m_{c_p} + m_{c_n}} E - \frac{R_c}{m_{c_p} + m_{c_n}} (-i_a - i_b)}_{V_b} - \underbrace{\frac{m_{a_p} - m_{a_n}}{m_{a_p} + m_{a_n}} E - \frac{R_a}{m_{a_p} + m_{a_n}} i_a}_{V_a}$$
$$- L_c \frac{d}{dt}(-i_a + i_{L_a}) - R_{L_c} \underbrace{(-i_a + i_{L_a})}_{i_{L_c}} \tag{8.55}$$

Clearly, ARRs (8.53)–(8.55) are not applicable in the case in which one switch in a half bridge is permanently off. For time intervals in which the other switch in that leg is also off, the denominator becomes zero. Therefore, the ARRs are reformulated. Let m_a, m_b, m_c denote the sum of the two discrete switch states in leg 'a', leg 'b' and 'c' respectively, i.e. $m_a := m_{a_p} + m_{a_n}$.

$$ARR_1': \quad 0 = m_b[(m_{a_p} - m_{a_n})E - R_a i_a] - m_a[(m_{b_p} - m_{b_n})E + R_b i_b]$$
$$- m_b \cdot m_a \left(L_a \frac{di_{L_a}}{dt} + R_{L_a} i_{L_a}\right) \tag{8.56}$$
$$ARR_2': \quad 0 = m_c[(m_{b_p} - m_{b_n})E - R_b i_b] - m_b[(m_{c_p} - m_{c_n})E - R_c(i_a + i_b)]$$
$$- m_c \cdot m_b \left[L_b \frac{d}{dt}(i_b - i_{L_a}) + R_{L_b}(i_b - i_{L_a})\right] \tag{8.57}$$
$$ARR_3': \quad 0 = m_a[(m_{c_p} - m_{c_n})E + R_c(i_a + i_b)] - m_c[(m_{a_p} - m_{a_n})E + R_a i_a]$$
$$- m_a \cdot m_c \left[L_c \frac{d}{dt}(-i_a + i_{L_a}) + R_{L_c}(-i_a + i_{L_a})\right] \tag{8.58}$$

The structure of ARRs (8.56)–(8.58) is indicated by the structural FSM in Table 8.3 in which the rows for the ON-resistance of the switches have been omitted. Their ON-resistance could be set to zero turning them into ideal switches.

Table 8.3 Structural FSM of the three-phase-inverter with sensors Df : i_a, Df : i_b, Df : i_{L_a}

Component	ARR_1	ARR_2	ARR_3	D_b	I_b
Sw : m_{a_p}	1	0	1	1	0
Sw : m_{a_n}	1	0	1	1	0
Sw : m_{b_p}	1	1	0	1	0
Sw : m_{b_n}	1	1	0	1	0
Sw : m_{c_p}	0	1	1	1	0
Sw : m_{c_p}	0	1	1	1	0
R : R_{L_a}	$m_a m_b$	0	0	$m_a m_b$	0
R : R_{L_b}	0	$m_b m_c$	0	$m_b m_c$	0
R : R_{L_c}	0	0	$m_a m_c$	$m_a m_c$	0
L : L_a	$m_a m_b$	0	0	$m_a m_b$	0
L : L_b	0	$m_b m_c$	0	$m_b m_c$	0
L : L_c	0	0	$m_a m_c$	$m_a m_c$	0
Se : E	1	1	1	1	0
Df : i_a	1	1	1	1	0
Df : i_b	1	1	1	1	0
Df : i_{L_a}	1	1	1	1	0

Possible faults in the inverter

In a paper on fault tolerant power inverter topologies used for alternating current (AC) motor drives [30], Welchko et al. distinguish the following faults inside the inverter.

1. Single inverter switch short circuit (the lower switch in a leg is permanently on)
2. Phase-leg short-circuit (both switches in a leg are permanently on)
3. Single inverter switch open-circuit (one switch in a leg is permanently off)
4. Single-phase open-circuit (both switches in a leg commutate in a complementary fashion but the phase line between the half-bridge and the AC motor is permanently disrupted.)

Consider that these faults happen in leg ‘a’. A short circuit failure in the lower switch is then taken into account by letting $m_{a_n} = \tilde{m}_{a_n} = 1$ as of some time point t_f. For $t > t_f$ potential V_a then reads

$$V_a = \frac{m_{a_p} - 1}{m_{a_p} + 1} E - \frac{R_a}{m_{a_p} + 1} i_a \tag{8.59}$$

In the case when both switches in leg ‘a’ are permanently on, the expression for potential V_a simplifies even further.

$$V_a = -\frac{R_a}{2} i_a \tag{8.60}$$

Suppose that the upper switch $\text{Sw} : m_{a_p}$ is switching. Whenever it is on (8.44) yields

$$V_a = E + R_a i_a \tag{8.61}$$

For time intervals in which both switches are off, (8.56) reduces to

$$0 = -m_b R_a i_a \tag{8.62}$$

In that case, the node with the potential V_a is not connected to the voltage $+E$ supply. As its capacitance C_a to ground is neglected, there is no current i_a in phase-line 'a' and $V_a = E$. That is, V_a does not switch between values $+E$ and $-E$ but retains the value $+E$. For the RL-load, phase line current $i_a = 0$ entails $i_{L_c} = i_{L_a}$ which means that two of the three inductors become dependent. That is, the number of state variables is mode-dependent and decreases to two when both switches in a leg are off. In the fourth case, the two switches in leg 'a' operate in a complementary fashion but phase line 'a' is interrupted so that the two load currents i_{L_c}, i_{L_a} are permanently equal.

Extension of the load model

The last two cases require an extension of the load model. The residual sinks introduced in Chap. 2, allow one to keep preferred integral causality when storage elements become dependent. In this application, a switched residual effort sink $\text{rSe} : \lambda$ is used that imposes an effort λ onto the two inductor elements $\text{I} : L_a$ and $\text{I} : L_c$ so that their output currents become equal. Figure 8.24 shows an extended BG that accounts for a mode dependent number of states.

The moduli $b_i(t) \in \{0, 1\}, i = 1, \ldots, 6$, account for various faults in leg 'a' by switching on or off voltage terms according to the fault that has occurred. From the bond graph in Fig. 8.24 the following set of equations is deduced.

$$V = b_1 V_a + b_2 E - b_3 E + b_5 \lambda + b_6 \frac{1}{2}(V_b + V_c) \tag{8.63}$$

$$\frac{d}{dt} i_{L_a} = \frac{1}{L_a}[V - R_{L_a} i_{L_a} - V_b - b_4 \lambda] \tag{8.64}$$

$$\frac{d}{dt} i_{L_b} = \frac{1}{L_b}[V_b - R_{L_b} i_{L_b} - V_c] \tag{8.65}$$

$$\frac{d}{dt} i_{L_c} = \frac{1}{L_c}[V_c - R_{L_c} i_{L_c} - V + b_4 \lambda] \tag{8.66}$$

$$i_a = i_{L_a} - i_{L_c} \tag{8.67}$$

$$i_b = i_{L_b} - i_{L_a} \tag{8.68}$$

$$i_c = i_{L_c} - i_{L_b} \tag{8.69}$$

The effort λ necessary to enforce $i_{L_a} = i_{L_c}$ is obtained by summing voltages at junctions 1_1 and 1_2.

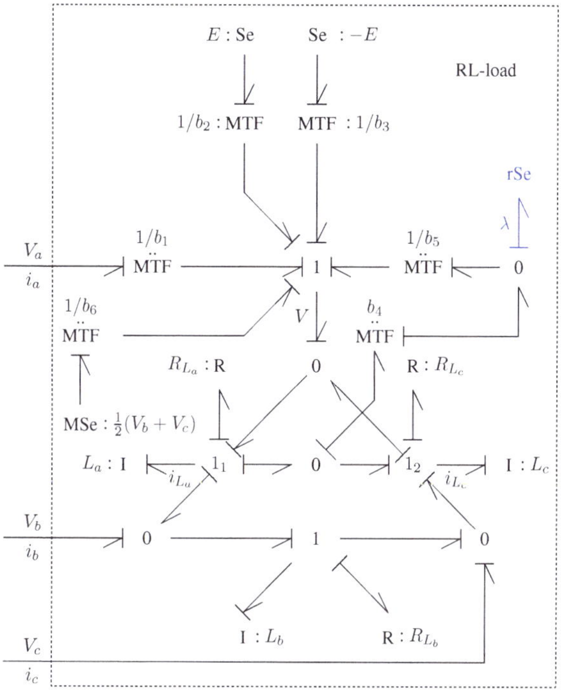

Fig. 8.24 Extended BG model of the RL-load in Fig. 8.20 accounting for various faults in leg 'a' [27]

$$E - V_b - R_{L_a} i_{L_a} - \lambda = V_c - E - R_{L_c} i_{L_a} + \lambda \tag{8.70}$$

Solving (8.70) for λ gives

$$\lambda = E - \frac{1}{2}(V_b + V_c). \tag{8.71}$$

if it is assumed that $R_{L_a} = R_{L_c}$.

When the residual effort sink is activated at time instant t_o, i.e. when the upper switch opens while the lower switch is permanently off, the two load currents i_{L_a} and i_{L_c} jump to a common value i_s so that $i_a = 0$. This jump at time instance t_o means that the common value i_s is to be determined and that numerical integration of the model equations must be re-initialised at t_o. When the upper switch abruptly closes at time instances t_c while the lower switch is still off, then both currents i_{L_a} and i_{L_a} start from their common value.

The magnitude δ of the jump is determined by the requirement that both currents must be equal at the time instant of the switching event.

$$i_{L_a}(t_o) + \delta = i_{L_c}(t_o) - \delta = i_s. \tag{8.72}$$

Hence,

$$\delta = \frac{1}{2}[i_{L_c}(t_o) - i_{L_a}(t_o)] \tag{8.73}$$

In the case of an interrupted phase line 'a', i.e. load currents i_{L_a} and i_{L_c} are permanently equal, $b_4 = 1$ while all other transformer moduli in the extended BG of the RL-load (Fig. 8.24) are set to zero. For computation of the ODEs for the load currents i_{L_a} and i_{L_c} the difference $V - \lambda$ must be known. This difference can be determined by summing efforts at junctions 1_1 and 1_2.

$$0 = V - R_{L_a} i_{L_a} - L_a \frac{d}{dt} i_{L_a} - V_b - \lambda \tag{8.74}$$

$$0 = V_c - R_{L_c} i_{L_c} - L_c \frac{d}{dt} i_{L_c} - V + \lambda \tag{8.75}$$

Assuming that $R_{L_a} = R_{L_c}$ and $L_a = L_c$, the equality of the two inductor currents i_{L_a}, i_{L_c} enforced by the residual effort sink rSe : λ gives

$$V - \lambda = \frac{1}{2}(V_b + V_c) \tag{8.76}$$

Simulation results for this case are presented in Sect. 8.2.2.

8.2.1 *Dynamic Behaviour of the Healthy Inverter*

The waveforms in Fig. 8.25 display the dynamic behaviour of the healthy system. The simulation results are well in agreement with the ones presented in [29, 31]. The parameters used for the simulation runs are listed in Table 8.4.

8.2.2 *Fault Scenario 1: Rupture of Phase Line 'a' for $t \geq t_1$*

Interruption of phase line 'a' means that $i_a = 0$ as of $t \geq t_1$. By consequence, $i_{L_a} = i_{L_b}$. That is, the number of states decreases to two so that the extended load model in Fig. 8.24 is to be used. Let `step(t,t1)` be the unit-step function that jumps from zero to one at $t = t_1$. Then, the residual effort sink in the load model is activated by choosing $b_4(t) = \text{step}(t, t_1)$. The other transformer moduli in (8.63) and (8.64) are $b_1(t) = 1 - \text{step}(t, t_1)$, $b_2(t) = \text{step}(t, t_1)$ and $b_3 = 0$ in this case.

The effect of a rupture of phase line 'a' for $t \geq t_1 = 0.05$ s on some currents is displayed in Fig. 8.26.

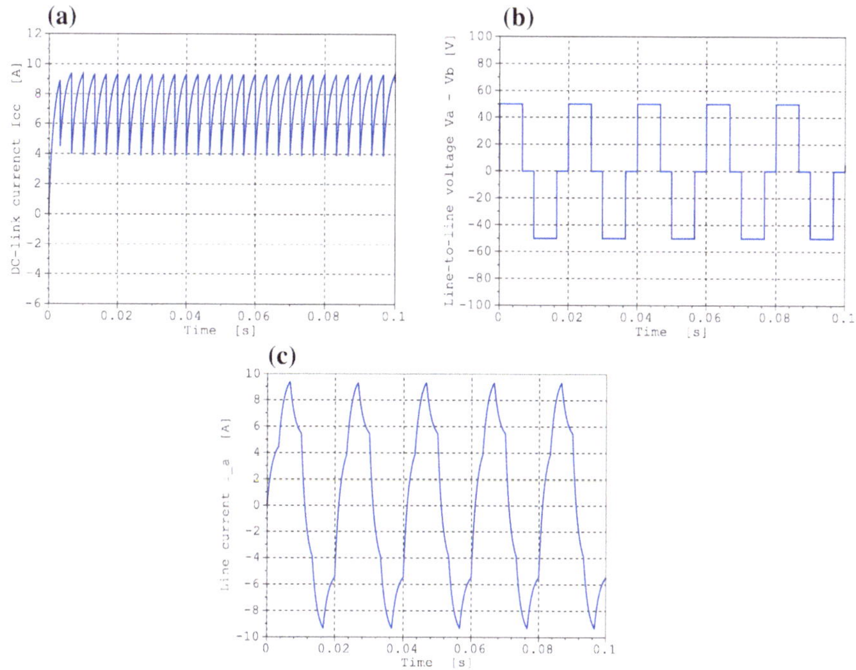

Fig. 8.25 Waveforms of some variables in the case of a healthy inverter. **a** DC-link current I_{cc}. **b** Line-to-line voltage V_{ab}. **c** Line current i_a

Table 8.4 Parameters of the three phase power inverter

Parameter	Value	Units	Meaning
E	25.0	V	Voltage supply
T	0.02	s	Period
Phase φ_1	0	rad	Phase of $\sin(2\pi/T + \varphi_1)$
Phase φ_2	$4\pi/3$	rad	Phase of $\sin(2\pi/T + \varphi_2)$
Phase φ_3	$2\pi/3$	rad	Phase of $\sin(2\pi/T + \varphi_3)$
Maximum	1	[V]	Parameters of the comparator
Minimum	0	[V]	
$R_a = R_b = R_c$	0.1	Ω	ON resistance of the switches
$L_a = L_b = L_c$	0.015	H	Load inductances
$R_{L_a} = R_{L_b} = R_{L_c}$	10	Ω	Load resistances

A vanishing phase line current i_a affects all three ARRs. Their residuals have been computed numerically by coupling a model of the faulty system to a model of the non-faulty system by means of residual effort sinks as depicted in Fig. 8.27.

Figure 8.28 displays the time evolution of the ARR residuals. Apart from the spikes at switching time points, which could be filtered, the values of all three residuals are

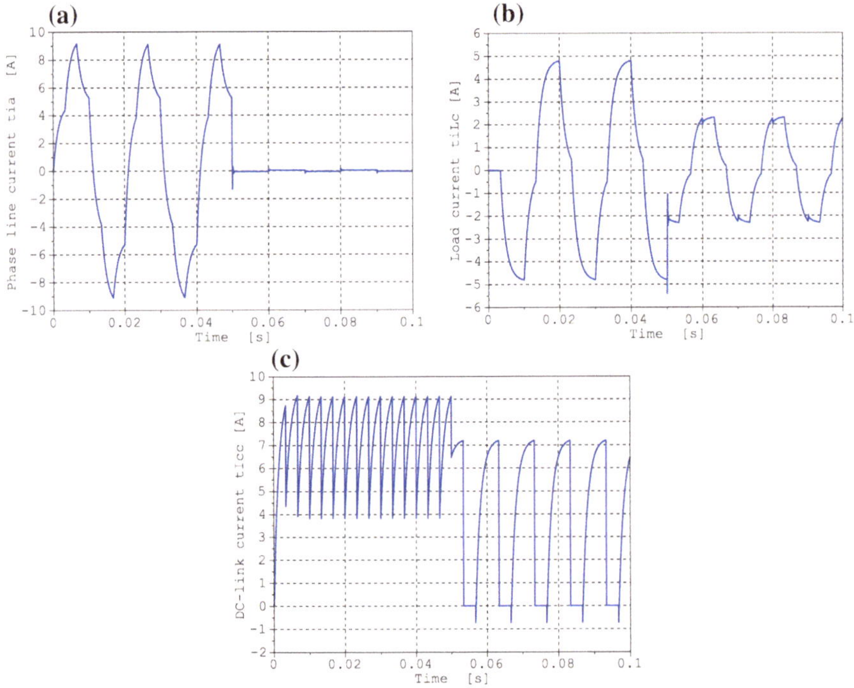

Fig. 8.26 Waveforms of some currents in the case of an interruption of phase line 'a' for $t \geq t_1 = 0.05$ s. **a** Phase line current $\tilde{i}_a$. **b** Inductor current $\tilde{i}_{L_c}$. **c** DC-link current $\tilde{I}_{cc}$

clearly out of any small parameter uncertainty bounds for $t \geq t_1 = 0.05$ s indicating that a fault has happened at $t = t_1$. For identification of the fault more sensors are needed.

8.2.3 *Fault Scenario 2: Abrupt Increase of Load Resistance R_{L_a} at $t = t_2$*

The previous fault scenario considers an open circuit fault in phase line 'a' which can be viewed as a structural change causing a change of the number of state variables. In the following, the effect of an abrupt increase of a load resistance is studied. It is assumed that at $t = t_2 = 0.04$ s load resistance R_{L_a} doubles. That is, the inverter is faultless, the number of state variables remains invariant but a parametric fault occurs in the load. According to the FSM in Table 8.3, this fault can be detected but not isolated. A deviation of ARR residual r_1 from zero could also be due to a fault in the inductance L_a. Figure 8.29a shows that an increase of load resistance R_{L_a} reduces the phase line current i_a.

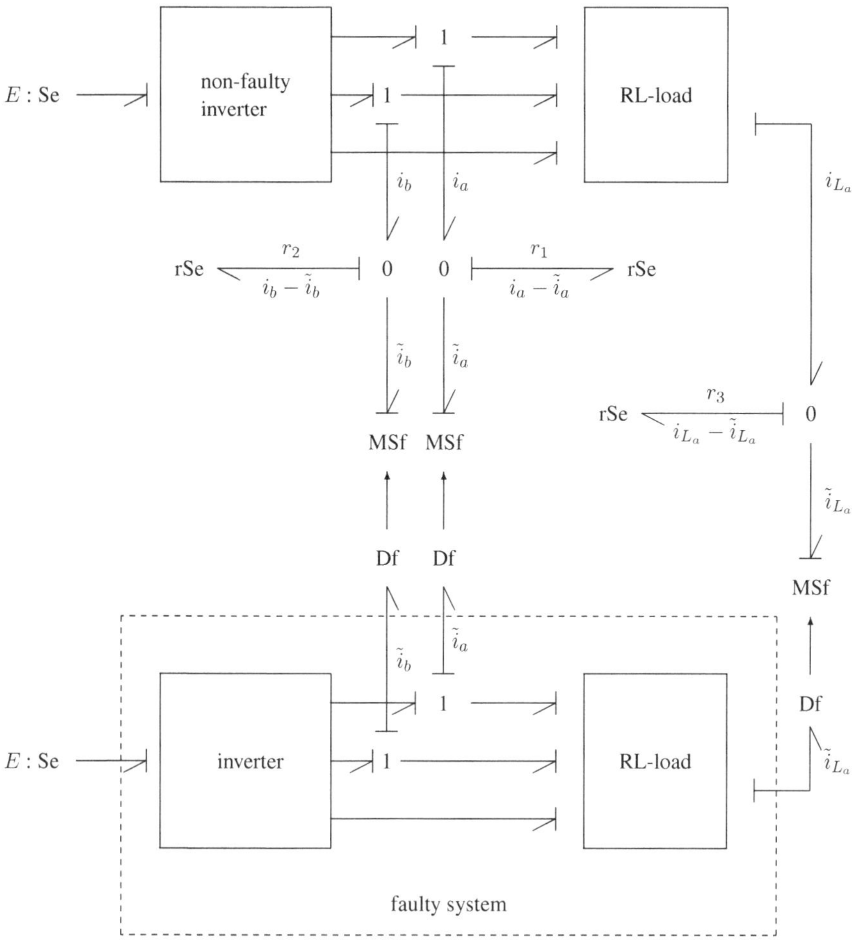

Fig. 8.27 Coupling of a faulty system model to a non-faulty system model for the numerical computation of ARR residuals

Whenever the upper switch in one of the half-bridges is closed, the phase line connected to the half-bridge contributes a current to the DC link current I_{CC}.

$$I_{CC} = m_{a_p} i_a + m_{b_p} i_b + m_{c_p} i_c \tag{8.77}$$

A change of the phase line current i_a thus affects the DC link current (Fig. 8.29b).

According to the FSM in Table 8.3, a parametric fault in resistance R_{L_a} only affects ARR residual r_1. This is confirmed by the waveforms of all three ARR residuals in Fig. 8.30. Apart from the spikes, ARR residual r_1 clearly deviates from zero while the other two residuals do not.

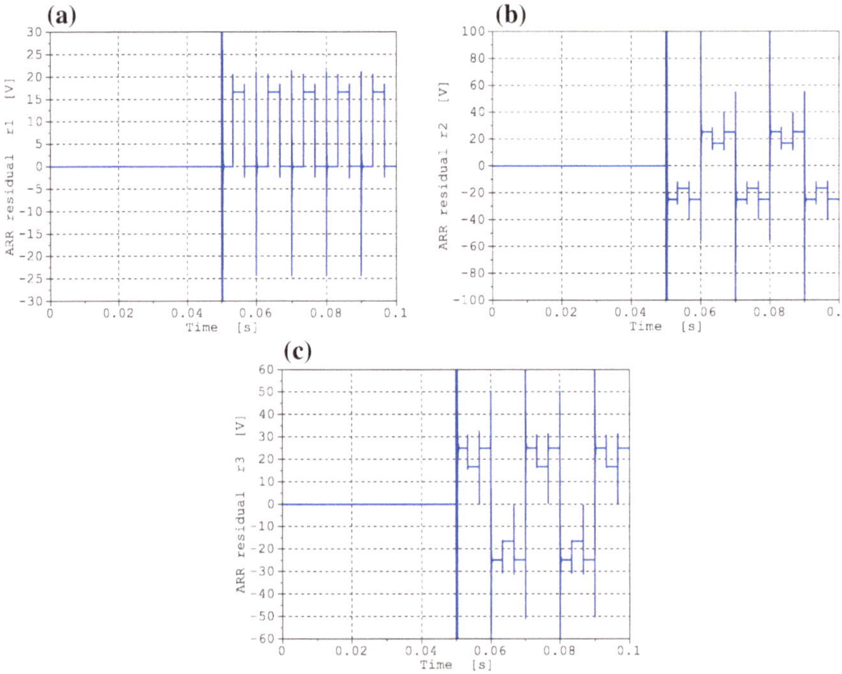

Fig. 8.28 Time evolution of ARR residuals in the case of an interruption of phase line 'a' for $t \geq t_1 = 0.05\,\text{s}$. **a** ARR residual r_1. **b** ARR residual r_2. **c** ARR residual r_3

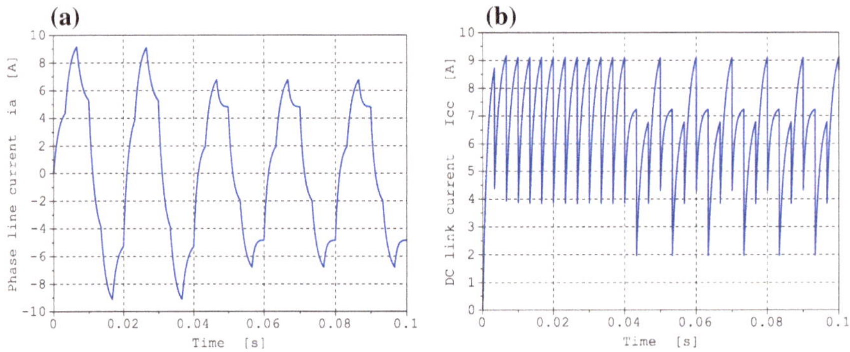

Fig. 8.29 Effect of an increase of load resistance R_{L_a} for $t \geq 0.04\,\text{s}$ on Phase line current i_a and on the DC link current I_{CC}. **a** Phase line current i_a. **b** DC link current I_{CC}

8.2.4 Fault Scenario 3: Open Circuit Fault in Diode D_4

This fault scenario considers the effect of an open circuit fault in the lower diode of leg 'a' and assumes that the upper diode is still perfectly switching. Accordingly,

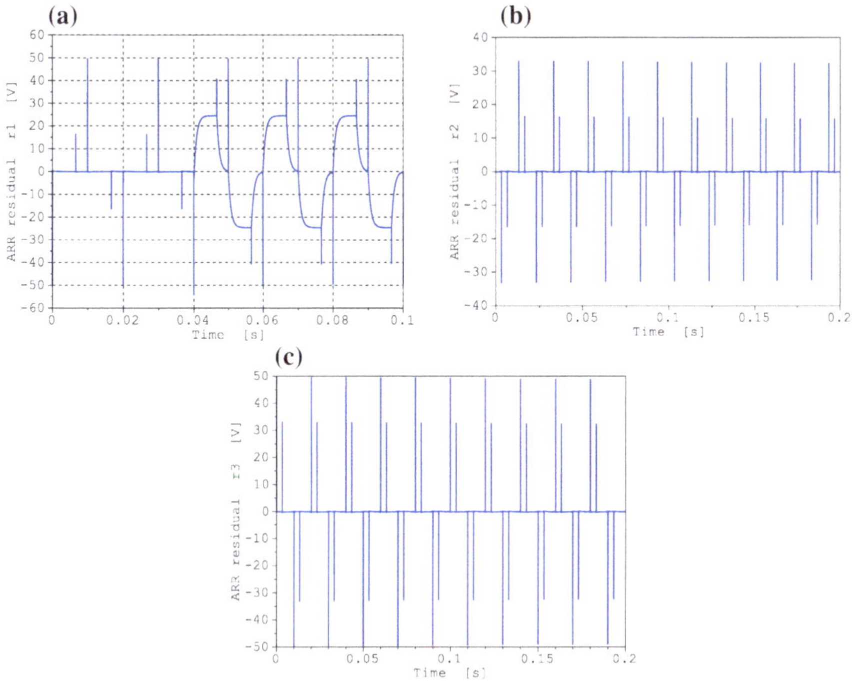

Fig. 8.30 Time evolution of ARR residuals in the case of an increase of load resistance R_{L_a} as of $t \geq t_1 = 0.04$ s. **a** ARR residual r_1. **b** ARR residual r_2. **c** ARR residual r_3

phase line 'a' is temporarily disconnected from voltage supply $+E$ whenever the upper switch S_{a_p} off so that line current i_a vanishes for these time intervals. As a result, the two inductor currents i_{L_a}, i_{L_c} must be equal. In order to preserve integral causality at the two I-elements I : L_a, I : L_c, a residual effort sink rSe : λ is switched on that enforces the outputs of the two I-elements to be equal.

8.2.4.1 Sinusoidal Pulse Width Modulation

In the following, widely used Sinusoidal Pulse Width Modulation (SPWM) [32] is adopted to control the switches of the voltage source inverter. This technique uses three sine waves of frequency f_s mutually phase shifted by 120° against each other, called reference signals. The magnitude of a sine wave is compared with that of a triangular carrier wave of much higher frequency $f_c = 2nf_s$ where $n \in \mathbb{N}$. Whenever the sine wave magnitude is greater than that of the triangular wave, the PWM signal controlling a switch is equal to one, otherwise it is zero. Figure 8.31 depicts a sine wave, a triangular wave with much higher frequency and the generated PWM signal.

In order to avoid undefined switching states and undefined AC output line voltages in the VSI, the switches of any leg in the inverter are controlled so that one switch is on

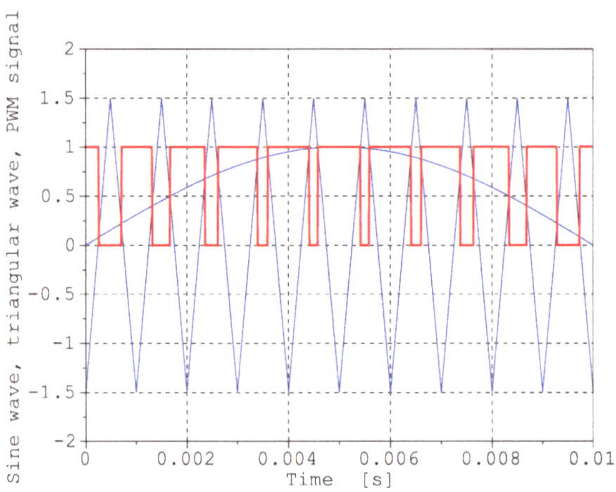

Fig. 8.31 SPWM generation

while the other is off. A healthy inverter controlled by SPWM signals generates three sinusoidal phase currents with a phase shift of 120° against each other. Figure 8.32 shows the inverter's phase line currents obtained by simulation.

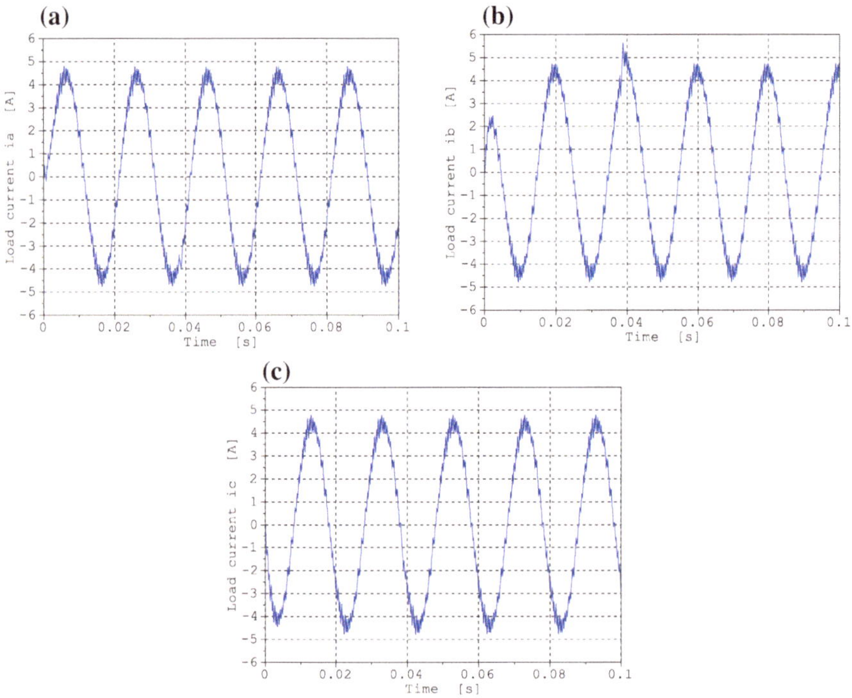

Fig. 8.32 Phase line currents of the healthy inverter. **a** Phase current i_a. **b** Phase current i_b. **c** Phase current i_c

8.2.4.2 Open Switch Fault Detection by Means of Park's Vector Method

One conventional method to detect open switch faults in voltage source inverters is to transform the three line currents i_a, i_b, i_c into a domain with two currents i_d, i_q and to study the shape of the trajectory i_q vs i_d. An advantage of the method is that no further sensors are required.

Let

$$i_a(t) = I_m \sin(\omega_s t + \varphi_0) \tag{8.78a}$$
$$i_b(t) = I_m \sin(\omega_s t - 2\pi/3 + \varphi_0) = I_m \sin(\omega_s t + 4\pi/3 + \varphi_0) \tag{8.78b}$$
$$i_c(t) = I_m \sin(\omega_s t + 2\pi/3 + \varphi_0) \tag{8.78c}$$

be the perfectly balanced three phase currents of a healthy inverter where I_m denotes the amplitude of the currents, $\omega_s = 2\pi f_s$, and φ_0 an initial phase angle. These three phase currents add up to zero, i.e.

$$0 = i_a(t) + i_b(t) + i_c(t)\ , \quad \forall\, t \geq 0 \tag{8.79}$$

The three phase currents i_a, i_b, i_c may be considered the components of a vector $\boldsymbol{i}_{abc}(t)$ which can then be mapped to a vector $\boldsymbol{i}_{dq}(t) = [i_d(t)\, i_q(t)]^T$ in a (d, q) two-phase reference frame by means of Park's transformation [5].

$$\begin{bmatrix} i_d \\ i_q \end{bmatrix} = \sqrt{\frac{2}{3}} \begin{bmatrix} 1 & -1/2 & -1/2 \\ 0 & \sqrt{3}/2 & -\sqrt{3}/2 \end{bmatrix} \begin{bmatrix} i_a \\ i_b \\ i_c \end{bmatrix} \tag{8.80}$$

The benefit of this transformation with regard to FDI is that an open circuit fault in one of the six switches of the inverter can be isolated. For a healthy inverter,

$$|\boldsymbol{i}_{dq}(t)| = \sqrt{i_d^2 + i_q^2} = \sqrt{\frac{3}{2}} I_m\ , \quad \forall\, t \geq 0 \tag{8.81}$$

That is, the trajectory $i_q(t)$ vs $i_d(t)$ is a circle. If there is an open circuit fault in one of the six switches the trajectory becomes a semicircle. The orientation of the axis that cuts the circle into two semicircles depends on the switch that is faulty [10].

For the healthy inverter, the circle in Fig. 8.33a is obtained after the phase line currents i_a, i_b, i_c in Fig. 8.32 have been averaged and transformed into currents i_d, i_q.

The radius of the circle in Fig. 8.33a is $|\boldsymbol{i}_{dq}| = \sqrt{3/2} \times I_m \approx \sqrt{3/2} \times 4.5 \approx 5.5$ in accordance with the amplitude of the phase currents in Fig. 8.32 and with (8.81).

The semicircle in Fig. 8.33b has been obtained by averaging the faulty phase currents in Fig. 8.34 and by applying the transformation in (8.80). The orientation of the diameter cutting the circle into two semicircles is in agreement with Fig. 3 in [10] (cf. also [7], Fig. 7).

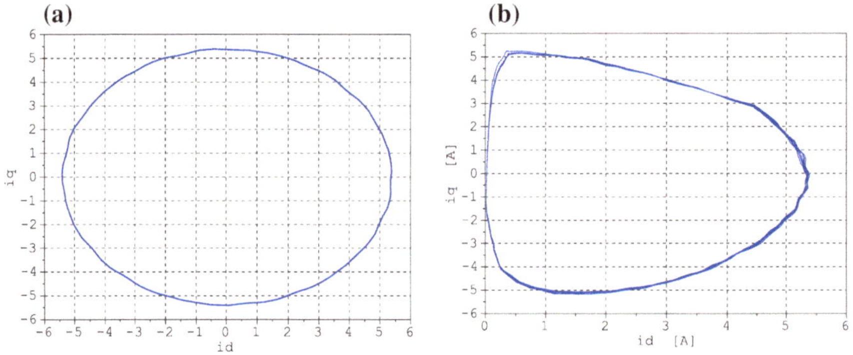

Fig. 8.33 Current space vector trajectory for a faultless inverter and in case of an open circuit fault. **a** Faultless inverter. **b** Open circuit fault in diode D_4

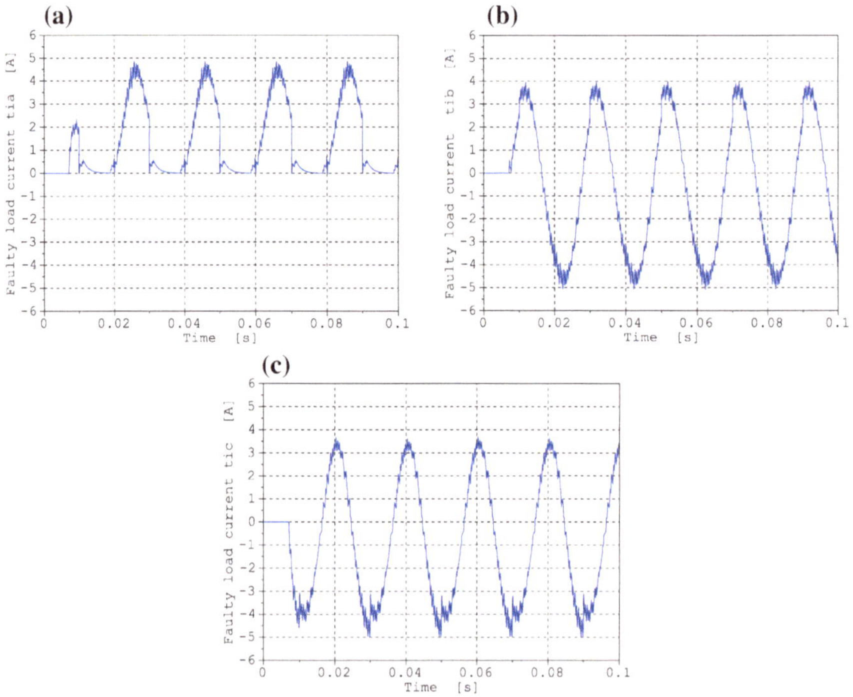

Fig. 8.34 Phase line currents in the case of an open circuit fault in diode D_4. **a** Phase current i_a. **b** Phase current i_b. **c** Phase current i_c

8.2.5 System Mode Identification

In a healthy three-phase inverter, the two switches in a half-bridge commutate in a complementary fashion. That is, 2^3 possible switch state combinations are to be considered. However, according to the controlling signals $U_1(t)$, $U_2(t)$, $U_3(t)$ (Fig. 8.21), the upper switches of the half-bridges are neither open altogether at some time nor is there a time interval for which all of them are closed. Accordingly, the number of active switch state combinations is six.

For a healthy system, denominators in ARRs (8.53)–(8.55) are equal to one so that the ARRs simplify.

$$
\begin{aligned}
ARR_1'': \quad 0 &= (2m_{a_p} - 1)E - R_a i_a - (2m_{b_p} - 1)E - R_b i_b \\
&\quad -L_a \frac{di_{L_a}}{dt} - R_{L_a} i_{L_a} \;=\; g_1(m_{a_p}, m_{b_p}) \qquad (8.82) \\
ARR_2'': \quad 0 &= (2m_{b_p} - 1)E - R_b i_b - (2m_{c_p} - 1)E + R_c(i_a + i_b) \\
&\quad -L_b \frac{d}{dt}(i_b + i_{L_a}) - R_{L_b}(i_b - i_{L_a}) \;=\; g_2(m_{b_p}, m_{c_p}) \qquad (8.83) \\
ARR_3'': \quad 0 &= (2m_{c_p} - 1)E + R_c(i_a + i_b) - (2m_{a_p} - 1)E - R_a i_a \\
&\quad -L_c \frac{d}{dt}(-i_a + i_{L_a}) - R_{L_c}(-i_a + i_{L_a}) \;=\; g_3(m_{c_p}, m_{a_p}) \qquad (8.84)
\end{aligned}
$$

A current system mode can be identified by evaluating the three ARR residuals for all six active system modes. Clearly, for a current system mode to be determined all three residuals must be close to zero. Figure 8.35a displays the waveforms of ARR residuals r_{11}, r_{r21}, r_{31} where the second subscript denotes mode 1. As can be seen, all three residuals are equal to zero for the time intervals $[5/6nT, nT]$, $n \in \mathbb{N}$. Figure 8.35b indicates that the three ARR residuals are equal to zero in mode 5 for the time intervals $[(n-1)T, (n-1)T + 1/6T]$. These results are in agreement with the waveform of system mode $m(t)$ depicted in Fig. 8.35c.

8.3 Three-Phase Full-Wave Rectifier

Figure 8.36 shows the circuit schematic of an uncontrolled three-phase full bridge rectifier with a capacitor on the dc-side and a resistive load used, for instance, in variable voltage variable frequency dc to ac inverters as a front end, or for battery charger applications [34]. Figure 8.37 displays the circuit diagram of a controlled three-phase rectifier.

Replacing passive diodes by transistors has the advantage that the latter devices can be turned on and off whenever this is required. As a result, a controller can switch on and off the six transistors so that the dc link voltage can be kept at a desired level. The voltage at the dc-side is measured, compared with a reference and the error is fed into the controller [35]. When the capacitor discharges, then the error signal causes the controller to generate signals for the gates of the transistors so that an increased current flows from the ac-side to the dc-side that reestablishes the desired dc voltage level. When the capacitor is overcharged, control signals for the transistors are generated so that power is returned to the ac-side and the capacitor is discharged.

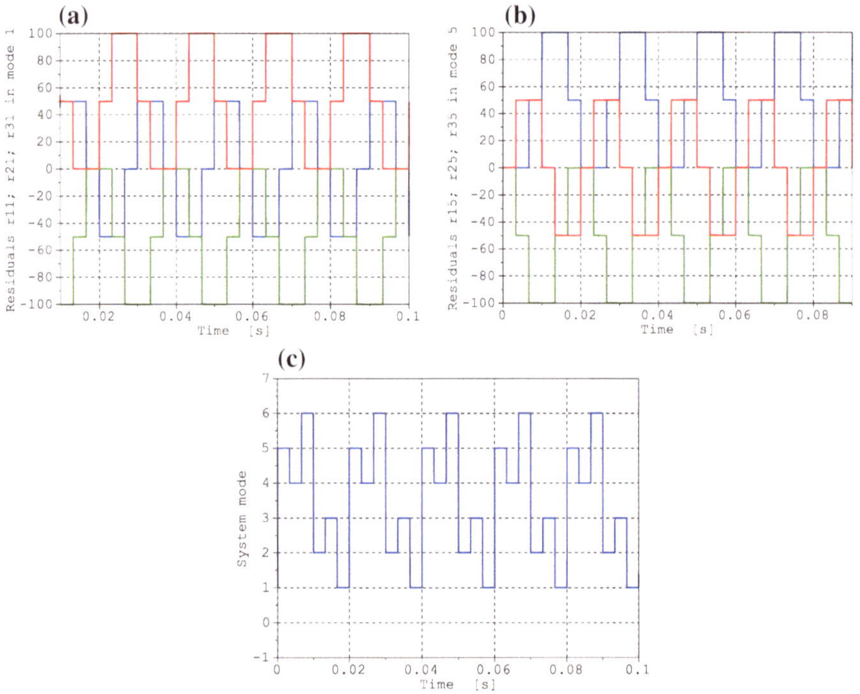

Fig. 8.35 Time evolution of ARR residuals in system modes 1 and 5 and waveform of system mode $m(t)$. **a** ARR residuals r_{11}, r_{21}, r_{31} in mode 1. **b** ARR residuals r_{15}, r_{25}, r_{35} in mode 5. **c** waveform of system mode $m(t)$

This ability of a bi-directional power transfer makes the controlled ac-dc converter suitable for a use between a synchronous machine driven by the combustion engine and the battery-bus in hybrid vehicles [36].

In the sequel, the uncontrolled three phase rectifier in Fig. 8.36 is considered. It is straightforward to convert its circuit schematic into a bond graph. Figure 8.38 displays a diagnostic bond graph with flow sensors Df : i_a, Df : i_b, and Df : i_c for the line currents and an effort sensor De : u_d for the load voltage.

Due to the fixed conductance causality of the switches the causality at junctions 0_a, 0_b, and 0_c remains undetermined. Therefore, C-elements in integral causality have been attached. Their constitutive ODE is formulated so that the capacitance can be set to zero.

Let V_s denote the knee voltage of the diodes. The transformer modulus in the piecewise linear model Sw : D_i of diode D_i, $i = 1, \ldots, 6$, then reads

$$m_i(t) = \begin{cases} 1 & u_i(t) \geq V_s \\ 0 & \text{otherwise} \end{cases} \tag{8.85}$$

where $u_i(t)$ denotes the voltage drop across the i-th diode.

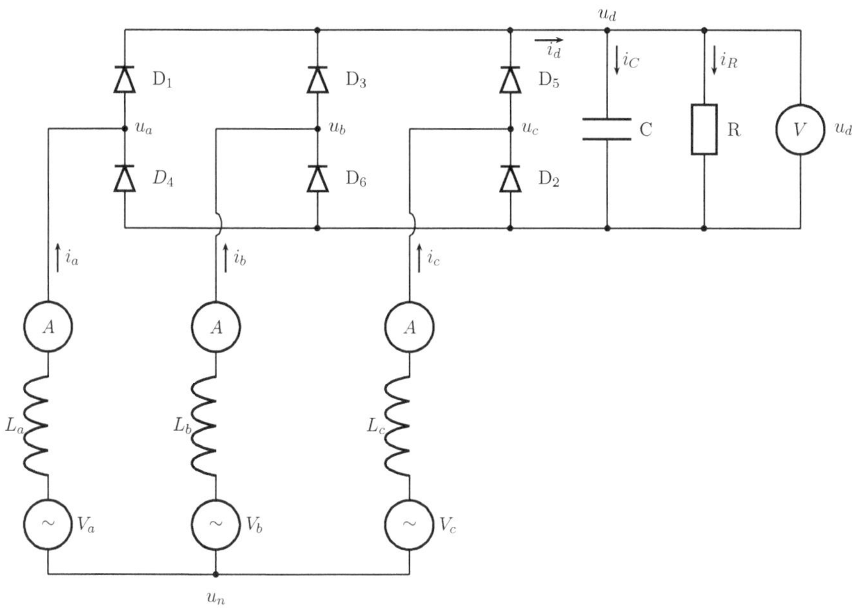

Fig. 8.36 Circuit schematic of an uncontrolled three-phase full bridge rectifier with sensors for the line currents and the load voltage [33]

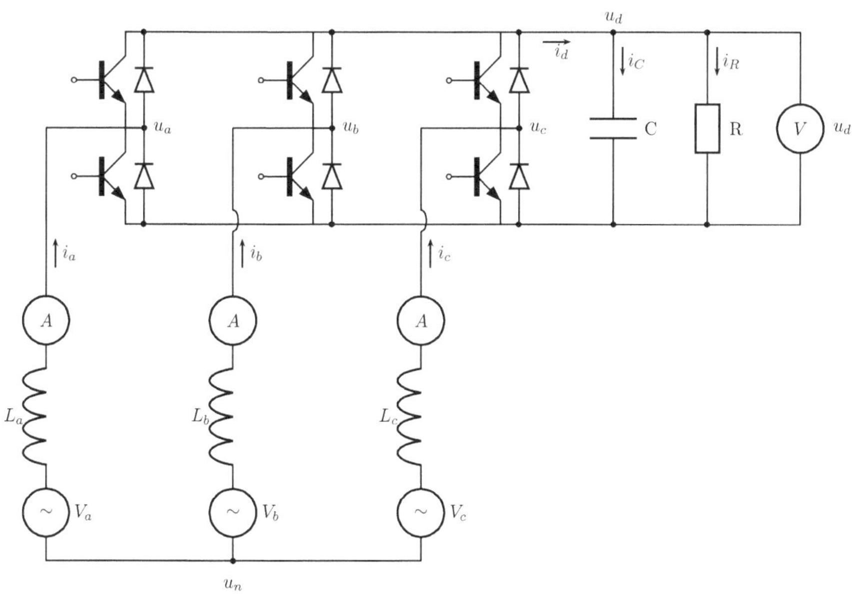

Fig. 8.37 Circuit schematic of a three-phase full bridge rectifier with sensors for the line currents and the load voltage

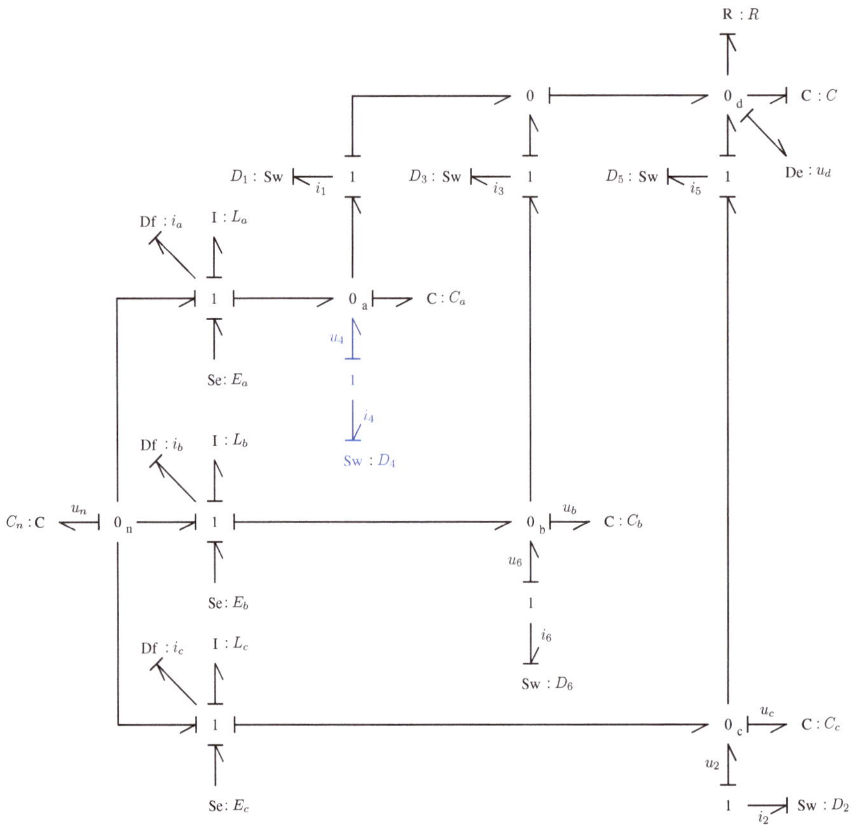

Fig. 8.38 Diagnostic bond graph of the three-phase rectifier in Fig. 8.36

Furthermore, let $g_i()$ denote the piecewise linear approximation of the nonlinear flow versus effort relation of the i-th diode according to Shockley's equation implemented by the switch model $\mathrm{Sw} : D_i$.

$$g_i(m_i, u_i) = \frac{m_i}{R_{\mathrm{on}}}(u_i - V_s) \tag{8.86}$$

Then the following equations can be deduced from the bond graph of Fig. 8.38 by summing efforts and flows respectively at the designated junctions.

$$1_a : \quad 0 = u_n + E_a - u_a - L_a \frac{di_a}{dt} = r_a \tag{8.87}$$

$$1_b : \quad 0 = u_n + E_b - u_b - L_b \frac{di_b}{dt} = r_b \tag{8.88}$$

$$1_c : \quad 0 = u_n + E_c - u_c - L_c \frac{di_b}{dt} = r_c \tag{8.89}$$

$$
\begin{aligned}
0_d: \quad 0 &= g_1(m_1, u_a - u_d) + g_3(m_3, u_b - u_d) \\
&= +g_5(m_5, u_c - u_d) - \frac{u_d}{R} - C\dot{u}_d = r_d
\end{aligned} \tag{8.90}
$$

$$
0_n: \quad 0 = -i_a - i_b - i_c \tag{8.91}
$$

$$
0_a: \quad 0 = i_a + g_4(m_4, -u_a) - g_1(m_1, u_a - u_d) \tag{8.92}
$$

$$
0_b: \quad 0 = i_b + g_6(m_6, -u_b) - g_3(m_3, u_b - u_d) \tag{8.93}
$$

$$
0_c: \quad 0 = i_c + g_2(m_2, -u_c) - g_5(m_5, u_c - u_d) \tag{8.94}
$$

Equations (8.87)–(8.94) relate the measured line currents i_a, i_b, i_c and the measured load voltage u_d to the unknown voltages u_a, u_b, u_c. The latter ones are determined by (8.91)–(8.94) which cannot be solved symbolically for the unknowns. That is, the ARRs cannot be obtained in symbolic form. However, (8.91)–(8.94) can be solved numerically for their unknown and the numerical results can be inserted in (8.87)–(8.94). If the diagnostic bond graph model is connected to a real process so that its evaluation takes place online, the line currents and the load voltage are provided by measurements from the real process. Their differentiation with respect to time in (8.87)–(8.90) must be performed in discrete time. This means that measurement uncertainties, noise, and the sampling time affect the numerical computation of the ARRs. The FSM in Table 8.5 captures the structural information of the implicit ARRs (8.87)–(8.94).

Because of the sum of flows at junctions $0_a, 0_b, 0_c$ the discrete states of both switches in a leg contribute to an ARR residual. For instance, m_1, m_4 contribute to r_a. In a healthy rectifier only one diode in a leg is conducting while the other one is blocking. Moreover, only one of the diodes D_1, D_3, D_5 may conduct at one time

Table 8.5 Structural fault signature matrix of the three phase rectifier with sensors Df : i_a, Df : i_b, Df : i_c and De : u_d

Component	r_a	r_b	r_c	r_d	D_b	I_b
Sw : m_1	1	0	0	1	1	0
Sw : m_2	0	0	1	1	1	0
Sw : m_3	0	1	0	1	1	0
Sw : m_4	1	0	0	1	1	0
Sw : m_5	0	0	1	1	1	0
Sw : m_6	0	1	0	1	1	0
L : L_a	1	0	0	0	1	1
L : L_b	0	1	0	0	1	1
L : L_c	0	0	1	0	1	1
C : C	0	0	0	1	1	0
R : R	0	0	0	1	1	0
Df : i_a	1	1	1	1	1	0
Df : i_b	1	1	1	1	1	0
Df : i_c	1	1	1	1	1	0
De : u_d	1	1	1	1	1	0

instant. The same holds for the lower diodes D_4, D_6, D_2. Faults in all components can be detected but only faults in the line inductors can be isolated with the given sensors. If the sensors can be assumed to be faultless their rows in the bottom part of the FSM can be omitted.

The components contributing to the ARR residuals can also be identified directly on the diagnostic bond graph in Fig. 8.38 by following causal paths from a sensor element in inverse causality into the bond graph and back to it. For instance, there is a causal path from and back to the flow detector Df : i_a.

$$\mathrm{Df} : i_a \rightarrow \mathrm{C} : C_a \rightarrow u_a \rightarrow \mathrm{Sw} : D_1 \rightarrow i_1 \rightarrow \mathrm{C} : C_a \rightarrow u_a \rightarrow \mathrm{Df} : i_a$$

That is, residual r_a depends on diode D_1. Likewise, a causal path shows that r_a also depends on diode D_4.

$$\mathrm{Df} : i_a \rightarrow \mathrm{C} : C_a \rightarrow u_a \rightarrow -u_4 \rightarrow \mathrm{Sw} : D_4 \rightarrow i_4 \rightarrow \mathrm{C} : C_a \rightarrow u_a \rightarrow \mathrm{Df} : i_a$$

Furthermore, causal paths can be identified indicating that residual r_b depends on diodes D_3 and D_6, while residual r_c depends on diodes D_5 and D_2. The following two causal paths show that r_d depends on diodes D_1, D_4.

$$\begin{aligned} &\mathrm{De} : u_d \rightarrow \mathrm{Sw} : D_1 \rightarrow i_1 \rightarrow \mathrm{De} : u_d \\ &\mathrm{De} : u_d \rightarrow \mathrm{Sw} : D_1 \rightarrow i_1 \rightarrow \mathrm{C} : C_a \rightarrow u_a \rightarrow -u_4 \rightarrow \mathrm{Sw} : D_4 \rightarrow i_4 \rightarrow \\ &\mathrm{C} : C_a \rightarrow u_a \rightarrow \mathrm{Sw} : D_1 \rightarrow i_1 \rightarrow \mathrm{De} : u_d \end{aligned}$$

There are further causal paths showing that r_d also depends on the remaining four diodes. Note that the residuals do not depend on the auxiliary capacitors as their parameter is set to zero.

8.3.1 Dynamic Behaviour of the Healthy Rectifier

In order to have a reference for the subsequent study of two fault scenarios, first, the dynamic behaviour of the healthy system is considered. To that end, the DAE system deduced from the diagnostic bond graph in Fig. 8.38 has been formulated as a Scilab script and has been computed by means of the dassl solver [37]. Simulation runs have used the parameters in Table 8.6. Alternatively, equations may be coded in the modelling language Modelica. The compiler Modelicac may then be used to generate an implicit Scicos block [38] in order to perform a simulation by means of the Scilab toolbox Scicos [39]. Scicos uses the solver DASKR [40].

Figure 8.39 shows the waveforms of line voltages u_a, u_b, u_c and the load voltage u_d. As to be expected for a full-wave rectification, the maximum value of the latter is $\sqrt{3}\,E \approx 1.73 \times 100\,\mathrm{V}$. Moreover, the frequency of the output voltage is three times the frequency of the line voltages.

Figure 8.40a displays the waveform of the scaled current $i_d = i_1 + i_3 + i_5$. The waveform of the scaled current through diode D_4 is depicted in Fig. 8.40b as an example of the six diode currents.

Table 8.6 Parameters of the uncontrolled three phase rectifier in Fig. 8.36

Parameter	Value	Units	Meaning
E	100	V	Amplitude of the line voltages
f	50	Hz	Frequency of the line voltages
φ_a	0	rad	Phase a
φ_b	$2\pi/3$	rad	Phase b
φ_c	$4\pi/3$	rad	Phase c
$L_a = L_b = L_c$	10^{-5}	mH	Phase line inductances
V_s	0.7	V	Knee voltage of the diodes
R_{on}	0.1	Ω	ON resistance of the diodes
C	50	μF	Capacitance
R	50	Ω	Load resistance

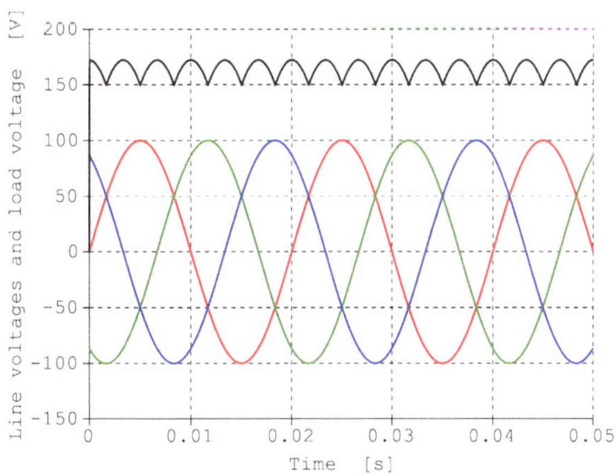

Fig. 8.39 Waveforms of the line voltages u_a, u_b, u_c and the load voltage u_d (*upper graph*)

The enlargement of Fig. 8.40b shows that the length of the time period diode D_4 is on corresponds to $2\pi/3$ while the diode is off for a time period corresponding to $4\pi/3$ (Fig. 8.41). Furthermore, in the case of a pure resistive load R, the maximum of a diode current equals $\sqrt{3}\,E/R \approx 1.73 \times 100\,\text{V}/50\,\Omega = 3.46\,\text{A}$. For a rectifier with an RC-load and a capacitance $C = 50\,\mu\text{F}$, this value is 4.2 A as Fig. 8.41 shows. If a simulation run is performed with $C = 5\,\mu\text{F}$ then the maximum of diode current i_4 equals 3.5 A.

In the following, two fault scenarios are studied. In both cases, the single fault hypothesis is adopted. Names of faulty quantities carry a tilde to distinguish them from names of their faultless counterparts. In figures, this is expressed by preceding names with the letter 't'.

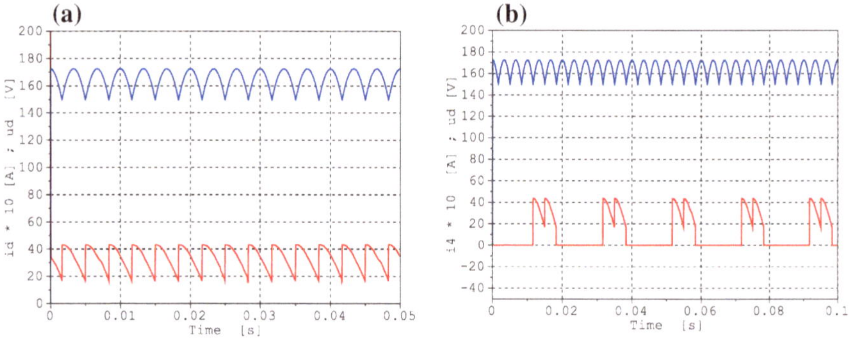

Fig. 8.40 Waveforms of the scaled current i_d, the diode current i_4, and the load voltage u_d. **a** Current i_d (*lower graph*), load voltage u_d (*upper graph*). **b** Current i_4 (*lower graph*), load voltage u_d (*upper graph*)

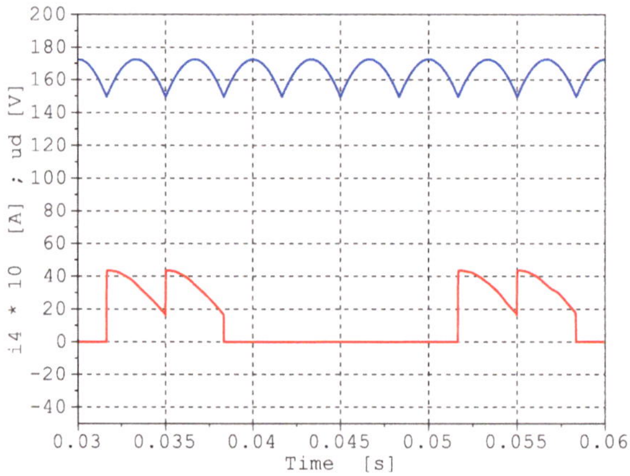

Fig. 8.41 Enlargement of Fig. 8.40b

8.3.2 Fault Scenario 1: Open-Circuit Fault in Diode 4

In this scenario, it is assumed that diode D_4 is no longer conducting for $t > t_2 = 0.02$ s. The simulation result in Fig. 8.42a indicates that, in fact, the diode is no more conducting for $t > t_2$. As can be seen, this failure has a significant impact on the output voltage u_d. Figure 8.42b shows the effect of this open-circuit fault on the current i_d.

Figure 8.43 depicts the impact of an open-circuit fault in diode D_4 on the phase-line currents. For $t > 0.02$ s, negative values of line current $\tilde{i}_a$ are missing and current $\tilde{i}_d$ drops to zero in those time intervals in which $\tilde{i}_a$ remains zero (Fig. 8.43a). From Fig. 8.43b it can be seen that positive values of line currents $\tilde{i}_b$, $\tilde{i}_c$ are affected as of $t_2 = 0.02$ s.

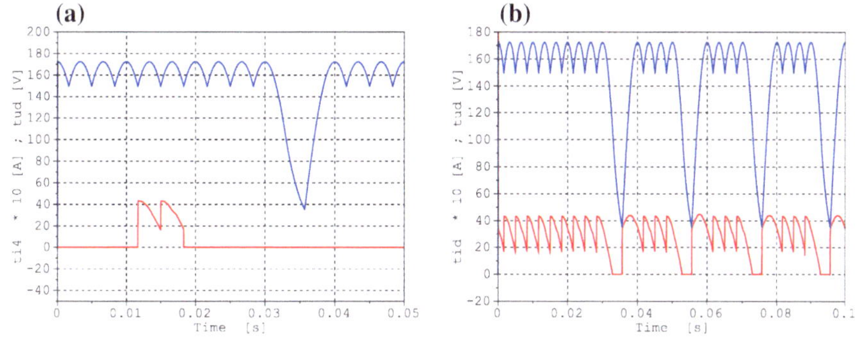

Fig. 8.42 Waveforms of the scaled faulty diode current $\tilde{i}_4$, the faulty current $\tilde{i}_d$, and the faulty output voltage $\tilde{u}_d$ in case of an open-circuit fault of Diode D_4. **a** Scaled faulty diode current $\tilde{i}_4$, faulty output voltage $\tilde{u}_d$. **b** scaled faulty diode current $\tilde{i}_4$, faulty current $\tilde{i}_d$

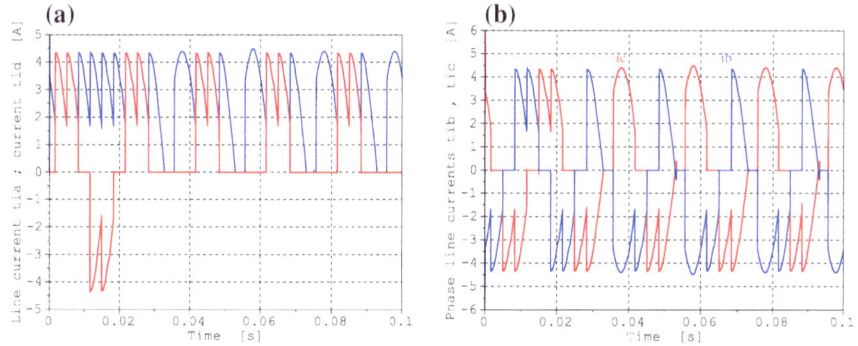

Fig. 8.43 Impact of an open-circuit fault in diode D_4 on the phase-line currents. **a** Phase-line current $\tilde{i}_a$, current $\tilde{i}_d$. **b** Phase line currents $\tilde{i}_b$, $\tilde{i}_c$

An open-circuit fault in diode D_4 for $t > t_2$ means that the current through the diode, $i_4(t) = g_4(m_4(t), -u_a(t))$, vanishes. This affects the node potential u_a. The latter one is the solution of the implicit algebraic Eq. (8.92) which in turn results in a residual $r_a \neq 0$ according to (8.87).

Figure 8.44a clearly shows that ARR residual r_a significantly deviates from zero whenever the current through diode D_4 is missing. In those time intervals in which diode D_4 is expected to be off, its failure is not detected.

According to the structural FSM in Table 8.5, a faulty diode D_4 also affects ARR residual r_d. However, the impact on r_d is negligible (Fig. 8.44b). A residual $r_a \neq 0$ does not permit to conclude that diode D_4 is faulty. It could also be the other diode D_1 in leg 'a' that causes residual r_a to produce values significantly different from zero. A fault in one of the two diodes in a leg of the rectifier can be detected but cannot be isolated with the given sensors.

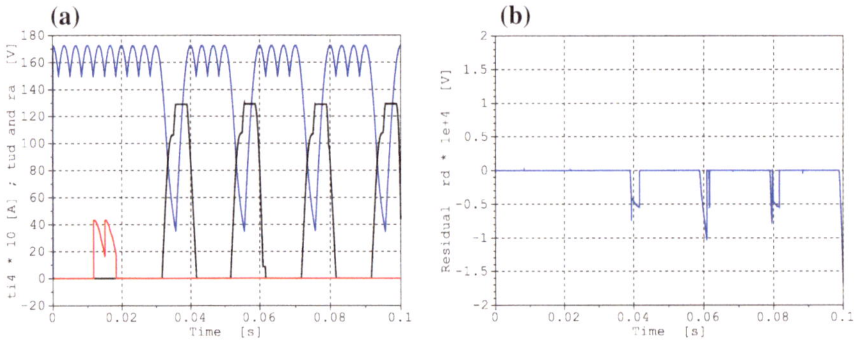

Fig. 8.44 Waveforms of the scaled faulty diode current $\tilde{i}_4$, the faulty output voltage $\tilde{u}_d$ and residual r_a and r_d in the case of an open-circuit fault of Diode D_4. **a** Residual r_a. **b** Residual r_d

The ARR residuals have been computed numerically. To that end, a bond graph model of the faulty rectifier has been coupled to a bond graph of the healthy system by means of residual sinks as displayed in Fig. 8.45.

8.3.3 Fault Scenario 2: Short-Circuit Fault in Diode 4

In this fault scenario, it is assumed that diode D_4 is permanently conducting as of $t_2 = 0.02\,\mathrm{s}$ and that its resistance is $1.2\,\Omega$. This means that there is a short-circuit

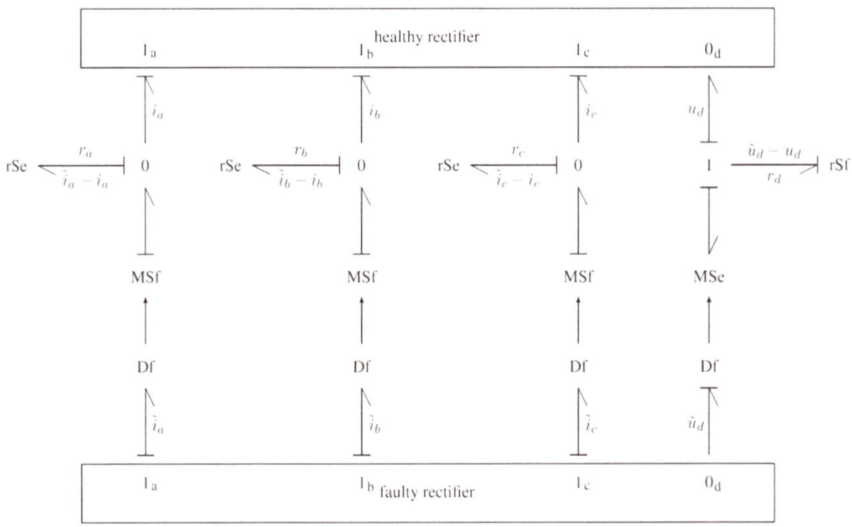

Fig. 8.45 Coupling of the bond graphs of the faulty rectifier and the faultless system by means of residual sinks

in phase leg 'a' whenever the upper switch is on. This mode is called shoot-through mode in the literature. In general, it is a forbidden mode in voltage source inverters (VSIs) as a shoot-through short circuit current would damage the inverter. Therefore, the upper and the lower switch of any of the three phase legs are never gated on at the same time. However, a switch may fail and become permanently conducting as of some time instant. A circuit that responds to a short-circuit in a phase leg and isolates the voltage source from the fault current is the Z-source DC circuit breaker proposed by Corzine and Ashton [41]. Its details are briefly considered in the next section.

Figure 8.46a displays the diode current i_4 in the healthy system and the diode current $\tilde{i}_4$ in the faulty system. As the two currents are identical when D_4 is on in the healthy system, the current i_4 has been scaled to better distinguish it from the faulty one. A comparison with Fig. 8.42 shows that a short-circuit fault in diode D_4 affects the output voltage u_d less than an open switch fault. The values of $\tilde{u}_d$ are lower by about 10 V whenever the faulty diode current $\tilde{i}_4$ is negative.

According to the FSM in Table 8.5 a fault in one of the two diodes in leg 'a' affects residuals r_a and r_d. As can been seen from Fig. 8.46a both residuals clearly deviate from zero whenever the diode D_4 should be OFF blocking negative currents.

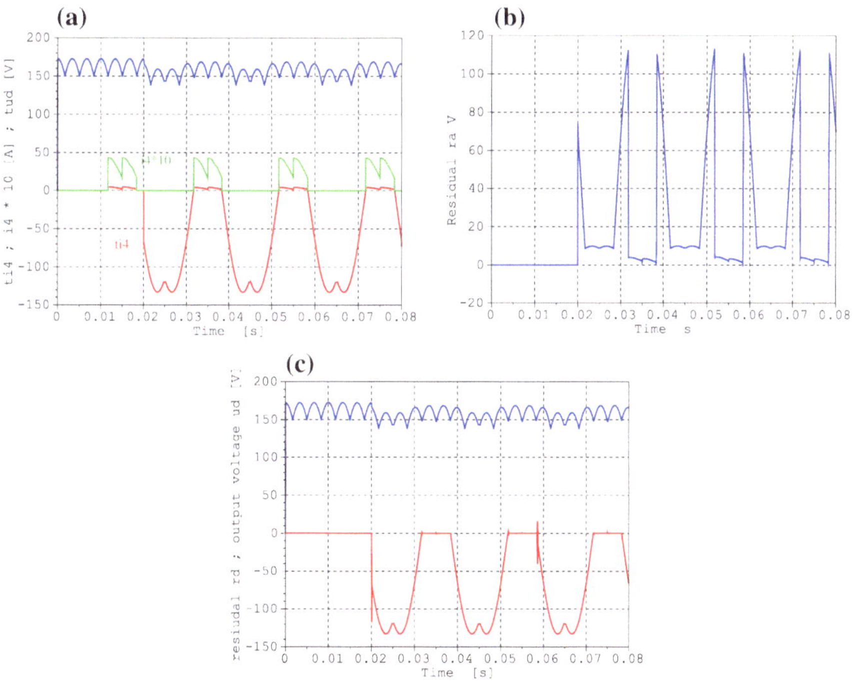

Fig. 8.46 Waveforms of the scaled diode current i_4, the faulty diode current $\tilde{i}_4$, and ARR residuals r_a and r_d in case of a short-circuit fault in diode D_4. **a** Diode current i_4, faulty diode current $\tilde{i}_4$. **b** Residual r_a. **c** Residual r_d

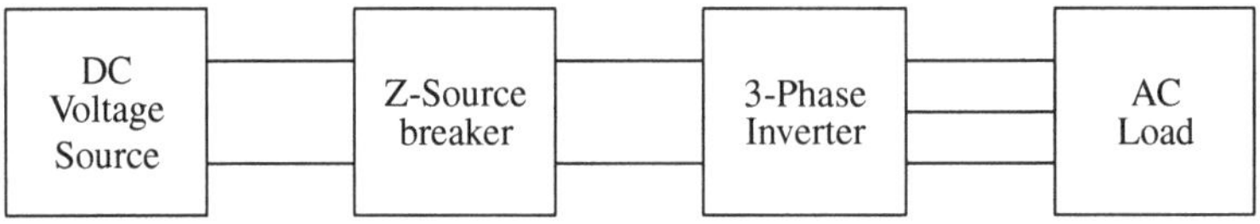

Fig. 8.47 Circuit breaker coupling a DC voltage source to a 3-phase inverter

8.3.4 Short-Circuit Fault in the Load of a Z-Source DC Circuit Breaker

An interesting feature of the Z-source DC circuit breaker is its ability to automatically respond to a short-circuit fault in its DC load and to protect the power source from the fault. A Z-source circuit breaker between a power source and an inverter (Fig. 8.47) allows to simultaneously close the upper and the lower power switch of an inverter phase leg. The combination of a Z-source circuit and an inverter is known as Z-source inverter (ZSI). If both switches of a phase leg are on, the inverter is said to be in shoot-through mode. That is, the Z-source inverter has another mode in addition to the six active modes and the two null modes of a traditional 3-phase inverter.

A number of research works on Z-source inverters has been reported in the literature (see, for instance, [42–44]). A bond graph model of a mono-phase Z-source inverter with a series RL load has been presented in [45]. The approach uses switched power junctions (SPJs) [20] and residual sinks [46].

In the following, a bond graph model of the Z-source breaker and a short-circuit fault in its load shall be briefly considered. The circuit design of the Z-source breaker proposed by Corzine and Ashton [41] has been inspired by the Z-source inverter introduced by Peng [26]. Figure 8.48 shows the circuit schematic.

Fig. 8.48 Schematic of the Z-source DC circuit breaker

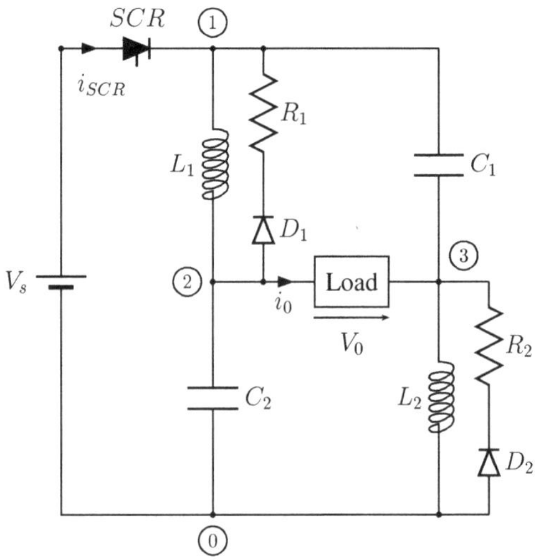

The Z-source DC circuit breaker basically consists of a silicon controlled rectifier (SCR) and two crossed L-C series connections. In case there is no fault, the SCR is on and the capacitors are charged by the voltage source. In steady state, the capacitor currents are zero, the voltages across the inductor vanish and a constant current flows through the series connection of inductors and load. Suppose that the resistances of the inductors can be neglected and that the load is the parallel connection of a load resistor R_L and a load capacitor C_L. Then steady-state values are:

$$u_{C_1} = u_{C_2} = V_s \tag{8.95}$$

$$V_o = V_s \tag{8.96}$$

$$i_o = \frac{V_s}{R_{\text{on}} + R_L} \tag{8.97}$$

where R_{on} denotes the ON-resistance of the SCR.

When the load resistance R_L drops to a very low value in a small time interval Δt then the Z-source output voltage $V_o = R_L i_o$ goes to zero and there there will be a fault current through the Z-source capacitors back to the DC-voltage source resulting in an increase of the capacitor current i_{C_1} and a drop of the SCR forward current i_{SCR} to zero during the time interval Δt. This will cause the SCR to commutate off. A control circuit can then remove the gate current from the SCR so that the SCR will remain off as of that time instant. As a result, the fault current is isolated from the voltage source and the latter one is protected against a large fault current after the voltage drop across the SCR, u_{SCR}, becomes negative. Clearly, when $V_o = 0$, the voltage drop across the inductor must be equal to the voltage drop across the capacitor in each of the two meshes.

Once the SCR is off, two series L-C branches connected to the load of low impedance and disconnected from the voltage source will start to resonate. When the inductor voltages become negative, the wheel diodes turn on. The capacitor currents go to zero and the stored energy will be dissipated in the two inductor-diode-resistor loops.

If the SCR is considered a non-ideal switch, conversion of the circuit diagram in Fig. 8.48 into a BG (Fig. 8.49) is straightforward. If there is no short-circuit fault in the load, modulus m of the MTF is equal to one. In the case of a short-circuit in the load evolving over a short time interval, the current through the SCR goes to zero. After this transition mode, a control circuit removes the gate current from the SCR so that the SCR remains off. This is modelled by the signal block that sets the MTF modulus to zero. As a result, the voltage drop across the SCR, u_{SCR}, is still computed but the current i_{SCR} remains zero. That is, the BG in Fig. 8.49 with invariant causalities captures the fault-free mode, the transition mode, and the shoot-through mode without using SPJs. Also, residual sinks are not needed.

From the BG in Fig. 8.49, the following ordered set of equations is easily deduced

$$u_{SCR} = Vs - u_{C_1} - u_{C_2} + V_o \tag{8.98}$$

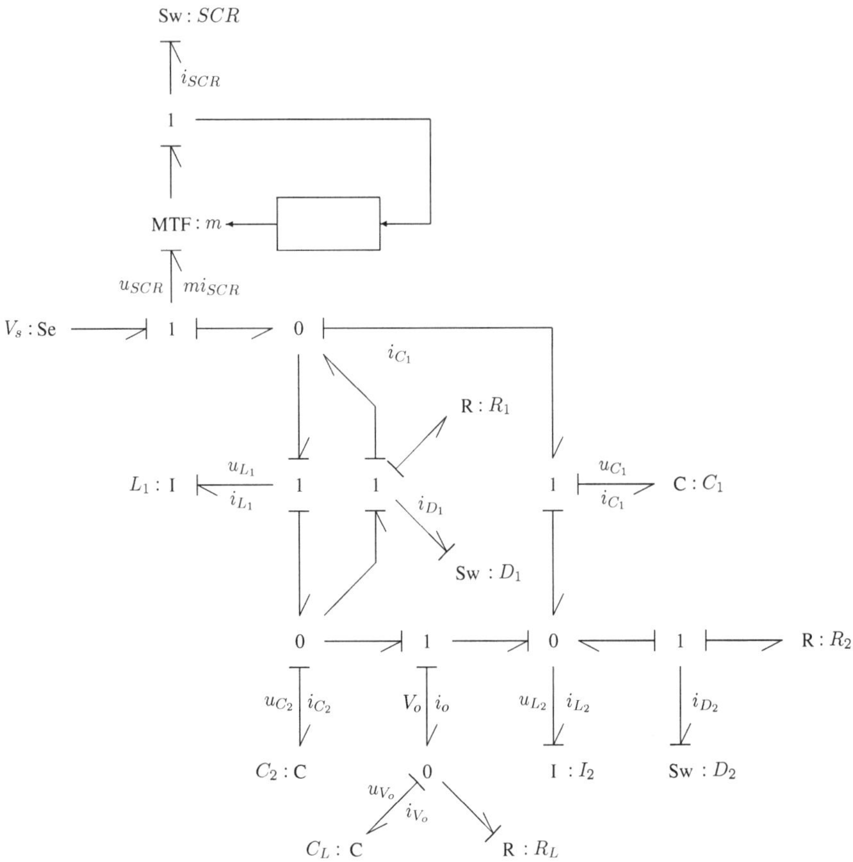

Fig. 8.49 Bond graph of the Z-source DC circuit breaker

$$i_{SCR} = \frac{\texttt{step}(u_{SCR}, 0)}{R_{\text{on}}} \cdot u_{SCR} \tag{8.99}$$

$$mi_{SCR} = m \cdot i_{SCR} \tag{8.100}$$

$$u_{L_1} = u_{C_1} - V_o \tag{8.101}$$

$$i_{D_1} = \frac{\texttt{step}(-u_{L_1}, 0)}{R_{\text{on}} + R_{D_1}} \cdot (-u_{L_1}) \tag{8.102}$$

$$u_{L_2} = u_{C_2} - V_o \tag{8.103}$$

$$i_{D_2} = \frac{\texttt{step}(-u_{L_2}, 0)}{R_{\text{on}} + R_{D_2}} \cdot (-u_{L_2}) \tag{8.104}$$

$$R_L = \texttt{ramp}(t, t_1, t_2, R_{L_0}, R_f) \tag{8.105}$$

$$i_o = i_{L_2} - mi_{SCR} + i_{L_1} - i_{D_1} - i_{D_2} \tag{8.106}$$

$$i_{C_1} = m i_{SCR} - i_{L_1} + i_{D_1} \tag{8.107}$$

$$i_{C_1} = i_{L_1} - i_o - i_{D_1} \tag{8.108}$$

$$i_{V_o} = i_o - \frac{V_o}{R_L} \tag{8.109}$$

$$\dot{u}_{C_1} = \frac{1}{C_1} i_{C_1} \tag{8.110}$$

$$\dot{u}_{C_2} = \frac{1}{C_2} i_{C_2} \tag{8.111}$$

$$\dot{V}_o = \frac{1}{C_L} i_{V_o} \tag{8.112}$$

$$\frac{d}{dt} i_{L_1} = \frac{1}{L_1} u_{L_1} \tag{8.113}$$

$$\frac{d}{dt} i_{L_2} = \frac{1}{L_2} u_{L_2} \tag{8.114}$$

where $y = \mathtt{step}(x, 0)$ denotes the unit step function. $R_L(t) = \mathtt{ramp}(t, t_1, t_2, R_{L0}, R_f)$ accounts for the decrease of the load resistance from an initial value R_{L_0} to the short-circuit impedance R_f during the time interval $\Delta t = t_2 - t_1$.

Simulation runs have used the parameters in Table 8.7. Parameter values have been chosen so that the resonance in shoot-through mode is visible. An ode-solver with a root finding capability has been used to detect the time point at which the SCR current becomes zero.

Figure 8.50a displays that, in shoot-through mode, the sum of currents $i_{L_1} + i_{C_1} - i_{D_1}$ in fact vanishes, i.e. the DC voltage source is isolated from the fault. In the faultless case, the steady state current flows through the SCR, the inductors and the load resistor. If the resistances of the inductors are neglected then the load

Table 8.7 Parameters of the Z-source DC circuit breaker in Fig. 8.48

Parameter	Value	Units	Meaning
V_s	240	V	Supply voltage
$C_1 = C_2$	200	μF	
$L_1 = L_2$	200	mH	
R_{on}	0.1	Ω	ON resistance of the SCR
$R_1 = R_2$	20	Ω	ON resistance of the wheel diodes
C_L	200	μF	Load capacitance
R_{L0}	60	Ω	Initial load resistance
R_f	10	mΩ	Load resistance in shoot-through mode
t_1	500	ms	Start time of the shoot-through fault
$\Delta t = t_2 - t_1$	10	ms	Fall time of the load resistance

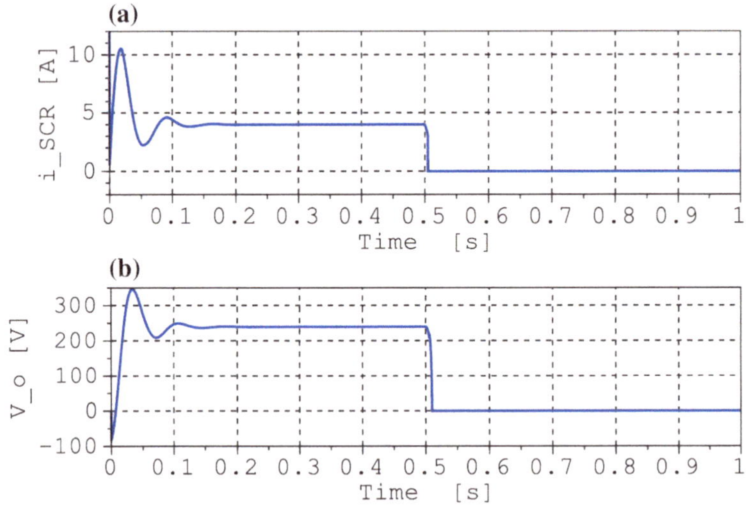

Fig. 8.50 Waveforms of the SCR current and the output voltage of the Z-source. **a** Waveform of the current $i_{SCR} = i_{L_1} + i_{C_1} - i_{D_1}$. **b** Output voltage V_o of the Z-source

resistance basically determines the value of the current, $i_{SCR} = V_s/(R_{\text{on}} + R_{L0}) = 240\,\text{V}/(0.1 + 60)\,\Omega = 3.993\,\text{A}$.

Figure 8.50b shows that the output voltage V_o of the Z-source drops to zero at the event of a short-circuit fault in the load. In the faultless case, the steady state value of the output voltage, V_o, equals the value of the supply voltage, $V_o = V_s = 240\,\text{V}$.

The short-circuit fault at $t_s = 0.5\,\text{s}$ causes a pulse in the waveform of u_{SCR}. After that event, $u_{SCR}(t) \rightarrow V_s$ for $t \rightarrow \infty$ as would be expected (Fig. 8.51). In case there is no short-circuit fault in the load, the steady state value of u_{SCR} is equal to $R_{\text{on}} \times i_{SCR} = 0.1 \times 3.993 \approx 0.4\,\text{V}$ as can be seen from the enlargement of Fig. 8.51 in Fig. 8.52.

Figure 8.53 indicates that, in fact, $u_{C_1} = u_{L_1}$ when the output voltage V_o drops to zero and that $u_{L_1} = 0$ in steady state.

Finally, Fig. 8.54 shows that during the resonance phase, the current flows through free wheel diode D_1 and resistor R_1 whenever the voltage u_{L_1} is negative. The same

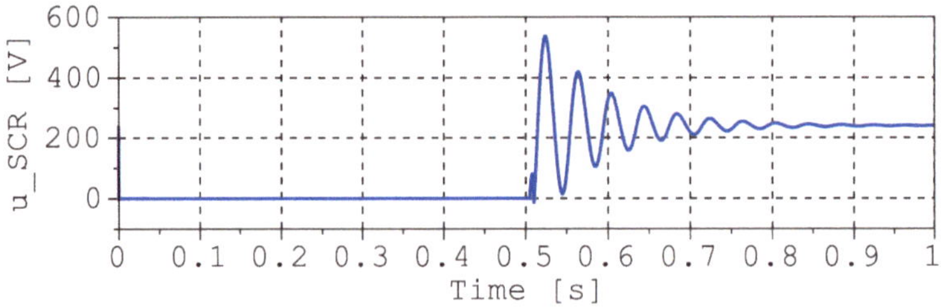

Fig. 8.51 Waveform of the voltage drop u_{SCR} across the SCR

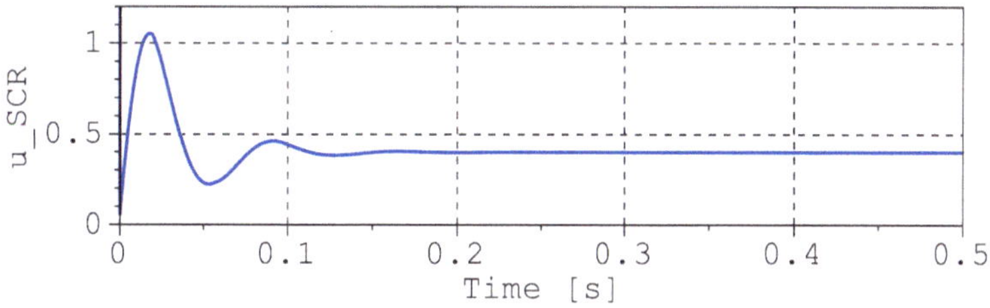

Fig. 8.52 Enlargement of Fig. 8.51

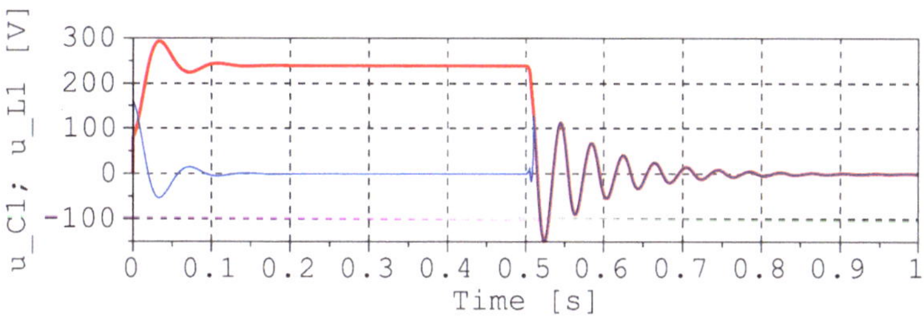

Fig. 8.53 Waveforms of voltage drops u_{C_1}, and u_{L_1} across capacitor C_1 and inductor L_1 respectively

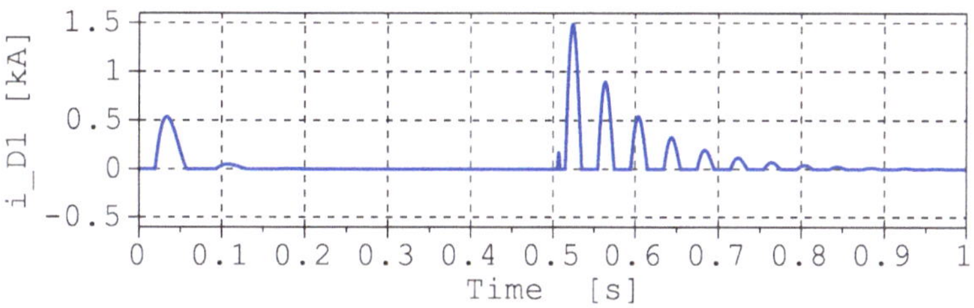

Fig. 8.54 Waveforms of current i_{D_1} through free wheel diode D_1

applies for diode D_2. That way, the energy is taken out of the Z-source after the short-circuit fault had happened.

8.4 Summary

Power electronic subsystems are widely used in a variety of electromechanical systems including motor drives with a mechanical load. They are built up by means of semiconductor switches that are commonly operated at high frequency. Their failure quite often give rise to faults in power electronic systems. For these reasons, mechatronic systems using power converters are well suited for an application of a bond graph model-based approach to FDI of system represented by a hybrid model.

The small example systems considered in this chapter are widely used basic components of power electronic systems. In power electronics, it is common to model the fast switching semiconductor devices that use various types of transistors, diodes, or thyristors simply as ideal or non-ideal switches although more sophisticated transistor models can be used and are used depending on the application and the purpose of a simulation study.

Representing switching devices by a non-ideal BG switch model with invariant conductance causality at its port, it is straightforward to convert a circuit schematic into a DBG. At least an initial not simplified BG closely resembles the circuit topology. In contrast to controlled junctions, a BG model of a switch makes the representation of switching devices in a BG explicit. For each switching cell in the circuit schematic a switch model can be found in the BG. The models of the inverter and of the rectifier respectively are pure switched resistor networks as any energy storage is neglected in the three legs of the inverter and the rectifier. This leads to causal conflicts at 0-junctions in the BG corresponding to the mid node of the legs. The causal conflicts have been resolved by adding auxiliary capacitors to the 0-junctions. For the three small example systems, equations have been deduced manually from the BG and have been coded in the script language of Scilab. During this process, auxiliary capacitances have been set to zero so that the result is a DAE system. Derivation and reformulation of equations could be performed automatically.

The structural information of the ARRs has been captured in an all-mode FSM. With the sensors used, all faults can be detected but cannot be isolated by analysing ARRs. It depends on how a system to be considered is built up whether further sensors can be added so that voltages across switches and currents through them can be measured. For instance, if there is a fault in one leg of the three-phase inverter or the rectifier, it can be due to a fault in one of the two switches in that leg. As to open circuit faults in switching cells, characteristic trajectories can be obtained after application of the dq-transformation that allow to isolate open switch faults. This technique not based on ARRs but often used in power electronics has been illustrated in Sect. 8.2.4 for an open circuit fault in the lower switch Sw_4 of inverter leg 'a'.

The studies of fault scenarios in the power electronic systems considered in this chapter focus on a single switch fault and its impact on the load. The switch fault is achieved by a modification of the control signal. It is assumed that a switch changes instantaneously from a healthy state into a faulty one. The latter one is accounted for by a modification of its discrete state variable.

Section 8.2.3 also considers the effect of an abrupt parametric fault in one phase of the inverter's load. In this case study, the load of the three-phase inverter is an RL-network in delta configuration often used in studies of three-phase PWM voltage source inverters. The study may be extended by replacing the RL-network by a sophisticated BG-model of an induction motor (see for instance [31, Chap. 8]). Studies of the three-phase diode bridge rectifier typically assume a resistive load in parallel to a filter capacitor.

ARR-based system mode identification has been illustrated for the boost converter with only two modes and for the three-phase diode bridge inverter with six active modes.

Section 8.1.5 applies the incremental bond graph based determination of ARR residual thresholds to the simple boost converter.

References

1. Mohan, N., Undeland, T. M., & Robbins, W. P. (2002). *Power Electronics, converters, application and design*. New York: Wiley.
2. Nagel, L.W. (1975). SPICE2: A computer program to simulate semiconductor circuits [*Ph.D. Thesis*]. University of California, Berkeley; 1975. Memorandum No. ERL-MS20.
3. Cadence. OrCAD 16.6 Demo Software (Capture and PSpice only). Retrieved from: http://www.cadence.com/.
4. Lu, B., & Sharma, S. K. (2009). A literature reviewof IGBT fault diagnostic and protection methods for power inverters. *IEEE Transction on Industry Applications, 45*(5), 1770–1777.
5. Ubale, M., Dhumale, R., & Lokhande, S. (2013). Open switch fault diagnosis in three phase inverter using diagnostic variable method. *IJRET: International Journal of Research in Engineering and Technology, 02*(12), 636–640. Retrieved from: http://www.ijret.org.
6. Meinguet, F., Sandulescu, P., Aslan, B., Lu, L., Ngac-Ky, N., Kestelyn, X., et al. (2012). A signal-based technique for fault detection and isolation of inverter faults in multi-phase drives. In *IEEE International Conference On Power Electronics, Drives and Energy Systems 2012 (PEDES 2012)*. (pp. 1–6). Bengaluru, India.
7. Im, W.-S., Kim, J.-S., Kim, J.-M., Lee, D.-C., & Lee, K.-B. (2012). Diagnosis methods for IGBT open switch fault applied to 3-phase AC/DC PWM converter. *Journal of Power Electronics, 12*(1), 120–127. Retrieved from: http://jpe.or.kr/archives/list_detail_articles.asp.
8. Khomfoi, S., Sae-Kok, W., & Ngamroo, I. (2011). An open circuit fault diagnostic technique in IGBTs for AC to DC converters applied in microgrid applications. *Journal of Power Electronics, 11*(6), 801–810, Retrieved from: http://jpe.or.kr/archives/list_detail_articles.asp.
9. Rothenhagen, K., & Fuchs, K. W. (2004). Performance of diagnosis methods for IGBT open circuit faults in voltage source active retifiers. In *Proceedings of the 2004 IEEE 35th Annual Power Electronics Specialists Conference* (Vol. 6, pp. 4348–4354).
10. Fuchs, K. W. (2003). Some diagnosis methods for voltage source inverters in variable speed drives with induction machines—A survey. In *Proceedings of the 29th Annual Conference of the IEEE Industrial Electronics Society, IECON'032* (pp. 1378–1385).
11. Buisson, J., Cormerais, H., & Richard, P.-Y. (2001). Bond graph modeling of power converters with switches commutating by Pairs. In J. J. Granda, & G. Dauphin-Tanguy (Eds.), *Proceedings 2001 International Conference on Bond Graph Modelling and Simulation (ICBGM'01)*. Simulation Series, (Vol. 33(1), pp. 179–184). San Diego: SCS.
12. Garcia-Gomez, J., Dauphin-Tanguy, G., & Rombaut, C. (1999). Average bond graph models of DC/DC power converters. In J. J. Granda, F. E. Cellier (Eds.), *Proceedings of the 1999 International Conference on Bond Graph Modeling and Simulation*. Simulation Series, (Vol. 31(1), pp. 338–343). San Diego, CA. SCS.
13. Edström, K., Strömberg, J.-E., Söderman, U., & Top, J. (1997). Modelling and simulation of a switched power converter. In J. J Granda, & G. Dauphin-Tanguy (Eds.), *Proceedings 1997 International Conference on Bond Graph Modeling and Simulation (ICBGM'97)*. Simulation Series, (Vol. 29(1), pp. 195–200). San Diego: SCS.
14. Allard, B., Morel, H., Lautier, P. h., & Retif, J. M. (1997). Bond graphs for averaged modeling of power electronic converters. In J J. Granda, & G. Dauphin-Tanguy (Eds.), *Proceedings 1997 International Conference on Bond Graph Modelling and Simulation (ICBGM'97)*. Simulation Series, (Vol. 29(1), pp. 201–206). San Diego: SCS.

15. Ducreux, J. P., Dauphin-Tanguy, G., & Rombaut, C. (1993). Bond graph modelling of commutation phenomena in power electronic circuits. In J J. Granda, & F. E. Cellier (Eds.), *International Conference on Bond Graph Modeling, ICBGM'93, Proceedings of the 1993 Western Simulation Multiconference*. Simulation Series (Vol. 25(2), pp. 132–136). San Diego, CA.: SCS Publishing. ISBN: 1-56555-019-6.
16. Dauphin-Tanguy, G., & Rombaut, C. (1993). Why a unique causality in the elementary commutation cell bond graph model of a power electronics converter. In *1993 IEEE International Conference on Systems, Man and Cybernetics*, (Vol. 1, pp. 257–263).
17. Aguilar-Justo, M., & de Luna-Ortega, C. (2011). FDI with hybrid bond graphs in the full bridge inverter. In *Proceedings 2011 8th International Conference on Electrical Engineering Computing Science and Automatic Control* (pp. 1–6). Merida City, Mexico: IEEE.
18. Alvari, M., Luo, M., Wang, D., & Zhang, D. (2011). Fault diagnosis for power electronic inverters: a model-based approach. In *Proceedinds 2011 IEEE International Symposium on Diagnostics for Electric Maschines, Power Electronics and Drives (SDEMPED)*. (pp. 221–228). Bologna, Italy: IEEE.
19. Kulkarni, C., Biswas, G., Koutsoukos, X., Goebel, K., & Celaya, J. (2010). Physics of failure models for capacitor degradation in DC-DC converters. In *The Maintenance and Reliability Conference*, MARCON. Knoxville, TN.
20. Umarikar, A. C., & Umanand, L. (2005). Modelling of switched systems in bond graphs using the concept of switched power junctions. *Journal of the Franklin Institute*, *342*, 131–147.
21. Scilab Enterprises. Scilab. 78000 Versailles, France. Retrieved from: http://www.scilab.org/.
22. Kulkarni, C., Biswas, G., & Koutsoukon, X. (2009). A prognosis case study for electrolytic capacitor degradation in DC-DC converters. In *Proceedings of Annual Conference of the Prognostics and Health Management Society*, San Diego, CA.
23. Carlson, E., Strunz, K., & Otis, B. (2010). A 20 mV input boost converterwith efficient digital control for thermoelectric energy harvesting. *IEEE Journal of Solid-state Circuits*, *45*(4), 741–750.
24. Rogers, E. (1999). Understanding boost power stages in switchmode power supplies. Texas Instruments. SLVA061, Retrieved from: http://www.ti.com/lit/an/slva061/slva061.pdf.
25. Erickson, R., & Maksimovic, D. (2001). *Fundamentals of power electronics*. Dordrecht: Kluwer Academic Publishers.
26. Peng, F. Z. (2003). Z-Source inverter. *IEEE Transactions on Industry Applications*, *39*(2), 504–510.
27. Borutzky, W. (2012). Bond graph modelling and simulation of fault scenarios in switched power electronic systems. *Proceedings IMechE Part I: J Systems and Control Engineering*, (pp. 1381–1393).
28. Controllab products. 20-sim the power in modeling, Retrieved from: http://www.20sim.com.
29. Junco, S., Diguez, G., & Ramirez, F. (2007). On commutation modeling in bond graphs. In J. Granda & F. Cellier (Eds.), *Proceedings of the 2007 International Conference on Bond Graph Modeling ICBM 2007*. Simulation Series (Vol. 39(1), pp. 12–19). SCS.
30. Welchko, B. A., Lipo, T. A., Jahns, T. M., & Schulz, S. E. (2004). Fault tolerant three-phase ac motor drive topologies: A comparison of features, costs, and limitations. *IEEE Transactions on Power Electronics*, *19*(4), 1108–1116.
31. Borutzky, W. (Ed.). (2011). *Bond graph modelling of engineering systems—theory, applications and software support*. New York: Springer.
32. Raju, N. I., Islam, Md S., & Uddin, A. A. (2013). Sinusoidal pwm signal generation technique for three phase voltage source inverter with analog circuit & simulation of pwm inverter for standalone load & micro-grid system. *International Journal of Renewable Energy Research*, *3*(3), 647–658, Retrieved from: http://www.ijrer.org/index.php/ijrer/article/view/771.
33. Borutzky, W. (2013). Bond-graph-based fault diagnosis in switched power electronic systems. In W. Borutzky (Eds.), *Proceedings of the 32nd IASTED International Conference on Modelling, Identification, and Control (MIC 2013). IASTED* (pp. 325–332). Innsbruck, Austria: Acta Press.
34. Schaefer, J. (1965). *Rectifier circuits: Theory and design*. New York: Wiley.

35. Dixon, J. (2010). Chapter 12: Three-phase controlled rectifiers. In M. Rashid (Ed.), *Power electronics handbook* (pp. 12-1–12-62). New York: Elsevier.
36. Visser, F. (2011). *Design and implementation of a bi-directional 3 phase converter for a 30 kW range extender application [Master thesis]*. Delft University of Technology.
37. Brenan, K. E., Campbell, S. L., & Petzold L. R. (1996) Numerical solution of initial-value problems in differential-algebraic equations. SIAM.
38. Najafi, M., & Nikoukhah, R. (2006). Modeling and simulation of differential equations in Scicos. In *Proceedings 5th International Modelica Conference 2006* (pp. 177–185). Arsenal Research, Vienna, Austria: The Modelica association.
39. Najafi, M., & Nikoukhah, R. (2005). SCICOS: A general purpose modeling and simulation environment. In G. Schmitz (Ed.), *Proceedings 4th International Modelica Conference* (pp. 367–374). Hamburg: The Modelica association.
40. Brown, P. N., Hindmarsh, A. C., & Petzold, L. R. (2006). Using Krylov methods in the solution of large-scale differential-algebraic systems. *SIAM Journal Scientific Computing*, *15*(6), 1467–1488.
41. Corzine, K. A., & Ashton, W. A. (2011). Structure and analysis of the Z-source MVDC breaker. In *2011 IEEE Electric Ship Technologies Symposium* (pp. 334–338). Alexandria, VA. IEEE.
42. Suresh Neve, V., Zope, P. H., & Suralkar, S. R. (2012). A literature survey on z-source inverter. *VSRD International Journal of Electrical, Electronics & Communication Engineering*, *2*(11), 889–896.
43. Miaosen, S., Joseph, A., Jin, W., Peng, F. Z., & Adams, D. J. (2007). Comparison of traditional inverters and Z-source inverter for fuel cell vehicles. *IEEE Transactions on Power Electronics*, *22*(4), 1453–1463.
44. Peng, F. Z., Joseph, A., Jin, W., Miaosen, S., Lihua, C., Zhiguo, P., et al. (2005). Z-source inverter for motor drives. *IEEE Transactions on Power Electronics*, *20*(4), 857–863.
45. Nacusse, M. A., & Junco, S. J. (2011). Simplifying switched bond graphs using residual sinks to enforce causality: application to modeling the zsource inverter. mecánica computacional, industrial applications (B). 2011;XXX(33):2533–2548. Retrieved online at: http://www.cimec.org.ar/ojs/index.php/mc/article/view/3931.
46. Borutzky, W. (1995). Representing discontinuities by means of sinks of fixed causality. In F. E. Cellier, J. J. Granda (Eds.), *1995 International Conference on Bond Graph Modeling, ICBGM'95, Proceedings of the 1995 Western Simulation Multiconference*. Simulation Series, (Vol. 27(1), pp. 65–72). SCS Publishing. ISBN: 1-56555-037-4.

Chapter 9
Failure Prognosis for Hybrid Systems Based on ARR Residuals

Besides abrupt faults considered in the case studies of the previous chapter, incipient faults due to tear and wear and ageing can occur and can lead to an increase of the failure rate and thus reduce a system's availability. A faulty valve in a process engineering system may lead to the leakage of a toxic gas, a crack in a blade of a turbine due to fatigue may lead to its failure, or corrosion due to humidity may lead to short circuits in a printed circuit board. One strategy is to monitor a system's behaviour and to *react* to a fault that has happened. An alternative that avoids unnecessary maintenance, thus reduces costs and improves scheduling of maintenance actions is so-called condition based maintenance (CBM), i.e. significant parameters of a system in operation are constantly monitored and sensor measurements are compared with thresholds and are assessed. *Preventive* or corrective actions are taken when abnormalities are observed, e.g. a faulty but still functioning component may be replaced when monitored feature values exceed admissible thresholds [1]. CBM thus comprises the collection of information about the health of a system, the processing of data in order to extract fault indicators and the decision making in order to issue maintenance recommendations. An even better strategy is *predictive failure prognosis*, i.e. to start from the diagnosis of a system's actual operating state, to detect the onset of a future failure, to project this state into the future and to estimate the operating time until a failure occurs, i.e. the time to failure (TTF). In the Prognostic and Health Management research community, the time to failure is commonly called remaining useful life (RUL). The results of a failure prognosis can then be used to take appropriate actions, e.g. by timely replacing a component so that foreseen failures do not occur and hamper the functionality of a system [2].

By using thresholds and a decision procedure, fault diagnosis can provide an indication of the onset of an incipient fault that may lead to a failure. If a model of the anticipated degradation of parameters is available, the time evolution of monitored parameters or features can be extrapolated. The time from the starting point of a fault until the projected parameter value intersects with a user defined alarm threshold is then an estimation of the remaining useful life. The prediction of the RUL is affected by uncertainties in the monitored parameters, by the choice of the degradation model, uncertainties in its parameters and uncertainties in the failure alarm thresholds. As

W. Borutzky, *Bond Graph Model-based Fault Diagnosis of Hybrid Systems*,
DOI 10.1007/978-3-319-11860-4_9

a result, the value of the RUL is a stochastic quantity with a mean within some confidence interval.

9.1 Prognosis Techniques

A brief survey of prognostic approaches is given in [3]. According to [2], prognosis techniques may be classified into three categories.

- Probability-based prognostic
- Data-driven prognostic
- Model-based prognostic

The first technique applies probabilistic methods to historical data from previous failures for a given class of systems. Parameters of probability functions are estimated from collected data. An advantage of probabilistic methods is that no detailed information is needed. However, the determination of parameters that faithfully describes an observed degradation phenomenon requires a significant amount of legacy statistical fault data. If these data are available, confidence limits can be determined that give an indication of the accuracy of a prediction.

Data-driven prognostic exploits methods from artificial intelligence such as neuronal networks. Measured data that captures the degradation evolution in a system is processed to extract features that are used to train a neural network so that it can assess the health of a system and can predict the RUL. That is, given a well designed monitoring system, the future evolution of a degradation can be predicted without having a mathematical model of the degradation.

Model-based prognostic starts from physical laws to develop an analytical dynamical model of a system under consideration including the evolution of degradation phenomena.

The RUL is predicted by determining failure thresholds and by numerically computing the time history of the degradation model as of the starting time point of an incipient fault obtained from a diagnostic system. The mathematical model of the degradation is the main advantage and at the same time the problem of this approach. On the one hand, it is based on an understanding of the physics of a degradation process and can provide accurate results. On the other side, a mathematical degradation model is difficult to obtain and holds only for a particular type of component, i.e. it cannot be used for other system components as well.

Failure prognosis is an essential part of condition-based management (CBM) and numerous papers on CBM have been published [1, 4, 5]. Some research work on failure prognosis recently reported in the literature may be found in [2, 6–9].

In the following, it is shown that bond graph model-based development of ARR residuals used for FDI can be extended to also serve model-based failure prognostic for hybrid systems. In [8, 9], residual-based failure prognosis uses continuous-time system models represented by bond graphs in integral causality, while reference [7] considers failure prognosis for systems modelled by hybrid bond graphs with controlled junctions.

9.2 Failure Prognosis Based on ARR Residuals Derived from a Bond Graph

The principle of the approach is easily outlined. Some issues residing in details are addressed later. As in previous chapters, first, a bond graph model of a system using nominal parameters is constructed and ARRs are deduced from the bond graph.

As to degradation phenomena in components, an attempt is made to describe them by a mathematical model assuming that degradation can be captured by a continuous drift of component parameters. Once a mathematical model with known parameters is available, this time dependency of parameters can be used to calculate the RUL of the system [7], or the trend of a time varying parameter can be incorporated into the ARRs together with outputs of the nominal behavioural system model causing an ARR residual sensitive to a certain parameter to drift away from zero as the system under consideration leaves the mode of normal behaviour [9]. That is, in online diagnosis, measurements from the real system are constantly compared with outputs of the model with nominal parameters by evaluating ARRs and by checking whether ARR residuals exceed predefined thresholds in order to detect the start of an occurring fault. Once the initiation of a fault has been detected, failure prognosis starts. The time difference between an initial time point indicating the start of an incipient fault provided by diagnosis and the time point the trend of the faulty parameter or a residual intersects with a predefined failure alarm threshold then is an estimate of the RUL. In the case of multiple simultaneous degradation phenomena a number of RULs for the components that are going to fail is obtained. Clearly, for the RUL of the system, the shortest component RUL is significant.

9.2.1 Degradation Models

Let $\boldsymbol{\Theta}$ denote the vector of component parameters $\Theta_1, \ldots, \Theta_p$, let $\tilde{\boldsymbol{y}}(t)$ be the vector of known measured outputs of the real system, $\boldsymbol{u}(t)$ the vector of inputs into the system and into the nominal model of the system and $\boldsymbol{m}(t)$ the vector of all switch states, i.e. $\boldsymbol{m}(t)$ denotes the system mode. Assume that fault diagnosis has detected and isolated Θ_1 as a single incipient fault starting at time instant t_0. Then

$$\exists\, j \quad 0 \neq r_j(t) = g_j(\boldsymbol{u}(t), \tilde{\boldsymbol{y}}(t), \Theta_1, \ldots, \Theta_p, \boldsymbol{m}(t)) \quad t > t_0 \tag{9.1}$$

for those ARRs in which Θ_1 is present. Let $j = j_1$. For given computed and stored values $r_j(t_n)$ at sample points t_n, (9.1) can be considered an implicit relation for the degradation profile of Θ_1 vs time. Assuming that the conditions of the implicit function theorem (cf. Appendix C.3) hold, then (9.1) can be solved for Θ_1. That is, there exists a real valued function ψ_1 such that

$$\Theta_1(t) = \psi_1(r_j(t), \Theta_2, \ldots, \Theta_p, \boldsymbol{u}(t), \tilde{\boldsymbol{y}}(t), \boldsymbol{m}(t)) \tag{9.2}$$

The type of the system under consideration and an understanding of the physics of the relevant degradation phenomena suggest to use formulae of certain form with unknown coefficients from a set of potential models for an approximation of the anticipated progression of an incipient fault. These mathematical models may be linear or nonlinear, may be an algebraic relation or an ODE. For a set of potential prescribed degradation models, the one that best matches the profile derived from collected data identifies the fault degradation. The unknown degradation model coefficients are obtained by means of parameter identification. Clearly, this matching with several potential degradation profiles can be performed in parallel.

For instance, suppose that the degradation trend $\Theta_1(t)$ can be approximated by the simple nonlinear ODE

$$\dot{\Theta}_1(t) \; = k\,\Theta_1^{3/2}(t), \quad \Theta_1(t_0) = \Theta_1^0 \tag{9.3}$$

with k being an unknown coefficient. Once k has been identified, the ODE can be solved.

$$\sqrt{\Theta_1(t)} = \frac{\sqrt{\Theta_1(t_0)}}{1 - \frac{k}{2}(t - t_0)\sqrt{\Theta_1(t_0)}} \tag{9.4}$$

With this solution replacing the nominal parameter Θ_1 in (9.1) residual $r_j(t)$ vanishes.

$$0 = g_j(\boldsymbol{u}(t), \tilde{\boldsymbol{y}}(t), \Theta_1(t), \Theta_2, \ldots, \Theta_p, \boldsymbol{m}(t)) \tag{9.5}$$

$j = 1, \ldots, N$ where N denotes the number of sensors.

In case multiple potential component faults may be the cause for the start of an abnormal behaviour of an ARR residual, then parameter estimation as part of fault isolation identifies the parameters that are going to deviate from their nominal values. For each of them, failure prognosis will have to identify a degradation model as a first step. This identification of degradation models can be performed in parallel. As a result, multiple faults may have different degradation profiles.

Degradation models for system components may be determined by accelerated life tests and then used after online fault diagnosis has identified the initiation of an incipient fault in order to assess the health of a system and to predict the RUL of the component with an incipient fault [8].

9.2.2 Estimation of the RUL

Given a failure alarm threshold and the starting point t_0 of an incipient fault, then the solution of the degradation model can be used to determine the time as of the starting point of the incipient fault until the degradation trend of the parameter intersects with the alarm threshold, which means that the RUL of the faulty component has been predicted. Let t_F denote the time point at which the degradation trend of $\Theta_1(t)$ intersects with the failure alarm threshold. The RUL is then simply obtained by solving (9.4) for $t_F - t_0$.

$$t_F - t_0 = \frac{2}{k}\left(\frac{1}{\sqrt{\Theta_1(t_0)}} - \frac{1}{\sqrt{\Theta_1(t_F)}}\right) \tag{9.6}$$

This way, the identified degradation model is directly used to predict the RUL of the faulty component. If multiple faults have been isolated, if for each of them a degradation model has been identified, and if for each fault a failure alarm threshold has been set, then the latter ones can be substituted into the solutions of the degradation models to obtain a RUL for each faulty component.

Alternatively, the RUL with regard to $\Theta_1(t)$ may be calculated by an evaluation of those residuals in which the slowly varying component parameter is present. If all parameters in (9.1) have nominal values and if the outputs of the nominal model, $\boldsymbol{y}(t)$, instead of measurements from the real system, $\tilde{\boldsymbol{y}}(t)$, are inputs into the function $g_j()$ then, clearly, its evaluation must give zero as a result. If, however, one nominal parameter is replaced by a function capturing its degradation trend then the ARR residual will leave a small interval around zero and with progressing time eventually will intersect with a prescribed failure alarm threshold. Suppose that the fault is identified by a unique component fault signature in the FSM. Then it is clear for a prognosis module which residuals are to be projected into the future. Figure 9.1 visualises the estimation of a component RUL based on the determination of a mathematical model of the degradation phenomenon and on an evaluation of ARRs.

9.2.2.1 Accounting for Degradation Phenomena in a Bond Graph

In [9], it is suggested to explicitly represent considered degradation phenomena in a nominal bond graph of the system once a mathematical model for them is available. However, if a mathematical model of a degradation process derived from first principles is not available, if a nominal parameter in the constitutive equation of an element is then replaced by a function of time so that measured data is fitted and can be extrapolated into the future, a bond graph representation might be a problem. If a resistance becomes time-varying as of sometime point t_0 then this may be simply captured by a nonlinear modulated resistor. If the nominal capacitance C_0 in the constitutive equation $q(t) = C_0 \cdot e(t)$ of a capacitor is replaced by a time-dependent capacitance $C(t)$ then

$$f(t) = \dot{q}(t) = C(t)\dot{e}(t) + \dot{C}(t)e(t) \tag{9.7}$$

Equation 9.7 might be represented by means of a C element in derivative causality and a nonlinear resistor (Fig. 9.2).

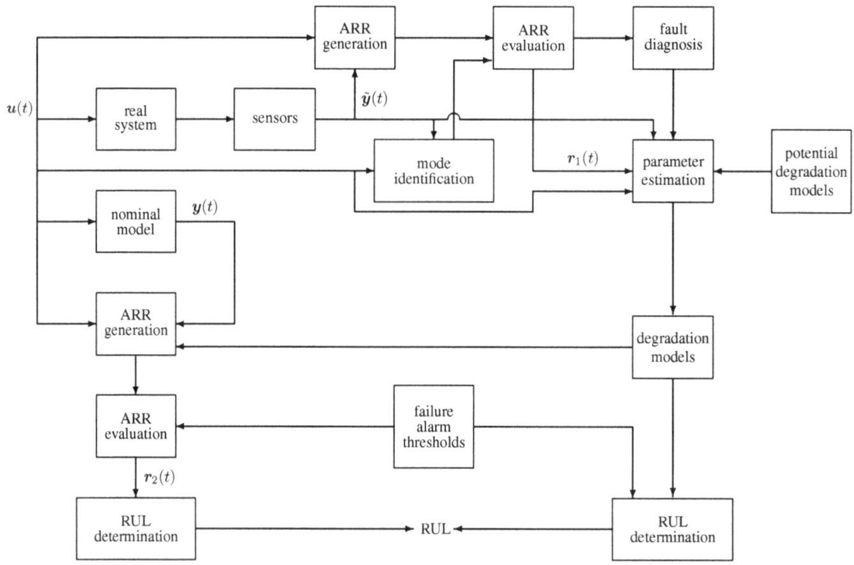

Fig. 9.1 Failure prognosis as an add-on to online fault diagnosis

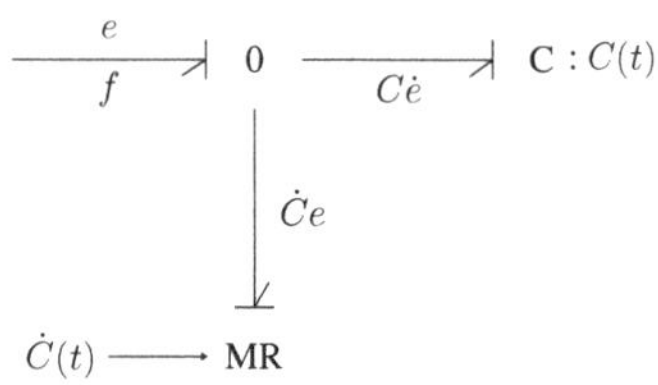

Fig. 9.2 Bond graph model of a capacitor accounting for a degradation of the capacitance

ARRs deduced from a bond graph then capture the degradation trend of physical parameters. As a result, their residuals will not stay close to zero. When time progresses they will eventually intersect with a user set failure alarm threshold. Finally, as failure prognosis starts from an initial state provided by the monitoring system, i.e. initial conditions are known, the bond graph may be in integral causality.

9.2.2.2 Degradation Not Described by Physical Parameters

In [7], Yu Ming extends the consideration of performance deterioration to components for which a model with physical parameters is not available. Such components may be sensors and actuators or monitored subsystems for which only measurements are known. In order to quantify the severity of nonparametric faults, he introduces an efficiency factor $0 \leq \beta \leq 1$, where $\beta = 0$ denotes a failure while $\beta = 1$ indicates no fault. Accordingly, $1 - \beta$ is a measure for the severity of a fault. For sensors, a fault is captured by multiplying the normal measurement by β, an actuator fault is taken into account by multiplying the input by β. These efficiency coefficients contribute to ARRs as parameters do. In the case of an incipient fault, their trend is unknown and is to be matched with a potential degradation profile in the same manner as for physical parameters.

9.2.3 Failure Prognosis for Hybrid Systems

If a system is described by a hybrid model, at least a subset of all ARRs is mode dependent. As a result, some faults may not be detectable in some modes. Accordingly, the starting point of an incipient fault cannot always be determined in the mode the fault had happened. If only the product of a switch state m_1 and a function of a parameter Θ_1, e.g. m_1/Θ_1 is present in some ARRs and if the degradation behaviour $\Theta_1(t)$ starts at a time point t_0 when m_1 is off ($m_1 = 0$) then this fault cannot be detected until the system enters into a new mode where m_1 is on ($m_1 = 1$). That is, there will be a delay between the time point of a fault occurrence and the time point of its detection. Data to be used for prognosis is collected as of the moment a fault is detected for the length of a certain observation window. Suppose that in the current mode fault detection reveals a coherence vector, say $\boldsymbol{c} = [1\ 1\ 0]$, that matches the fault signature of two parameters Θ_1, Θ_2 then one of them or both may be the cause for the abnormal behaviour of residuals r_1, r_2. This may be decided by means of ARR residual based parameter estimation. If there is a third parameter Θ_3 with a signature $[0\ 1\ 0]$ and if it cannot be decided whether one the parameters has started in the previous mode to deviate from its nominal value then the third parameter must be considered a potential fault candidate. It may be possible that the faulty behaviour of Θ_3 started already in the previous mode and would have contributed to an abnormal behaviour of residual $r_1(t)$ in that mode if a switch state premultiplying a function of Θ_3 in ARR_1 would have been on in the previous mode. Hence, it is not sufficient to just take an obtained coherence vector and to look for matches with component fault signatures in the FSM of the current mode. Therefore, once a fault condition has been detected, Wang et al. [6] suggest to continue monitoring the system at least until the next mode change following the fault condition detection happens. At that time point a new coherence vector can be detected. Comparing this vector with the rows of the FSM and observing the result of the previous fault identification, a new set of possible fault candidates can be determined from which a fault that was not detectable in the previous mode but is detectable in the current mode can be identified.

Finally, components of a hybrid system may exhibit a different degradation behaviour in different modes of operation. Accordingly, for different system modes, different sets of mathematical models may be used for matching a degradation profile. As a result, degradation models are mode-dependent and so is the RUL of faulty components.

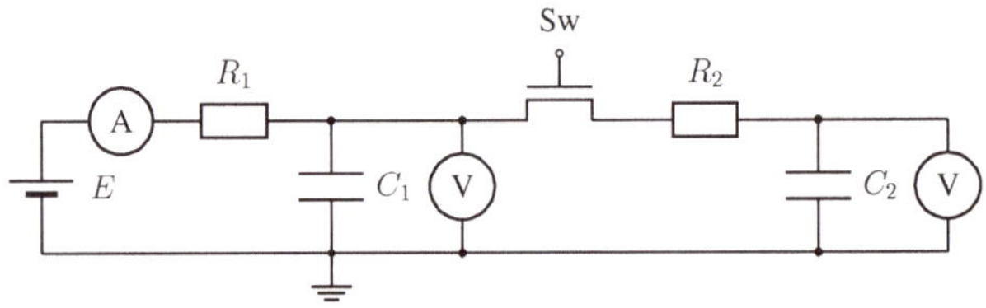

Fig. 9.3 Simple network with one switch and three sensors

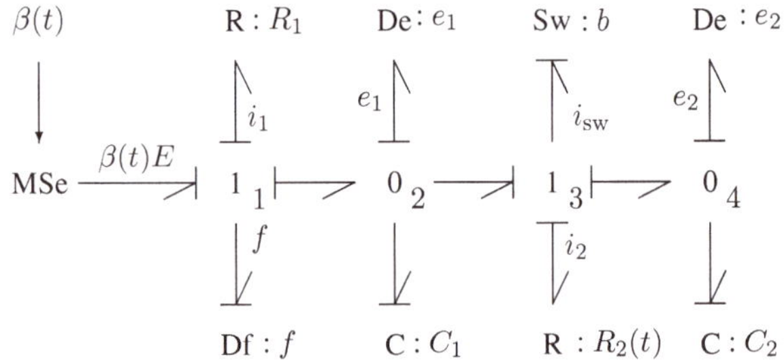

Fig. 9.4 Diagnostic bond graph of the network in Fig. 9.3

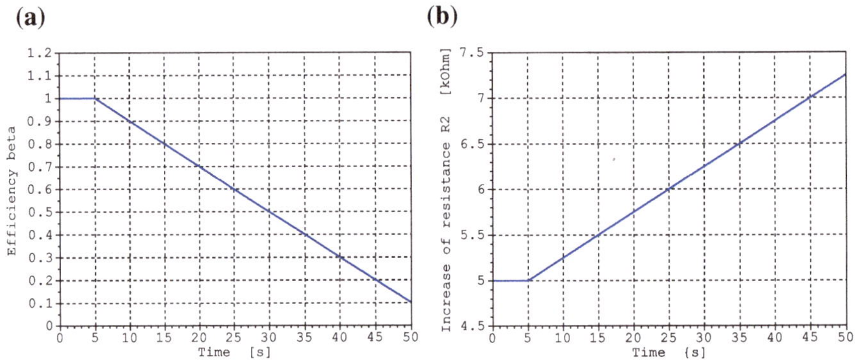

Fig. 9.5 Degradation profile of the incipient fault in the efficiency β and in the resistance R_2. **a** Decrease of β. **b** Buildup of R_2

Example

For illustration, a simple passive network with one switch and three sensors depicted in Fig. 9.3 is considered.

The DBG of the network in Fig. 9.4 accounts for a degradation of the voltage supply indicated by the efficiency $\beta(t)$ and an incipient fault in resistance R_2.

Let both phenomena start at $t = 5$ s and assume that their degradation profile may be approximated by linear models as depicted in Fig. 9.5. The study considers an initially open switch that closes at time instant $t = 10$ s (Fig. 9.6).

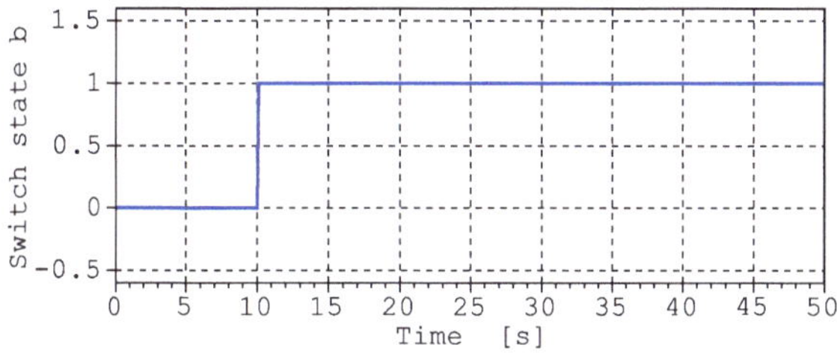

Fig. 9.6 Switch state b(t)

Table 9.1 Mode-dependent FSM for the switched network in Fig. 9.3

Component	Parameter	r_1	r_2	r_3	D_b
Switch state	b	0	1	1	1
Voltage source	β	1	0	0	1
R: R_1	R_1	1	0	0	1
R: $R_2(t)$	$R_2(t)$	0	b	b	$b = \begin{cases} 0 \\ 1 \end{cases}$
C: C_1	C_1	0	1	0	1
C: C_2	C_2	0	0	1	1

From the DBG in Fig. 9.4 three structured ARRs can be deduced.

$$1_1: \quad \mathrm{ARR}_1: \quad 0 = \beta(t)E - R_1 f - e_1 \tag{9.8}$$

$$0_2: \quad \mathrm{ARR}_2: \quad 0 = f - C_1\dot{e}_1 - \frac{b}{R_{\mathrm{on}} + R_2(t)}(e_1 - e_2) \tag{9.9}$$

$$0_4: \quad \mathrm{ARR}_3: \quad 0 = \frac{b}{R_{\mathrm{on}} + R_2(t)}(e_1 - e_2) - C_2\dot{e}_2 \tag{9.10}$$

The structure of the ARRs is captured by the FSM in Table 9.1. It is assumed that the sensors are faultless. Therefore, their rows are omitted from the FSM. Also, the ON-resistance of the switch is considered a faultless parameter that has been removed from the FSM. Simulation runs used the parameters in Table 9.2.

Figure 9.7a displays the time history of the capacitor voltages u_1, u_2 and their degradated waveforms $\tilde{u}_1(t), \tilde{u}_2(t)$ due to the incipient faults in the voltage supply and in resistance R_2.

In the healthy system, voltage u_2 builds up after the switch has closed at $t = 10$ s. For $0 < t < 10$ s, only capacitor C_1 is charged. At time instant $t = 10$ s the switch is closed. Part of the charge of capacitor C_1 is transferred to the empty capacitor C_2. Accordingly, voltage u_1 temporarily slightly drops. As both capacitors are charged for $t > 10$ s their voltages are finally close to the supply voltage value of 5 V.

In the faulty system, the supply voltage is not stable but decreases as of $t = 5$ s and so do the capacitor voltages. The effect of the incipient faults on the resistor

Table 9.2 Parameters of the switched circuit in Fig. 9.3

Parameter	Value	Units	Meaning
E	5	V	Voltage supply
$\beta(t = 0)$	1	–	Initial value of efficiency β
R_1	500	Ω	Resistor R: R_1
$R_2(t = 0)$	5	kΩ	Initial value of Resistor R: R_2
R_{on}	0.1	Ω	ON resistance of the switch
C_1	5,000	μF	Capacitor C: C_1
C_2	1,000	μF	Capacitor C: C_2

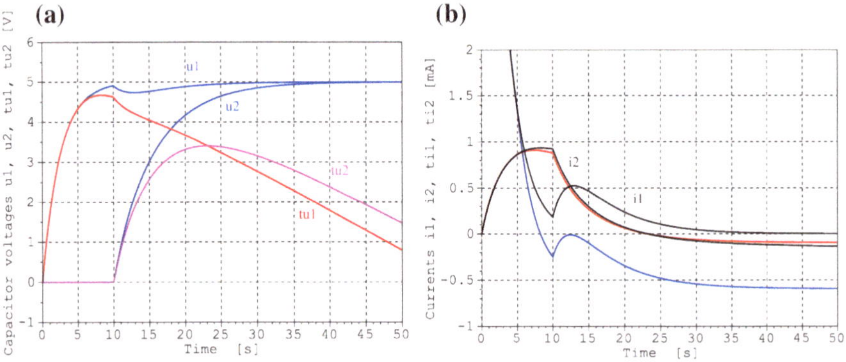

Fig. 9.7 Capacitor voltages u_1, u_2 and resistor currents i_1, i_2 and $\tilde{u}_1, \tilde{u}_2, \tilde{i}_1, \tilde{i}_2$ due to the incipient faults. **a** Voltages $u_1, u_2, \tilde{u}_1, \tilde{u}_2$. **b** Currents $\tilde{i}_1, \tilde{i}_2, \tilde{i}_1, \tilde{i}_2$

currents i_1, i_2 is less significant. In accordance with the waveforms of the capacitor voltages in the healthy system, the resistor currents decrease to values close to zero.

Figure 9.8 displays the time history of the ARR residuals. As Fig. 9.8a indicates, residual r_1 starts to deviate from zero at $t = 5$ s. Its waveform is dominated by the decrease of the supply voltage.

According to the profile depicted in Fig. 9.5b, in this offline simulation, it is known that the buildup of resistance $R_2(t)$ also starts at $t = 5$ s but residuals r_2, r_3 being sensitive to a fault in R_2 deviate from zero not before the mode change at $t = 10$ s happens. As can be seen from Fig. 9.8b, the absolute values of residuals r_2, r_3 are orders of magnitude smaller than the one of r_1. The degradation profile $R_2(t)$ has little effect on the dynamics of the circuit. This is also displayed by the waveforms of the capacitor voltages if $R_2(t)$ is replaced by its initial value $R_2(t = 0)$.

Assume that online monitoring and fault detection produce a coherence vector $\boldsymbol{c} = [1\,0]$ in the time interval $0 < t < 10$ s when the circuit is in mode $b = 0$. A comparison of the coherence vector with the FSM in Table 9.2 reveals efficiency β and resistance R_1 as possible fault candidates. Parameter estimation can identify β as a true fault. However, the incipient fault in R_2 starting simultaneously with the fault in β at $t = 5$ s cannot be detected. Residual r_1 is not sensitive to a fault in R_2 and the term $b/(R_{on} + R_2(t))$ in ARR_2 and ARR_3 is cancelled out by the switch state b

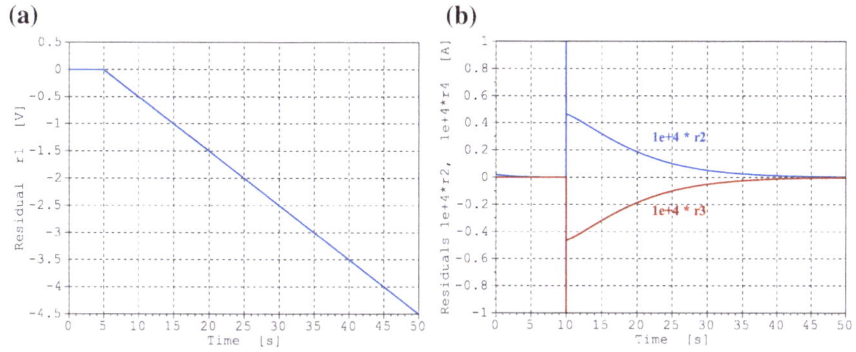

Fig. 9.8 Waveforms of the ARR residuals r_1, r_2, r_3. **a** Residual r_1. **b** Residuals r_2, r_3

equal to zero in that mode. If monitoring of the circuit continues then fault detection may deliver a new coherence vector $\boldsymbol{c} = [1\ 1]$ at $t = 10\,\text{s}$. By looking into the FSM in Table 9.2 and by observing that β has already been identified as a fault, the new fault candidates could be $\beta \wedge R_2$, $\beta \wedge C_1$. Parameter estimation will exclude C_1 from being a faulty parameter so that, after all, R_2 is identified as a faulty parameter.

Assume that the degradation trend of R_2 has been obtained by collecting measured data and by evaluating ARR$_2$ or ARR$_3$, that the profile can be approximated by a linear model and that its parameters have been identified. Let the graph in Fig. 9.5b display the resulting approximation. Further assume that an increase of the value of R_2, say by 20 %, sets a failure alarm threshold then the RUL of R_2 can be easily determined to be 20 s.

9.3 Summary

This brief chapter shows that beyond fault diagnosis and system mode identification ARRs deduced from a DBG of a hybrid model can also support model-based failure prognosis. When fault diagnosis has detected and isolated an incipient fault, data obtained from monitoring a system and collected over a certain time window as of the fault initiation time instant can be inserted into an ARR that is sensitive to the detected fault. By this way, the trend of the faulty component parameter vs time can be obtained. The degradation profile is then to be matched with a set of appropriate mathematical degradation models. Nonlinear least square minimisation will single out the best fitting model together with its parameters. Once such a model is available, the health of the system as of the time point of the fault initiation can be assessed and the RUL of the component can be predicted by inserting appropriate failure alarm thresholds into the degradation model and solving for the time the degradation profile intersects with a threshold. Alternatively, the mathematical degradation model for the faulty parameter may be inserted into an ARR together with the outputs of the nominal system model. Then the time point of an intersection of the ARR residual with the set failure alarm threshold is searched.

For systems described by a hybrid model at least some ARRs are mode-dependent. As a result, incipient faults may start when the system is in a mode that does not enable an ARR based approach to detect the fault because its contribution to an ARR or to a set of ARRs is cancelled out by switch states being equal to zero in that mode. An approach adopted from [6] and illustrated by a simple passive network with one switch is to allow for a delay time in fault isolation at least until the next mode change has happened so that faults that cannot be detected in the current mode can be detected and isolated after a mode change.

Failure prognosis based on models derived from physical laws has its advantages over other approaches such as data-driven methods. Degradation models are obtained from an understanding of degradation mechanisms to be taken into account and may provide accurate results. Their parameters have a physical meaning. Moreover, once a good understanding of the physics of a degradation process has been achieved, the model may be used also for other systems in which the same kind of degradation may take place. Only the parameters of the degradation model are to be adapted. However, an appropriate mathematical degradation model is also a weakness of model-based approaches as it may be difficult to be obtained. A physical model depends on the application domain, the system component under consideration and the kind of degradation phenomenon. On the contrary, a degradation profile obtained from measurements may suggest the use of a linear or an exponential characteristic with parameters to be fitted by least squares minimisation. Prognosis is a problem that has received increasing attention during past decades and is subject of ongoing research.

References

1. Jardine, A. K. S., Lin, D., & Banjevic, D. (2006). A review on machinery diagnostic and prognostics implementing condition-based maintenance. *Mechanical Systems and Signal Processing*, *20*(7), 1483–1510.
2. Vachtsevanos, G., Lewis, F. L., Roemer, M., Hess, A., & Wu, B. (2006). *Intelligent fault diagnosis and prognosis for engineering systems*. Hoboken, New Jersey: Wiley.
3. Dragomir, O. E., Gouriveau, R., Dragomir, F., Minca, E., & Zerhouni, N. (2009). Review of prognostic problem in condition-based maintenance. In *Proceedings of the European Control Conference ECC'09* (pp. 1585–1592), Budapest, New Jersey. Retrieved from: http://hal.archives-ouvertes.fr/docs/00/41/87/61/PDF/ECC09_0917_FI-1.pdf.
4. Peng, Y., Dong, M., & Zuo, M. J. (2010). Current status of machine prognostics in condition-based maintenance: A review. *International Journal of Advanced Manufacturing Technology*, *50*(1–4), 297–313.
5. Kothamasu, R., Huang, S. H., & VerDuin, W. H. (2006). System health monitoring and prognostics—A review of current paradigms and practices. *International Journal of Advanced Manufacturing Technology*, *28*, 1012–1024.
6. Wang, D., Yu, M., Low, C., & Arogeti, S. (2013). *Model-based health monitoring of hybrid systems*. New York: Springer.

7. Ming, Y. (2012). *Fault diagnosis and prognosis of hybrid systems using bond graph models and computational intelligence*, [PhD thesis]. School of Electrical and Electronic Engineering. Nanyang Technological University, Singapore. Retrieved from: https://dr.ntu.edu.sg/handle/10356/50480.
8. Medjaher, K., & Zerhouni, N. (2013). Hybrid prognostic method applied to mechatronic systems. *International Journal of Advanced Manufacturing Technology*, *69*(1–4), 823–834.
9. Medjaher, K., Zerhouni, N. (2009). Residual-based failure prognostic in dynamic systems. In *7th IFAC International Symposium on Fault Detection, Supervision and Safety of Technical Processes, SAFE PROCESS'09* (pp. 716–721). Barcelona, spain. Retrieved from: http://hal.inria.fr/docs/00/40/29/59/PDF/Medjaher_Safeprocess09.pdf.

Chapter 10
Overall Conclusion and Discussion

Since bond graphs were devised by Professor Paynter [1] more than five decades ago, they have been mostly used for the development of continuous time models of multidisciplinary engineering systems in order to study the dynamic behaviour of a system. In addition, intensive research has been carried out and various approaches have been proposed how to account for the abstraction of discontinuities in a bond graph framework and how to extend bond graph methodology to systems suitably described by a hybrid model. More recently, bond graphs have also been used for model-based FDI of systems described by a continuous time model. So far, none of the reported approaches has attained common usage.

This book shows that standard bond graph modelling can also support model-based

- fault detection and isolation,
- system mode identification, and
- failure prognosis

for systems described by a hybrid model. Devices with fast state transitions such as diodes, transistors, hydraulic valves, are modelled as non-ideal switches. Following a proposal of Ducreux et al. [2], they are represented in bond graphs by an MTF controlled by a modulus $m(t) \in \{0, 1\} \; \forall \, t \geq 0$ and an ON-resistor in fixed, mode-independent conductance causality. Causal conflicts that may occur at junctions are resolved by attaching an auxiliary element. Equations deduced from a bond graph are formulated such that the parameter of the auxiliary elements can be set to zero. Recently, a different approach to FDI for hybrid systems using controlled junctions has been proposed by Wang et al. [3].

If the abstraction of instantaneous state transitions is adopted then a switching device such as a mechanical clutch causes a change in the model structure. As a result, storage elements may become dependent. That is, the number of state variables become mode-dependent. One approach to this problem is to detect discrete system mode changes while simulating the dynamic system behaviour, to use a different mathematical model for the dynamics in each mode and to re-initialise numerical integration at the event of a discrete mode change when necessary. In order to keep

W. Borutzky, *Bond Graph Model-based Fault Diagnosis of Hybrid Systems*,
DOI 10.1007/978-3-319-11860-4_10

integral causality at storage ports independent of the system mode, residual sinks may be used that are switched on and off. When two storage storage elements become dependent due to a switch state change they supply an output that forces the two storage elements in integral causality to produce a joint output.

With regard to a simulation of the dynamic behaviour of a system represented by a hybrid model, this approach offers the following advantages.

- A BG with static causalities can be developed that holds for all system modes, i.e. causalities do not change with the commutations of switches.
- The standard SCAP can be applied without any modifications.
- A single set of mode-dependent equations of motions can be derived from a BG.
- Existing software could be used for the generation of equations.

Equations can be formulated so that parameters of auxiliary elements can be set to zero. Also, the ON-resistance of switches can be set to zero turning them into ideal switches so that small time constants and thus a set of stiff model equations can be avoided. In the case residual sinks are used, which is similar to the use of Lagrange multipliers, the resulting mathematical model is a DAE system of index 2.

As to FDI, mode-dependent ARRs can be deduced from a DBG with storage elements in derivative causality and sensors in inverted causality. Their structure is captured in an all-mode FSM. An evaluation of ARRs yields residuals that are used as fault indicators. In general, ARRs relate time derivatives of known variables. The necessary differentiation is carried out in discrete time.

Alternatively, for off-line simulation, a behavioural model of a real system subject to faults and a reference model with nominal parameters can be coupled by residual sinks. Their outputs being ARRs residuals force the nominal model to adapt to the behaviour of the faulty system model. In this approach, the two coupled bond graphs are in integral causality. Advantage of this approach are that

- all kinds of faults can be deliberately introduced into the behavioural model of the real process without any risk,
- ARR residuals can be numerically computed as unknowns of a mode-dependent DAE system. There is no need for ARRs in closed symbolic form.

For online simulation, the outputs of the behavioural model are to be replaced by measured outputs from the real system. As measurements carry noise they must be appropriately filtered.

ARR residuals serving as fault indicators should be distinguishably sensitive to true faults but little sensitive to noise and parameter faults in order to avoid false alarms on the one hand side and to make sure that fault detection does not miss any faults. Therefore, appropriate thresholds for ARR residuals are to be set. As the dynamic behaviour of hybrid systems can be quite different in different modes, predefined bounds of constant value may not be suitable. In this book, the incremental bond graph approach [4] has been briefly recalled and applied to hybrid system models to deduce adaptive mode-dependent ARR residual thresholds that account for parameter uncertainties.

If component parameters in a system mode share a fault signature in the FSM then a fault can be detected but not isolated by simple inspection of the FSM. In case online fault detection provides a coherence vector that matches with more than one row in the FSM in a system mode, the result is a set of potential fault candidates. One way to identify multiple faults is to perform parameter estimation by Gauss-Newton least squares output error minimisation. Bond graph modelling can support this approach to multiple fault isolation by providing ARRs. Their residuals are used in the functional to be minimised. This has been discussed in Chap. 6.

As ARRs derived from a hybrid model are mode dependent, it is important for an online FDI to know the current mode in order to use the correct values of the discrete states in the ARRs so that ARR residuals can serve as indicators of fault in the current system mode. It turns out that an evaluation of all ARRs or a subset of ARRs in case the previous mode is known can be used to identify the current system mode. The task can be performed in parallel in order to minimise the response time.

For illustration, in Chap. 8, the proposed bond graph model-based approach to FDI for hybrid systems has been applied to small often used power electronic systems. As those systems contain semiconductor power switches that are usually commutated at high frequency and are subject to typically open circuit and short circuit faults, they are well suited for FDI case studies. Furthermore, power electronic systems are part of mechatronic systems such as hybrid vehicles, or wind turbines which suggests to apply bond graph modelling. As failures in a power electronic subsystem affect the overall dynamic behaviour of a system and its performance it is important to apply FDI methods in order to avoid malfunctions or damages in the load of power electronic system. Therefore, it is not surprising that FDI in power electronic systems, especially in motor drives, has been a subject of many publications. The majority of them, however, does not use bond graph modelling. Clearly, the bond graph model-based approach to FDI proposed in this book is not confined to power electronic systems. The case study of the switched three phase power inverter, for instance, could be extended in further research. The RL-load could be replaced by an elaborated model of an induction motor and its mechanical load.

Finally, in comparison to fault diagnosis which has been a subject of a lot of research and has become established in industry, failure prognosis is yet a relatively young field. Fault detection, isolation and fault accommodation react to faults that had happened, while failure prognosis aims at assessing the current health of a system and to predict the remaining useful life or time to failure. This is of importance for maintenance in order to reduce costs while increasing reliability, availability and security of a system, of machinery or of equipment. Chapter 9 shows that ARRs derived from a bond graph can also support model-based failure prognosis for hybrid systems. If fault diagnosis has detected and isolated an incipient parametric fault then the time evolution of an ARR residual over a certain window and known inputs into the ARR define a degradation profile of the faulty parameter that is to be matched with a mathematical degradation model from a set of potentially appropriate models. The unknown parameters in this model can be determined by nonlinear least square parameter estimation. Once the parameters of the degradation model are available, the RUL of the faulty component can be determined by computing the time from the starting point of the incipient fault until the trend of the parameter intersects with a suitably defined failure alarm threshold. Chapter 9 considers a simple network with one switch for illustration. It is expected that bond graph model-based failure prognosis will be subject for further research.

The focus of this book has been on the presentation of a bond graph model-based approach to FDI and failure prognosis for hybrid systems. It turns out that ARRs derived from a bond graph play a key role in all tasks that have been considered, in FDI, in system mode identification and in failure prognosis. Simulation results have been obtained by using the dassl solver as part of the open source software Scilab [5].

As hybrid models include discrete state events, i.e. instantaneous state discontinuities, numerical multistep integration methods seem not to be particularly suited for their solution. The time instant of such discrete events must be located by constantly evaluating zero-crossing functions and by adjusting the integration step size. Once the time instant of a discrete event is located, numerical integration must be re-initialised. The necessary computational time may be of relevance in simulations of fault scenarios. Diagnostic models for generating ARRs use derivative causality at storage ports. The necessary differentiation with respect to time is usually performed in discrete time. Discontinuities in the inputs of the model, however, result in pulses.

DEVS simulation using state quantisation instead of time discretisation as outlined in Chap. 1 seems to be better suited for handling instantaneous state discontinuities. In comparison to the evolution of sophisticated numerical integration methods for DAE systems, the use of quantised state integration and the application of the DEVS methodology to hybrid system models is a rather young approach that still needs some problems to be addressed. However, the use of QSS and DEVS simulation for FDI in hybrid systems may be another subject of further research.

A topic that hasn't been considered in this book but is closely related to fault diagnosis is fault tolerant control (FTC) [6]. A 2008 bibliographical review on reconfigurable fault-tolerant control systems may be found in [7]. The aim is to react to a fault that has been detected and isolated so that the system can continue its operation in the presence of a component fault and to ensure safety at the same time. Fault tolerant strategies are categorised into two classes depending on whether they use a passive or an active approach [8]. Passive FTC relies on a robust controller of fixed structure that enables to cope with a set of faults taken into account at the stage of system design. A bond graph approach to passive fault-tolerant control of systems described by continuous-time model has been presented in [9]. Active FTC (AFTC) is a challenging task because the parameters of a control algorithm or even the algorithm is to be changed online by a supervision system. The integration of FDI and reconfigurable control for active fault-tolerant control systems has been addressed in [10]. Figure 10.1 displays the general scheme of active FTC.

In any case, FDI is a prerequisite also for this task and bond graph based ARR residual generation can provide the information needed by fault diagnosis. Chapter 11 of reference [11] addresses fault tolerant control of systems represented by continuous time model and related issues such as system inversion. In [12], a bond graph approach to diagnosis and FTC has been recently presented and applied to an intelligent autonomous vehicle. FTC of hybrid systems has been considered for instance

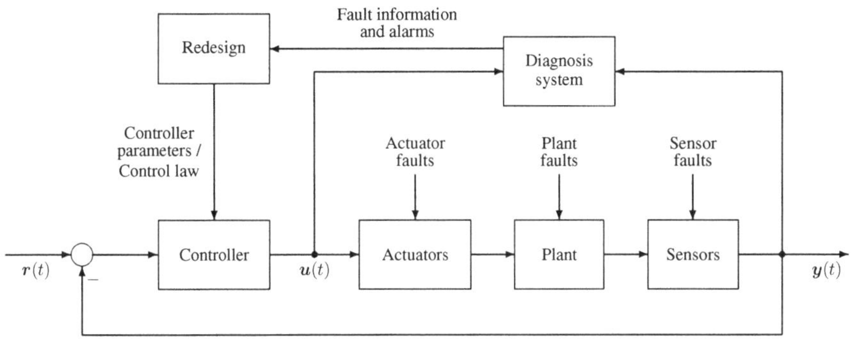

Fig. 10.1 General scheme of active FTC

in [13] without using bond graphs. A fault tolerant control approach for switched LTI systems has been proposed in [14] that uses bond graphs for mode identification and observer-based reliable state feedback control focusing on the time delay between FDI and fault accommodation during which still the original control law is applied to the faulty system.

References

1. Paynter, H. M. (1961). *Analysis and Design of Engineering Systems*. Cambridge: M.I.T. Press.
2. Ducreux, J. P., Dauphin-Tanguy, G., Rombaut, C. (1993). Bond graph modelling of commutation phenomena in power electronic circuits. In: J. J. Granda & F. E. Cellier (Eds.), *International Conference on Bond Graph Modeling, ICBGM'93, Proceedings of the 1993 Western Simulation Multiconference*. Simulation Series, 25(2) (pp. 132–136). San Diego: SCS Publishing. ISBN: 1-56555-019-6.
3. Wang, D., Yu, M., Low, C., & Arogeti, S. (2013). *Model-based health monitoring of hybrid systems*. New York: Springer.
4. Borutzky, W. (Ed.). (2011). *Bond graph modelling of engineering systems - theory, applications and software support*. New York: Springer.
5. Scilab Enterprises. 78000 Versailles, France. Available from: http://www.scilab.org/.
6. Blanke, M., Frei, C., Kraus, F., Patton, R., Staroswiecki, M. (2000). What is fault-tolerant control? Aalborg University, Department of Control Engineering. Available from: http://www.iau.dtu.dk/secretary/pdf/safeprocess_02h.pdf.
7. Zhang, Y., & Jiang, J. (2008). Bibliographical review on reconfigurable fault-tolerant control systems. *Annual Reviews in Control*, *32*, 229–252.
8. Patton, R. J. (1997). Fault-tolerant control systems: the 1997 situation. In *Proceedings of 3rd IFAC Symposium on Fault Detection, Supervision and Safety for Technical Processes* (pp. 1033–1055). Available from: http://hull.ac.uk/control/downloads/safepr.pdf.
9. Nacusse, M., Junco, S. J. (2011). Passive fault tolerant control: a bond graph approach. In A. Bruzzone, G. Dauphin-Tanguy, S. Junco, M. A. Piera (Eds.), *Proceedings of 5th International Conference on Integrated Modelling and Analysis in Applied Control and Automation* (IMAACA 2011) (pp. 75–82). Rome, Italy: Diptem Universit Di Genova.
10. Zhang, Y., Jiang, J. (2006). Issues on integration of fault diagnosis and reconfigurable control in active fault-tolerant control. In *Fault Detection, Supervision and Safety of Technical Processes* (pp. 1437–1448). Vol 6, Part 1. P.R. China: Tsinghua University.

11. Samantaray, A. K., & Ould Bouamama, B. (2008). *Model-based process supervision - A bond graph approach.*, Advances in Industrial Control London: Springer.
12. Loureiro, R. (2012). Bond graph model based on structural diagnosability and recoverability analysis: Application to intelligent autonomous vehicles [PhD thesis]. L' Université Lille 1.
13. Ocampo-Martinez, C., & Puig, V. (2009). Faul-tolerant model predictive control within the hybrid systems framework: application to sewer networks. *International Journal of Adaptive Control and Signal Processing.*, *23*(8), 757–787.
14. Hao, Y. A. N. G., Ze-Hui, M. A. O., & JIANG, Bin. (2006). Model-based fault tolerant control for hybrid dynamic systems with sensor faults. *Acta Automatica Sinica.*, *32*(5), 680–685.

Appendix A
Some Definitions

A.1 Notions Used in FDI

A number of key terms have been used throughout this book. This appendix provides a list of definitions that are in accordance with the outcome of a standardisation effort of the IFAC SAFEPROCESS Technical Committee. The definitions have been taken from [1, 2].

Definition A.1 (*Fault*) A fault is an unpermitted deviation of at least one characteristic property or parameter of the system from an acceptable, usual, or standard condition.

Definition A.2 (*Failure*) A failure is a permanent interruption of a system's ability to perform a required function under specified operating conditions.

Definition A.3 (*Malfunction*) A malfunction is an intermittent irregularity in the fulfillment of a system's desired function.

Definition A.4 (*Symptom*) A symptom is a change of an observable quantity from normal behaviour.

Definition A.5 (*Fault detection*) Fault detection: determination of the faults present in a system and the time of detection.

Definition A.6 (*Fault isolation*) Fault isolation: determination of the kind, location and time of detection of a fault.

Definition A.7 (*Fault identification*) Fault identification means the determination of the size- and time-variant behaviour of a fault.

Definition A.8 (*Fault diagnosis*) Fault diagnosis encompasses the determination of kind, size, location and time of detection of a fault by evaluating symptoms.

Definition A.9 (*Diagnostic model*) A set of static or dynamic relations which link specific input variables—the symptoms—to specific output variables—the faults.

W. Borutzky, *Bond Graph Model-based Fault Diagnosis of Hybrid Systems*,
DOI 10.1007/978-3-319-11860-4

Definition A.10 (*Analytical redundancy*) Use of two, not necessarily identical ways to determine a quantity where one way uses a mathematical process model in analytical form.

Definition A.11 (*Disturbance*) A disturbance is an unknown and uncontrollable system input.

Definition A.12 (*Residual*) Fault indicator based on deviations between measurements and model equation based calculations.

A.2 Hybrid Systems

Section 1.1 informally explains the notion of a hybrid system. The following formal definition has been given by Provan [3].

Definition A.13 (*Hybrid system*) A hybrid system is defined as $\Phi = (Q, X, \Sigma, Q_0, E, f, G)$ where

- Q is the set of discrete states or modes of the system,
- $X \subseteq \mathbb{R}^n$ is the continuous state space,
- Σ is a finite set of transition labels or events,
- $Q_0 \subseteq Q \times X$ is the set of initial conditions,
- $E \subset Q \times \Sigma \times Q$ is the transition relation, which defines the set of (controlled and autonomous) discrete transitions,
- $f : \mathbb{R} \times Q \times X$ is the flow condition for every mode defined by a differential equation,
- and $G : E \rightarrow 2^X \times \pi$ is a partial function that associates a guard condition (represented as a subset of X) with each autonomous transition, given a probability π.

Proven remarks:

The probability π introduces randomness into the transitions, which is important for transitions to failure states, which we assume occur randomly.

A state of a hybrid system is described by the pair $(q; x)$, where $q \in Q$ and $x \in X$.

A.3 DEVS Models

Definition A.14 (*Atomic DEVS model*) An atomic DEVS model is defined as the following tuple

$$M = \langle X, Y, S, \delta_{\text{int}}, \delta_{\text{ext}}, \lambda, ta \rangle$$

where

- X is the set of input event values,
- Y is the set of output event values,
- S is the set of state values,
- $\delta_{\text{int}} : S \rightarrow S$ is the internal transition function that defines how a state of the system changes internally when the elapsed time reaches the lifetime of the state,
- $\delta_{\text{ext}} : S \times \mathbb{R}_0^+ \times X \rightarrow S$ is the external transition function that defines how an input event changes a state of the system,
- $\lambda : S \rightarrow Y$ is the output function,
- $ta : S \rightarrow \mathbb{R}_0^+$ is the time advance function used to determine the lifespan of a state.

Remark A.1

1. For each state $s \in S$ function ta determines a time advance $ta(s)$ which is a non-negative real number that indicates how long the system remains in the current state in the absence of input events.
2. Let the system state at time t_1 be s_1. Then at time $t_1 + ta(s_1)$ an *internal* transition from s_1 to a new state $s_2 = \delta_{\text{int}}(s_1) \in S$ takes place.
3. When the system state changes from s_1 to s_2 an output event with the value $y_1 = \lambda(s_1) \in Y$ is produced.
4. Let the system state at time t_2 be s_2 and assume that an input event with the value x_3 arrives at time $t_2 + e_i < ta(s_2)$, i.e. while the system is in state s_2. Then the system changes instantaneously into a new state $s_3 = \delta_{\text{ext}}(s_2, e_i, x_3) \in S$ at time $t_2 + e_i$. At the advent of this externally triggered transition no output event is produced. □

Atomic DEVS models can be coupled by using input and output ports and by converting output events of one DEVS model into input events of another DEVS model. The DEVS formalism guarantees that the coupling of atomic DEVS models is a new DEVS model. A formal definition of a *coupled* DEVS model captures the names of the ports, the *internal* connections between input ports of one subsystem and output ports of another subsystem, external input connections from the input ports of a system and to input ports of subsystems, and external output connections from output ports of subsystems to output ports of the system. Let M_a denote the DEVS model of the static block f and M_b the DEVS model of the quantised integrator in the block diagram of Fig. 2.23. Then Fig. A.1 displays the coupled DEVS model N of the block diagram.

Given the enumeration of ports in Fig. A.1, the oriented connections can be represented by ordered pairs of tuples consisting of the name of the DEVS model and the port number, e.g. $((M_a, 0), (M_b, 0))$, $((M_b, 0), (M_a, 1))$, or $((N, 0), (M_a, 0))$.

Definition A.15 (*Coupled DEVS model*) A coupled DEVS model is defined as the following tuple

$$CM = \langle X, Y, D, \{M_d | d \in D\}, EIC, EOC, IC, select \rangle$$

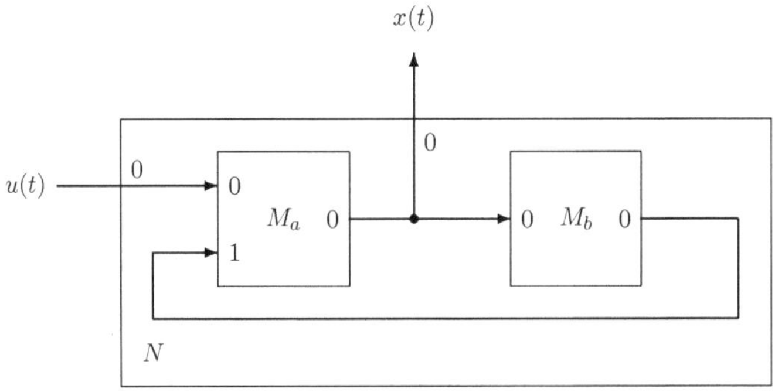

Fig. A.1 Coupled DEVS model of the block diagram in Fig. 2.23

where

- $X = \{(p, v)|p \in IPorts, v \in X_p\}$ is the set of input event values,
  $IPorts$ the set of input ports,
  X_p the set of values for the input ports,
- $Y = \{(p, v)|p \in OPorts, v \in Y_p\}$ is the set of output event values,
  $OPorts$ the set of output ports,
  Y_p is the set of values for the output ports,
- D is the set of component names and $d \in D$,
- M_d is a DEVS model,
- EIC is the set of external input couplings,
- EOC is the set of external output couplings,
- IC is the set of internal couplings,
- $select$ is the tie-breaking function $select : 2^D \to D$ that defines which event is to executed first in the case of of simultaneous events.

(see also [4, 5])

Appendix B
Short Introduction into Bond Graph Modelling

Bond graphs were devised by Professor Henry Paynter[1] at Massachusetts Institute of Technology (MIT), Cambridge, Massachusetts, USA as early as 1959 [6]. The concept was elaborated into a formal physical modelling methodology for multidisciplinary systems by his former Ph.D. students Professor Dean Karnopp and Professor Donald Margolis (University of California at Davis, California) and Professor Ronald Rosenberg (Michigan State University, East Lansing, Michigan). Since then this modelling approach has spread all over the world and is used in academia as well as in industry by many individuals. Software programs supporting bond graph modelling as well as a number of textbooks in various languages have emerged. The very first bond graph program for bond graph modelling and simulation was ENPORT™ developed by R. Rosenberg. Other programs are, for instance, 20sim [7], SYMBOLS [8], or the bond graph preprocessor CAMP-G [9] for Matlab®/Simulink® [10]. Some of the textbooks written in English besides the well known textbook of Karnopp, Margolis, and Rosenberg [11] are referenced in a list not meant to be exhaustive at the end of this appendix [12–18]. There are also some short introductions to bond graph modelling available [19–21]. In the following, some fundamentals of bond graph modelling are recalled so that the use of bond graph modelling for FDI and prognosis in this book can be followed more easily.

B.1 Basic Concepts

Bond graph based physical systems modelling starts from considering the exchange of energy between real or conceptual subsystems, components or elements of a multidisciplinary engineering system in which energy may be present in different forms and conversions from one form into another may take place. Like any other graph, bond graphs consist of nodes and edges. Nodes may represent models of subsystems, system components or elements. In the nodes, energy may be temporarily

[1] 1923–2002.

W. Borutzky, *Bond Graph Model-based Fault Diagnosis of Hybrid Systems*,
DOI 10.1007/978-3-319-11860-4

stored, transformed from one form into another, especially irreversibly into heat, or power may be distributed to other nodes. In bond graphs, the latter ones have got so-called *power ports* where energy may enter or leave a node. Of course, nodes may have more than one power port. They are called *multiports*. The edges of a bond graph are connections between the power ports of different nodes. They are called power bonds, or just *bonds*. They represent the transfer of power between ports and may be associated with physical links between real systems such as a shaft between a motor and a mechanical load, or a hydraulic line between hydraulic components if it is assumed that energy is neither stored nor dissipated into heat in a physical link. Otherwise, a physical model is to be developed for the physical link.

B.1.1 Power Variables and Energy Variables

Bond graph methodology assumes that the current amount of power $P(t)$ at time instant $t > 0$ transferred between two power ports can be expressed equally in all energy domains as the product of two power conjugated variables called *effort*, e, and *flow*, f, respectively.

$$P(t) = e(t) \cdot f(t) \tag{B.1}$$

Thus, the transfer of power between two subsystem ports can be depicted as indicated in Fig. B.1. The *half arrow* attached to the bond denotes the *reference* direction of the energy flow.

As a bond in a bond graph represents the energy exchange between two power ports of different nodes, all edges of a bond graph carry two power variables. Bonds may be annotated by the names of these power variables. It is a convention to place the name of an effort above a horizontal bond and the flow below the bond (cf. Fig. B.1). For vertical bonds the convention is to place the effort to the left and the flow to the right of the bond. For inclinations different from a multiple of 90°, a more sophisticated convention is to consider that variable as flow that is on the same side as the half arrow.

Effort and flow can be identified as force and linear velocity in translational mechanical engineering. In electrical engineering, the product of the voltage drop across the two pins of an electrical port and the joint current in both pins is the

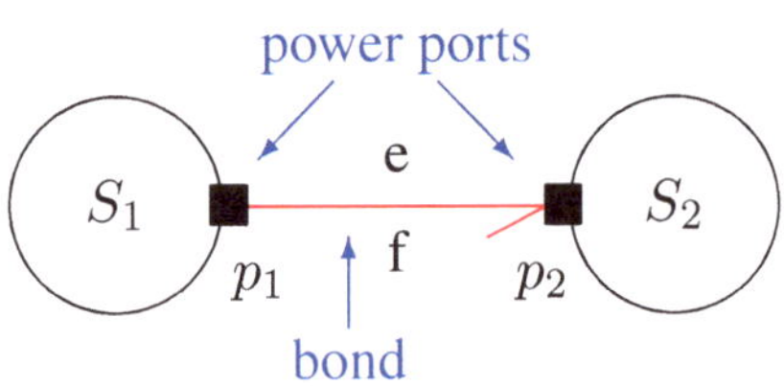

Fig. B.1 Power bond connecting two power ports of two subsystem models

Table B.1 Power and energy variables in various energy domains

Energy domain	Effort e	Flow f	Generalised momentum p	Generalised displacement q
Translational mechanics	Force F (N)	Velocity v (m/s)	Momentum p (N s)	Displacement x (m)
Rotational mechanics	Angular moment M (Nm)	Angular velocity ω (rad/s)	Angular momentum p_ω (Nms)	Angle θ (rad)
Electrical domain	Voltage u (V)	Current i (A)	Linkage flux λ (Vs)	Charge q (As)
Hydraulic domain	Total pressure p (N/m^2)	Volume flow rate Q (m^3/s)	Pressure momentum p_p (N/m^2 s)	Volume V_c (m^3)
Thermodynamic domain	Temperature T (K)	Entropy flow rate $\dot{S}$ (J/K/s)	–	Entropy S (J/K)
Chemical domain	Chemical potential μ (J/mole)	Molar flow $\dot{N}$ (mole/s)	–	Molar mass N (mole)

instantaneous amount of electrical power at this port. Correspondingly, in the thermal domain, these variables are the absolute temperature and the rate of change of the entropy. Table B.1 lists the effort and flow variables in the various energy domains.

The second and the third column of Table B.1 list the effort and flow variables in the various energy domains. The variables in the fourth column of Table B.1 are the time integral of the efforts and the variables in the fifth column are the time integral of the flows. They are called *energy variables* because they quantise the amount of energy in the energy storage elements of a model.

$$p(t) := p(t_0) + \int_{t_0}^{t} e(\tau)d\tau \tag{B.2}$$

$$q(t) := q(t_0) + \int_{t_0}^{t} f(\tau)d\tau \tag{B.3}$$

where $p(t_0)$, $q(t_0)$ are initial values.

B.1.2 Analogies

The power variables effort and flow play an equal role with respect to each other. They are just characterised by the fact that they are a factor in the power product. Given a pair of power variables, it is a matter of preference which of them is chosen as effort and which consequently serves as the flow. This gives rise to two possible *analogies*. One choice could be to relate a mechanical force, or a moment to an electrical voltage drop. Consequently, velocities, or angular velocities correspond to electrical currents. This analogy has been widely used for a long time. Therefore, occasionally, it is referred to as the *classical* analogy. The other possible association of a velocity with a voltage drop also makes perfectly sense from the point of *measurements*. It has been introduced by Firestone around 1933 and is called *mobility* analogy. If two modelling approaches just differ with respect to the analogy that is used, the resulting bond graph models will look different as well as the mathematical models derived from the graphs. However, as to the numerical evaluation of the mathematical models, simulation results should be the same.

B.1.3 Hierarchical Bond Graph Models

As other graphical modelling formalisms, bond graph methodology supports a hierarchical modelling approach. Bond graph models may be developed in a hierarchical combined top-down and bottom-up approach by using component models or elements from model libraries. For each hierarchy level, the structure of the model may be represented by a bond graph. While in iconic diagrams, or in electrical or hydraulic networks application-specific icons are used for the nodes of the graph, in bond graphs, nodes are presented by words enclosed by an optional ellipsis. For that reason, the notion of a *word bond graph* is common in the process of a bond graph-based model decomposition approach.

At the lowest hierarchy level, bond graph nodes represent basic energetic processes, that is, the delivery or storage of energy, the irreversible transformation of energy into heat, or the power conservative distribution of power. For these fundamental energetic processes, specific type names for nodes are used. For instance, the storage of electrical energy in a capacitor, or the storage of potential energy in a mechanical spring is represented by a node of type C. Of course, as with other graphical representations, user defined nodes may be introduced, e.g., a node labelled `DisplacementPump` representing a mathematical model of a hydraulic displacement pump.

As bond graph modelling starts from considering the energy exchange between system components and since this exchange is associated with physical quantities such as momentum, mass, electrical charge, or entropy, bond graph models should comply with physical conservation laws. In contrast, signal processing blocks in block diagrams may represent any functional relation between signals.

B.2 Bond Graph Elements

At the lowest hierarchy level, bond graph modelling uses a set of nine basic conceptual elements for representation of the fundamental energetic processes. They can be grouped into five categories.

B.2.1 Supply and Absorption of Energy

The supply of energy into a system is modelled by source elements. The absorption of energy flowing out of a system into its environment can be represented by sinks, which can be considered negative sources. As a power port has two variables, two kinds of sources exist. Sources may impose either an effort or a flow onto a system. For instance, a battery serving as a constant voltage source can be modelled by an effort source, while a hydraulic pump providing a constant volume flow rate can be modelled as a flow source. In bond graphs, sources and sinks respectively are denoted by the character S (Source). The type is naturally indicated by adding either the characters e or f respectively (Se or Sf). Sources may have more than one power port. Moreover, as there exist, e.g. stabilised voltage sources, or controlled hydraulic pressure pumps, sources may also have a signal port for feedback control. At a signal port, the amount of power is negligible small. In this case, the node type identifier is prefixed by the character M standing for *modulated* source (MSe, or MSf). (Conventionally, the signal port is located on the M side of the element.)

B.2.2 Energy Storage

The generalised momentum and the generalised displacement are not just the integral of an effort and a flow respectively, but have a physical meaning in most energy domains as indicated in Table B.1. There is no generalised momentum in thermodynamics. In chemical engineering and in hydraulics, the use of a pressure momentum is rather uncommon. Except in the thermodynamic and in the chemical domain, the generalised momentum and the generalised displacement respectively can be related to a power variable resulting in the constitutive relation of a 1-port storage element. That is, one of the two power variables of the port is the rate of change of a so-called *conserved*, or stored quantity, also called a *state*, while the other power variable is an equilibrium determining variable. As either the effort or the flow can be the rate of change of the conserved quantity, two types of energy stores can be distinguished.

In a C energy storage element, the flow f is integrated and the resulting generalised displacement q is related to the conjugate effort of the port.

$$q(t) = q(t_0) + \int_{t_0}^{t} f(\tau)d\tau \tag{B.4a}$$

$$q(t) = \Phi_C(e(t)) \tag{B.4b}$$

where Φ_C is a one-to-one function $\Phi_C : \mathbb{R} \to \mathbb{R}$ that has a unique single valued inverse Φ_C^{-1}.

For the second type of energy store, the I energy storage element, the role of effort and flow is just interchanged. The effort is integrated and the resulting generalised momentum is related to the conjugate flow.

$$p(t) = q(t_0) + \int_{t_0}^{t} e(\tau)d\tau \tag{B.5a}$$

$$p(t) = \Phi_I(f(t)) \tag{B.5b}$$

where Φ_I is a one-to-one function $\Phi_I : \mathbb{R} \to \mathbb{R}$ that has a unique single valued inverse Φ_I^{-1}.

In this sense, both storage types are dual to each other. For instance, an electrical capacitor, or a mechanical spring can be modelled as a C-type energy store, while a rigid body storing kinetic energy, or a coil storing magnetic energy can be modelled by an I type storage element. Energy stores can be multiport elements. Note that, in contrast to sources, modulation of storage elements would violate the principle of energy conservation.

B.2.3 Irreversible Transformation of Energy into Heat

The irreversible transformation of energy into heat, e.g. in an electrical resistor, or due to friction in mechanical and hydraulic systems, is often modelled as a loss of *free* energy. In bond graphs, it is represented by an R element (resistive element). If the production of entropy is taken into account, the two-port RS element introduced by Thoma [22] is used (cf. Fig. B.2). The character S (Source) indicates the thermal port and expresses the entropy production.
The RS element conserves power.That is,

$$e \cdot f = T \cdot \dot{S} \tag{B.6}$$

where T denotes the temperature and S the entropy.

While the relation between the power variables of the non-thermal port may be linear, for the thermal port it is always nonlinear. If a linear constitutive equation

Fig. B.2 Irreversible transformation into heat [22]

$$e(t) = R \cdot f(t) \tag{B.7}$$

with a resistance R is assumed for the non-thermal port, then

$$\dot{S}(t) = \frac{R \cdot f^2(t)}{T(t)} \tag{B.8}$$

According to the second law of thermodynamics, entropy production must be positive. Consequently, the graph of the constitutive relation must be within the first and third quadrant.

Like sources, resistive elements may be modulated. For instance, variable hydraulic orifices in spool valves controlled by the displacement of the spool may be modelled by displacement controlled R elements.

B.2.4 Reversible Transformation of Energy

In this kind of process, entropy is neither stored nor produced. Consequently, it is power conservative. There are two types of bond graph elements representing this kind of process. They are denoted by the acronyms TF and GY respectively. In the simplest case they are 2-port elements. Power conservation means that the instantaneous power at one port equals the instantaneous power at the other port. An element of type TF relates the efforts at the ports and separately relates the flows, while an element of type GY relates the effort of one port to the flow of the other port and vice versa. In the constitutive relations of both elements, a variable is multiplied by either a constant or by a function of time. In the second case, the elements must have a signal port in addition to the power ports. This is pointed out by prefixing their acronyms with the character M (MTF, MGY). Physical components that may be modelled by a TF element are electrical transformers, mechanical gear boxes, or hydraulic displacement pumps. The GY element may be a simple model for the transformation of electrical energy into mechanical energy in an electrical motor. Transformers as well as gyrators can be multiport elements.

B.2.5 Power Conservative Distribution of Energy

Modelling of energy flows in a system means that energy is supplied by sources and conveyed and distributed between the conceptual elements of the model. As energy storage and irreversible transformation of energy have already been taken into account by energy storage elements and by resistors, distribution of energy between elements can be considered power conservative. There are two types of multiport interconnection elements, called *0-junction* and *1-junction*. The two of them distribute power and have linear constitutive relations. For a 0-junction, the

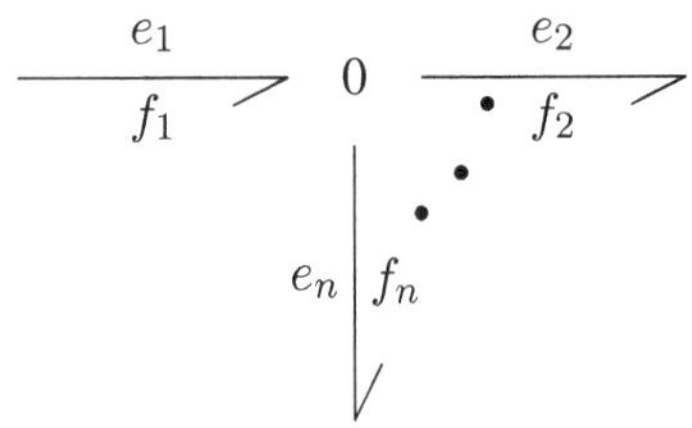

Fig. B.3 0-junction with n ports

efforts of all power ports are the same and all power conjugated flows sum up to zero taking into account their signs. The sign is determined by the orientation of the half arrow for the energy reference direction. If a half arrow is oriented towards a port, the flow is taken positive, otherwise the flow has a negative sign. The 1-junction plays the dual role. That is, the flows of all bonds incident the node are the same and all conjugate efforts sum up to zero by taking into account their signs. A 0-junction corresponds to an interconnection node in a network. Such a node has an effort (voltage, pressure) and according to the generalisation of Kirchhoff's current law, all flows in the branches incident to the node sum up to zero. In contrast, a 1-junction in a bond graph has no node as a counterpart in networks. The constitutive equation of a 1-junction relating all efforts is embodied implicitly in networks as it corresponds to the generalisation of Kirchhoff's voltage law.

According to the orientations of the half arrows in Fig. B.3, the constitutive equations of the 0-junction read

$$e_1 = e_2 = \ldots = e_n \tag{B.9a}$$

$$0 = f_1 - f_2 - \ldots - f_n \tag{B.9b}$$

Definition B.1 (*Junction structure*) The subgraph of a bond graph that contains only nodes of type 0, 1, (M)TF, (M)GY is called a *junction structure*.

B.3 Systematic Construction of Bond Graphs

One of the advantages of the bond graph approach is that the topological connectivity of components in a system can guide the systematic construction of a bond graph. Two procedures can be formulated, one for the construction of bond graphs for mechanical subsystems and one for the construction of bond graphs for subsystems in energy domains other than the mechanical domain (non-mechanical subsystems).

B.3.1 Mechanical Subsystems (Translation and Fixed-Axis Rotation)

1. Identify distinct inertial velocities and angular velocities; represent them by a 1-junction. Zero absolute velocities may be represented by a 0-junction also.
2. Insert C- and R-ports via a 0-junction between a proper pair of two 1-junctions. (The 0-junction represents a difference of velocities and, at the same time, a force or a moment. A spring or a dashpot reacts to a difference of velocities at its terminals.)
3. Insert TF- and GY-elements between appropriate pairs of 1-junctions representing either absolute or relative velocities.
   (A TF element relates a velocity at one port to a velocity at another port. A GY element relates a velocity at one port to a force or moment at another port. Velocities may be either inputs at all ports of a GY element, or they all must be outputs. Consequently, TF- and GY-elements are connected to 1-junctions.)
4. Attach inertia 1-port elements to their respective 1-junction.
5. Attach 1-port sources and 1-port sinks to appropriate 1-junctions.
6. Assign a reference direction for the energy flow to each bond (half arrow).
7. Remove all 1-junctions representing a velocity or angular velocity $\equiv 0$ along with all incident bonds and simplify the bond graph.

B.3.2 Non-mechanical Subsystems

1. Identify distinct efforts (potentials of nodes in electrical networks, absolute pressures in hydraulic and acoustic systems, absolute temperatures in thermodynamic systems); represent them by a 0-junction.
2. Insert the non-mechanical power port of a source, energy store, dissipator, transformer or gyrator via a 1-junction between two proper 0-junctions.
   (In bond graphs of electrical systems, the 1-junction represents the voltage drop across the port and, at the same time, the current through the port. In case of an electrical transformer, the 1-junctions at both ports of the associated TF element represent the currents through the coils of the transformer. In bond graphs of hydraulic systems, C elements are inserted via a 1-junction between the 0-junction of an absolute pressure and the 0-junction of the atmospheric pressure. In bond graphs of thermal systems, the thermal port of a C element is attached directly to the 0-junction of an absolute temperature.)
3. Add half arrows to all bonds.
4. Choose a potential as a reference; eliminate its corresponding 0-junction along with all incident bonds. If two sub-circuits of an electrical network are connected via an isolating transformer, a reference potential must be chosen in each sub-circuit.
5. Simplify the bond graph.

For hydraulic subsystems, it is common to choose the atmospheric pressure as reference. After elimination of its associated 0-zero junction along with all incident bonds, 0-junctions represent gage pressures. This results in a simplification of the construction of bond graphs for hydraulic systems. Gage pressures are represented by 0-junction, C elements are attached directly to a proper 0-junction. TF elements in bond graphs of hydraulic systems relate a pressure, p, to its associated mechanical force, F, and a volume flow rate, $\dot{V}$, of incompressible fluid flow to its associated translational velocity v.

$$F = A \cdot p \tag{B.10a}$$

$$\dot{V} = A \cdot v \tag{B.10b}$$

where A is the cross-section area perpendicular to the direction of a one-dimensional fluid flow. That is, the hydraulic port of a TF element is connected to a 0-junction of a gage pressure, while its mechanical port is connected to the 1-junction of a velocity.

B.3.3 Assignment of Power Reference Directions

There are intuitive rules for the assignment of power reference directions to the ports of all types of bond graph elements. For sources, the half arrow points away from the element's port, for storage elements and resistors, the orientation of the incident bonds is towards the element. This is quite intuitive as sources supply energy, energy storage elements store energy temporarily and resistors transform the absorbed energy irreversibly into heat. TF and GY elements adopt a 'through direction' of the reference direction of the energy flow because energy passes through these elements without storage and without entropy production. In essence, energy reference directions are from the sources through the junction structure into energy stores, resistors and sinks.

There should a *difference* of power variables at 0-junctions in bond graphs of mechanical systems and at 1-junctions in bond graphs of non-mechanical systems because springs and dampers react to a velocity difference and in electrical circuits, it is a voltage drop, viz. a difference of potentials, across a two-terminal element. The through direction of half arrows at the 1-junction corresponds to the reference direction of the current through the two-terminal element.

B.4 The Concept of Computational Causality at Power Ports

As each bond connecting two power ports of different nodes A and B carries two power variables, one of the two may be determined by one of the two submodels, while the other is determined by the other model. In other words, from

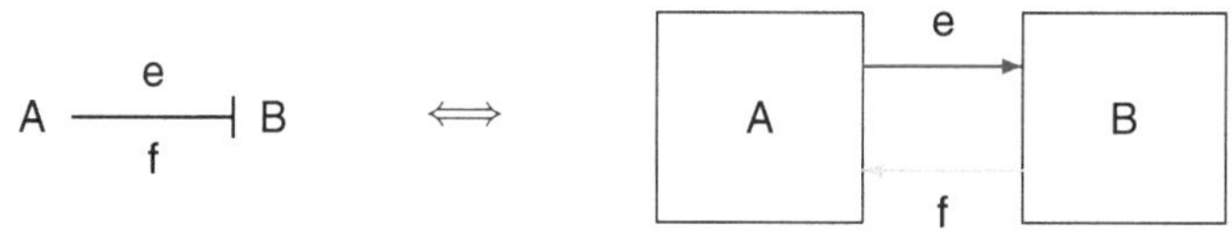

Fig. B.4 Computational causality: indicating the direction of effort and flow

a computational point of view, the effort could be computed by the evaluation of model A, while the flow is computed in model B. It could be the other way around as well. The decision as to local input-output pairs is indicated by a perpendicular stroke attached to the bond and is called the assignment of *computational causality* (Fig. B.4).

The end of a bond without perpendicular stroke indicates the model in which the effort is computed and in which the conjugate flow must be known. Another possible and common view is to consider both power variables of a bond as signals of *opposite* direction. That is, the perpendicular stroke indicates the signal direction of the effort, which implies that the end without the perpendicular stroke displays the direction of the conjugate flow. The perpendicular stroke is called a *causal* stroke. A bond graph is called a *causal* bond graph if a causal stroke has been added to each bond. Note that the half arrow and the causal stroke are orthogonal concepts. That is, there are four possible pattern of half arrow and causal stroke. Assigning causalities in a bond graph means that the bond graph is overlayed with a block diagram representing the *computational* structure of a model.

If sensors and instruments are included in the bond graph modelling of an engineering system, then the power conveyed between two ports of different components can be neglected, if the sensing of signals is of primary concern. This means that one of the two conjugate power variables associated with a bond can be dropped, turning the bond into a so-called *activated* bond or reducing it to a conventional signal arrow. As a result, the ports linked by a bond that has become a signal arrow turn into signal ports.

The hierarchical development of a bond graph model and the connection of component submodels according to the topological structure of a system implies that bond graphs of component models must be non-causal. The decision, which of the two power variables of a power port plays the role of an input signal, forcing the conjugate variable to be an output variable, is determined by the nature of component models. That is, causal strokes cannot be assigned before the hierarchical development of an overall system model is finished and the hierarchy has been 'flattened'. The latter means that each node representing a submodel is to be replaced recursively by a bond graph until the overall system bond graph only contains standard bond graph elements. Causal strokes, or computational causalities, at the ports of one and the same component sub-model can be different depending on the component sub-models it is connected to. Equations derived from a non-causal bond graph should be expressed initially in implicit form.

B.4.1 Rules for Computational Causalities at Power Ports

At each power port of a component model, it can be decided which one of the two power variables is computed in the component model, or in other words, which of the two power variables is an outgoing signal, or an output variable in one of the constitutive relations. However, these decisions cannot be made completely arbitrarily.

For sources there is no choice. For an effort source, the output is the effort, for a Sf source, it is the flow.

For 2-port TF and GY elements causal patterns must be as displayed in Fig. B.5. A TF relates efforts. That is, if the effort at one port is an input, the effort at the other port must be an output. Since the constitutive equations of GY elements relate the effort of one port to the flow of the other one, both causal strokes must either point to the element or away from it.

At 0-junctions the effort at all incident bonds is the same. Consequently, one causal stroke can point to the junction, while all others must point away from it (cf. Fig. B.5). For the dual 1-junctions, the role of effort and flow is interchanged. That is, at one bond the causal stroke may be pointing away from the junction. At all other bonds it must point towards the 1-junction. This causal pattern reflects the fact that one effort is equal to the sum of all other efforts and simultaneously it indicates that

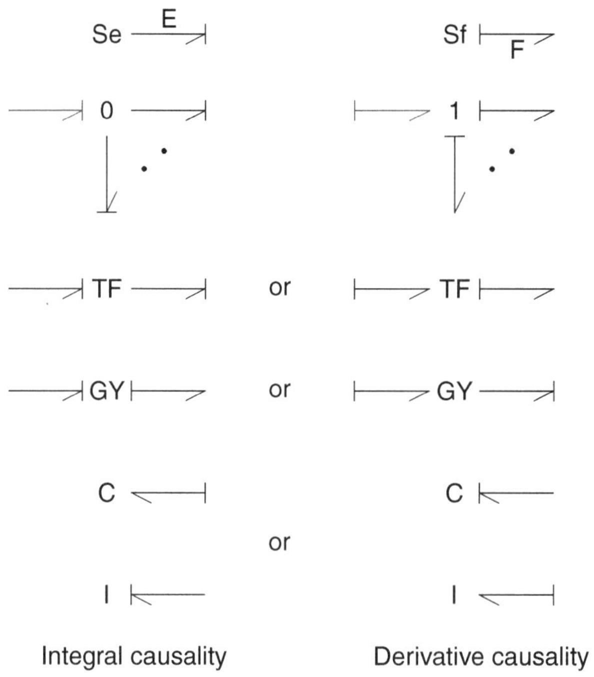

Fig. B.5 Admissible causal patterns at bond graph element ports

all flows are the same. One flow may be input to the junction, while all others are outputs.

For storage elements so-called *integral* causality is preferred. The power variable that is integrated with respect to time is the input variable. The conjugate power variable, related to the state of the energy store, is taken as output variable. For a C energy store, this is the effort. Consequently, the causal stroke points away from the port of the C element. For the dual I element, the effort of the power port is taken as an input variable. Consequently, the causal stroke points to the port of the I element. The effort is integrated with respect to time. The flow related to the resulting state of the I energy store is its output. If, for a C energy store, the causal stroke is on the side of the power port, or if the casual stroke points away from the port of an I energy store, then this means that the output is obtained by differentiation of the conjugate power variable. In this case, so-called *derivative* causality has been assigned to the port.

For resistors with linear constitutive relations, causal strokes may be either on the side of a port (conductance causality), or pointing away from the port (resistance causality). In some cases, however, there is no such choice. For a resistor representing dry friction, only the effort (force) can be the output.

Finally, causal patterns other than the admissible ones are termed *causal conflicts*. They give clear, valuable indications to consequences of modelling assumptions and may give rise to changes of the model.

B.4.2 Sequential Assignment of Computational Causalities

Given the previously discussed rules for the assignment of computational causalities to power ports, the question is in which order causalities are assigned to ports and how this information added to a bond graph is propagated through the graph, in other words, how does the choice of computational causality at power port affects the ports of adjacent elements. The step by step assignment of causalities follows the Sequential Causality Assignment Procedure (SCAP) introduced by Karnopp and Rosenberg [11, 23]. This procedure has become a standard and is widely used. A modification with regard to so-called causal bond graph loops has been discussed by van Dijk in [24]. In order to support the derivation of ARRs from hybrid bond graphs that hold for all system modes, Low et al. [25] recently proposed another modification of the SCAP by introducing preferred causalities for controlled junctions.

Sequential Causality Assignment Procedure (SCAP)

1. Assign causality to one of the sources according to its type and *propagate* this causal information into the bond graph through its junction structure as far as possible by observing causality rules at element ports.
2. Repeat step 1 until all ports of sources are assigned an appropriate causality.
3. Assign preferred integral causality to a port of an energy store and propagate this causal information into the bond graph as far as possible. Propagation of the

causality of a storage port may lead to derivative causality at power ports of other energy stores and often entails an assignment of causality at resistor ports. If, by this way, a nonlinear resistor port receives a causal stroke not in accordance with the resistor's constitutive equations, the causality assignment must reversed at this port. This, however, may result in a causal conflict at the junction the resistor port is attached to. This conflict at the junction must be resolved before preferred integral causality can be assigned to the next storage port. For instance, if an I element and a 1-port resistor representing Coulomb friction are attached to a 1-junction, then the I element must take derivative causality.

4. If there are any resistor ports left without causality after causality has been assigned to all storage ports, then the procedure continues with assignment of causality to resistors with characteristics that do not have a unique inverse, to ensure their correct formulation.
5. Finally, if there are still resistor ports or internal bonds in the junction structure without causality, one resistor port or an internal bond must be chosen. Causality is arbitrarily assigned and propagated through the junction structure. This step is repeated until no causally unassigned bonds are left.
   Note, if this last step is needed, then algebraic loops, viz. implicit algebraic equations, will be part of the mathematical model. Many of today's software programmes supporting bond graph modelling are able to cope with algebraic loops. They just issue a warning or process the model silently.

B.5 Derivation of Equations from Causal Bond Graphs

Once causal strokes have been added to the bonds of a bond graph, a mathematical model can be derived in a systematic manner. First, it must be decided for which unknowns a set of mathematical relations is to be derived. An obvious choice is the set of states of energy storage elements with integral causality at their ports. These variables determine the energetic state of a system in the sense that they quantify the content of each energy store at all time instances $t \geq 0$. As the output variable of a 1-port energy storage element with integral causality, also called *co-energy* variable, is related to its state, it can be chosen as an alternative unknown. This choice is adopted in this book. Note that energy stores with derivative causalities do not contribute to the system's state. Their output variable algebraically depends on the output variables of energy stores with integral causalities.

Software supporting bond graph modelling such as 20sim®, Symbols™, or CAMPG mentioned at the beginning of this appendix can automatically deduce model equations from a causal bond graph. For small to medium scale models, equations can also be deduced manually in a systematic manner. For that purpose, a procedure is given in the following that uses the notion of a *causal path*.

Definition B.2 (*Causal path*) A sequence of bonds from one power port of an element to a power port of another element is called a causal path if there is no 2-port gyrator in between and if all bonds have their causal stroke at the same end.

A cascade of bonds between two power ports with a gyrator in between is called a causal path if all bonds on one side of the gyrator have there causal stroke at the same end, while all bonds on the other side of the gyrator have their causal stroke on the opposite end. That is, the gyrator switches the direction of efforts on one of its sides. □

Remark B.1 An essential feature of bond graphs is that, once assignment of causalities has been completed, conclusions can be drawn with regard to the form of mathematical models that can be derived from the graph by looking for causal paths in a causal graph. There is no need to know the actual form of possibly nonlinear constitutive relations nor to establish and to reformulate any equations. For instance, if there are no energy storage elements with derivative causality, no causal paths between two ports of different resistors and no closed causal paths in the junction structure, then a mathematical model in the form of an explicit state space model can be derived from the graph. More details may be found, for instance, in [14, 24]. □

One straightforward way towards the formulation of a mathematical model in a modelling language, well suited for automation, is to write the constitutive equations for all nodes of the bond graph and to have all redundancies removed symbolically. If the aim is to perform a simulation, the equations can be sorted and transformed into a programming language. If the equations are linear and if the aim is to come up with the matrices of a linear state space model in symbolic form to be processed by a mathematical program, e.g., the open source software Scilab [26], then, clearly, all algebraic equations must be eliminated. Auxiliary variables can be eliminated when equations are derived from the causal bond graph by walking back causal paths until the input, e.g. into a resistor can be expressed by the outputs of sources and storage elements in integral causality. For bond graphs that are not too large, this can be done manually in a systematic manner. The derivation of an ordered set of equations is guided by the following procedure [27, 28].

B.5.1 Procedure for Manually Deducing Equations from a Causal Bond Graph

1. Write the constitutive equations for all independent sources. Their outputs are given functions of time.
2. In contrast, the output of a controlled source is algebraically related to its input. If the latter is not an output of an independent source or an energy store with integral causality, then it can be expressed by means of such outputs by back propagation of causal paths in the junction structure and by eliminating intermediate variables.

3. The outputs of resistors depend algebraically on their inputs. By back propagation along causal paths through the junction structure, their outputs can be expressed by outputs of sources either independent, or controlled ones and outputs of energy stores. The outputs of dependent sources do not need to be eliminated, since they have already been determined in the previous step.
4. For storage ports, the derivative with respect to time of an output is a function of the input(s). By working back causal paths, the inputs can be expressed by outputs of other energy storage elements, of resistors, or sources.

Note, that if there are causal paths between resistor ports then implicit, algebraic equations will result. This means, that the output of the resistor port at one end of the causal path cannot be computed without knowing the output of the resistive port at the other end of the causal path. In this case, intermediate variables cannot be eliminated and expressed by system inputs and state variables by back propagation along causal paths. The mathematical model will be of the form of a DAE system that can be transformed into an ODE system if the system of coupled algebraic equations can be solved symbolically.

Implicit algebraic equations also result, if there are closed causal paths in the junction structure. If there are storage elements that must accept derivative causality, then their output variables are algebraically dependent on the outputs of the storage elements in preferred integral causality and the number of states is smaller than the number of storage elements.

B.5.2 Illustrating Example

The systematic manual derivation of equations from a causal bond graph shall be illustrated by considering the simple circuit depicted in Fig. B.6.

The input-output behaviour of the operational amplifier is captured by the simple commonly used model depicted in Fig. B.7. That is, the example circuit contains a dependent controlled source.

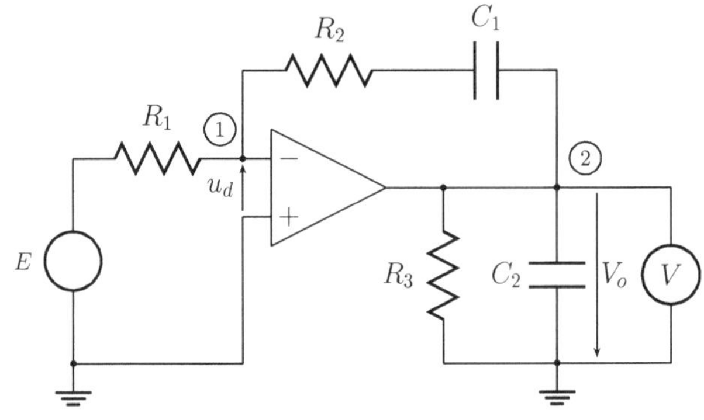

Fig. B.6 Example circuit

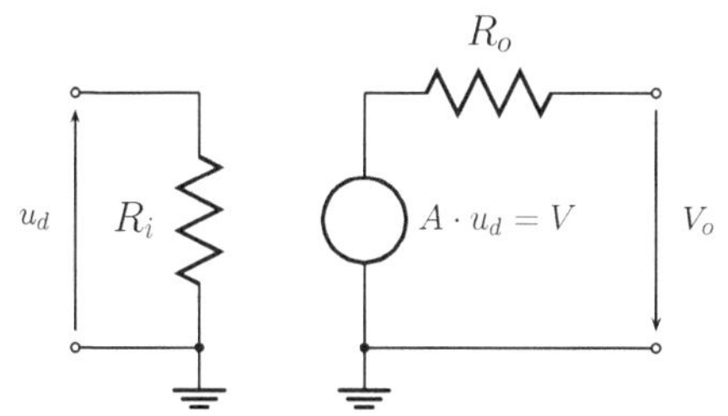

Fig. B.7 Behavioural model of the operational amplifier

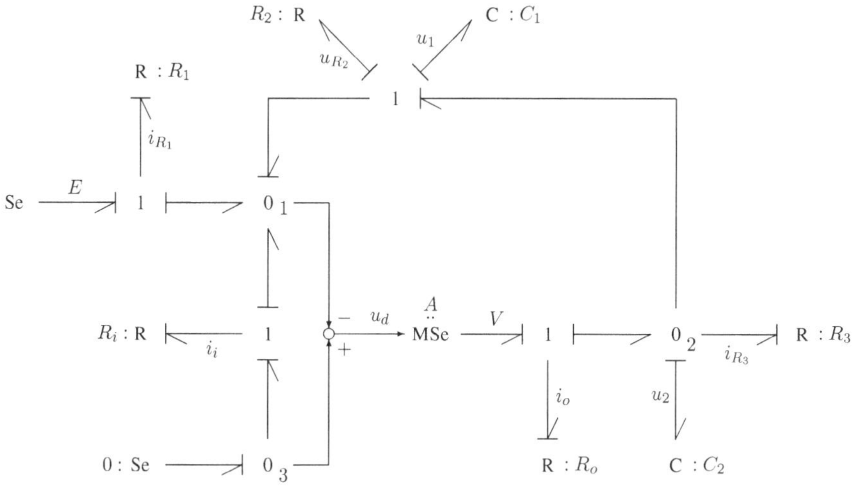

Fig. B.8 Bond graph of the example circuit in Fig. B.6

Application of the procedure for non-mechanical subsystems results in the causal bond graph of Fig. B.8.

The causally completed BG in Fig. B.8 indicates that the two capacitors C : C_1, C : C_2 are in integral causality and that there are causal paths from R : R_2 to R : R_1, from R : R_2 to R : R_i, and from R : R_2 to R : R_o via the controlled source MSe. These causal paths share bonds. The output V of the controlled source is algebraically dependent on its input u_d. The latter one depends algebraically on the capacitor voltages u_1, u_2, and the resistor voltage u_{R_2}.

This means that the mathematical model to be deduced will consist of two explicit differential equations including resistor currents and a set of coupled implicit algebraic equations for the outputs of the resistors and the input of the controlled source. As the model is linear, these algebraic equations could be solved symbolically turning the DAE system into an explicit ODE system. Alternatively, the DAE system could be directly formulated in the Scilab script language and evaluated by the DASSL solver.

From the bond graph of the example circuit the following equations can be deduced.
Independent sources:

$$E = f_E(t) \tag{B.11}$$

Dependent sources:

$$u_d = -u_2 + u_1 + u_{R_2} \tag{B.12}$$
$$V = A \cdot u_d \tag{B.13}$$

Resistors:

$$i_{R_1} = \frac{1}{R_1}(E - u_d) \tag{B.14}$$
$$i_i = \frac{1}{R_i}u_d \tag{B.15}$$
$$u_{R_2} = R_2(-i_i - i_{R_1}) \tag{B.16}$$
$$i_o = \frac{1}{R_o}(V - u_2) \tag{B.17}$$
$$i_{R_3} = \frac{1}{R_3}u_2 \tag{B.18}$$

Storage elements:

$$\dot{u}_1 = \frac{1}{C_1}(-i_i - i_{R_1}) \tag{B.19}$$
$$\dot{u}_2 = \frac{1}{C_2}(i_o + i_i + i_{R_1} - i_{R_3}) \tag{B.20}$$

The set of coupled linear algebraic equations (B.12)–(B.18) deduced from the bond graph in Fig. B.8 can be easily solved manually for the input u_d of the modulated source MSe.

$$\begin{aligned} u_d &= -u_2 + u_1 + u_{R_2} \\ &= -u_2 + u_1 - R_2(i_i + i_{R_1}) \\ &= -u_2 + u_1 - R_2\left(\frac{1}{R_i}u_d + \frac{1}{R_1}(E + u_d)\right) \\ \left(1 + \frac{R_2}{R_1} + \frac{R_2}{R_i}\right)u_d &= -u_2 + u_1 - \frac{R_2}{R_1}E \end{aligned} \tag{B.21}$$

Once u_d is known, the currents of the resistors can be computed so that the right hand side of the two state equations is known. In the case of a larger set of coupled linear algebraic equations, software such as Matlab's Symbolic Math Toolbox™ could be used to solve the subsystem of linear algebraic equations symbolically.

If there were no causal paths between resistor ports then their outputs could be expressed by the two state variables u_1, u_2 and the input E by back propagation of causal paths. The result would be an ordered set of equations that could be computed in that order. Clearly, if state space matrices are needed, the outputs of the resistors could be inserted into the constitutive equations of the storage elements.

B.6 Bond Graphs: A Core Model Representation

Bond graph modelling cannot only support the generation of state space models for the simulation of the dynamic behaviour of a system. Bond graphs can serve as core model representation from which various forms of mathematical models can be directly generated such as

- DAE systems,
- Lagrange equations,
- transfer functions,
- equations of a partially inverse model with respect to a given pair of variables,
- canonical form of state space equations or the standard interconnection form both used for robust control.

Moreover, causal bond graphs can support

- the analysis of structural observability and structural controllability,
- model based control,
- parameter sensitivity analysis,
- the analysis of a direct LTI model for structural invertibility,
- model based fault detection and isolation (FDI) for systems represented by a hybrid model which is the subject of this book.

B.7 Summary

Bond graph modelling of multidisciplinary systems comprising various energy domains starts from considering energy flows between ports of subsystems or components and is based on physical principles. An essential feature of the methodology is that modellers do not need to start with equations but may focus in a qualitative way on physical phenomena a model should capture. The formulation of equations is postponed to a later stage of the model development process. Moreover, the path down from a conceptional view, from a system schematic that may not be formalised down to a mathematical model is rule based. Rules guide the systematic development of a bond graph, the assignment of causalities, and the generation of equations.

A causal bond graph clearly reflects modelling assumptions and displays the topological structure, i.e. the connectivity of components or elements like a circuit diagram and furthermore indicates the computational structure of a model such as a block diagram does.

Bond graph modelling allows for a representation that is uniform over all energy domains, does interface with block diagrams and is formalised so that it can be processed by software. If software supporting bond graph modelling is not available, causal bond graphs can be systematically converted into a block diagram, or the equations of a time domain model deduced from a causal bond graph can be processed by a numerical solver for ODE or DAE systems.

References

[1] Isermann, R. (2006). *Fault-Diagnosis systems an introduction from fault detection to fault tolerance*. Berlin: Springer.

[2] Blanke, M., Kinnaert, M., Lunze, J., & Staroswiecki, M. (2006). *Diagnosis and fault-tolerant control*. Berlin: Springer.

[3] Provan, G. (2009). Model abstractions for diagnosing hybrid systems. *Proceedings of the 20th international workshop on principles of diagnosis, DX-09* (pp. 321–328). Stockholm, Sweden. Retrieved from http://www.cs.ucc.ie/ccsl/GP-papers/2009/Provan_DX_2009_2.pdf.

[4] Wainer, G. A. (2009). *Discrete-Event modeling and simulation—a practitioner's approach*. Boca Raton, FL, USA: CRC Press, Taylor & Francis Group.

[5] Zeigler, B., Kim, T. G., & Praehofer, H. (2000). *Theory of modeling and simulation* (2nd ed.). New York: Academic Press.

[6] Paynter, H. M. (1961). *Analysis and design of engineering systems*. Cambridge, Massachusetts, USA: M.I.T. Press.

[7] Controllab Products. *20-sim the power in modeling*. Retrieved from http://www.20sim.com.

[8] HighTec Consultants. *SYMBOLS Sonata*™. Retrieved from http://www.htcinfo.com/.

[9] Cadsim Engineering. *CAMP-G*. Retrieved from http://www.bondgraph.com.

[10] The Mathworks. *MATLAB®/Simulink®*. Retrieved from http://www.mathworks.com/products/simulink/.

[11] Karnopp, D. C., Margolis, D. L., & Rosenberg, R. C. (2012). *System dynamics: Modeling, simulation, and control of mechatronic systems* (5th ed.). Wiley: New York.

[12] Merzouki, R., Samantaray, A., Pathak, P., & Ould Bouamama, B. (2013). *Intelligent mechatronic systems*. Berlin: Springer.

[13] Borutzky, W. (Ed.). (2011). *Bond graph modelling of engineering systems—Theory, applications and software support*. NY, USA: Springer.

[14] Borutzky, W. (2010). *Bond graph methodology—Development and analysis of multidisciplinary dynamic system models*. London, UK: Springer. ISBN : 978-1-84882-881-0.

[15] Das, S. (2009). *Mechatronic modeling and simulation using bond graphs*. Boca Raton: CRC Press.
[16] Samantaray, A. K., & Ould Bouamama, B. (2008). *Model-based process supervision—A bond graph approach. Advances in industrial control*. London: Springer.
[17] Mukherjee, A., Karmakar, R., & Samantaray, A. K. (2006). *Bond graph in modeling, simulation and fault identification*. New Delhi, India: I.K. International Publishing House. ISBN: 81-88237-96-5.
[18] Damić, V., & Montgomery, J. (2003). *Mechatronics by bond graphs: An object-oriented approach to modelling and simulation*. Berlin: Springer.
[19] Borutzky, W. (2009). Bond graph modelling and simulation of multidisciplinary systems—An introduction. *Simulation Modelling Practice and Theory, 17*(1), 3–21.
[20] Broenink, J. F. (1999). Introduction to physical systems modelling with bond graphs. Enschede, Netherlands: University of Twente. Retrieved from http://www.ce.utwente.nl/bnk/papers/BondGraphsV2.pdf.
[21] HighTec Consultants. (2013). *Bondgraphs.com—Everything about bondgraphs*. Retrieved from: http://bondgraph.org/.
[22] Thoma, J. U. (1975). Entropy and mass flow for energy conversion. *Journal of The Franklin Institute, 299*(2), 89–96.
[23] Karnopp, D. C., & Rosenberg, R. C. (1968). *Analysis and simulation of multiport systems—The bond graph approach to physical system dynamics*. Cambridge, MA: MIT Press.
[24] van Dijk, J. (1994). *On the role of bond graph causality in modelling mechatronic systems*. [Ph.D. thesis]. University of Twente. Enschede, The Netherlands.
[25] Low, C. B., Wang, D., Arogeti, S., & Zhang, J. B. (2010). Causality assignment and model approximation for hybrid bond graph: Fault diagnosis perspectives. *IEEE Transactions on Automation Science and Engineering, 7*(3), 570–580.
[26] Scilab Enterprises. Scilab; 78000 Versailles, France. Retrieved from http://www.scilab.org/.
[27] Borutzky, W. (2005). Exchange and reuse of bond graph models based on XML. In J. J. Granda & F. E. Cellier (Eds.), *International conference on bond graph modeling, and simulation (ICBGM 2005), Simulation Series* (pp. 129–136, Vol. 37(1)). SCS Publishing. ISBN: 1-56555-287-3.
[28] Wellstead, P. E. (1979). *Introduction to physical system modelling*. London: Academic Press.

Appendix C
Some Mathematical Background

C.1 Index of a DAE System

Definition C.1 (*DAE index*) (cf. [1, 2]) Let $\boldsymbol{F}$ be a vector valued function that is sufficiently differentiable and

$$\boldsymbol{F}(\boldsymbol{x}(t), \dot{\boldsymbol{x}}(t), \boldsymbol{y}(t), \boldsymbol{u}(t), \boldsymbol{p}, t) = \boldsymbol{0} \tag{C.1}$$

a set of implicit DAEs where the vectors have the following meaning.

$\boldsymbol{x}$: dynamic variables

$\boldsymbol{y}$: output variables

$\boldsymbol{u}$: input variables

$\boldsymbol{p}$: vector of constant parameters

t : time

Then (C.1) has a *differentiation index* ν if ν is the minimal number of analytical differentiations such that analytical manipulations of the set of equations

$$\boldsymbol{0} = \boldsymbol{F}(\ldots) \tag{C.2}$$

$$\boldsymbol{0} = \frac{d}{dt}\boldsymbol{F}(\ldots) \tag{C.3}$$

$$\vdots \tag{C.4}$$

$$\boldsymbol{0} = \frac{d^{\nu}}{dt}\boldsymbol{F}(\ldots) \tag{C.5}$$

allow to establish an explicit ODE for $\boldsymbol{x}$. □

W. Borutzky, *Bond Graph Model-based Fault Diagnosis of Hybrid Systems*,
DOI 10.1007/978-3-319-11860-4

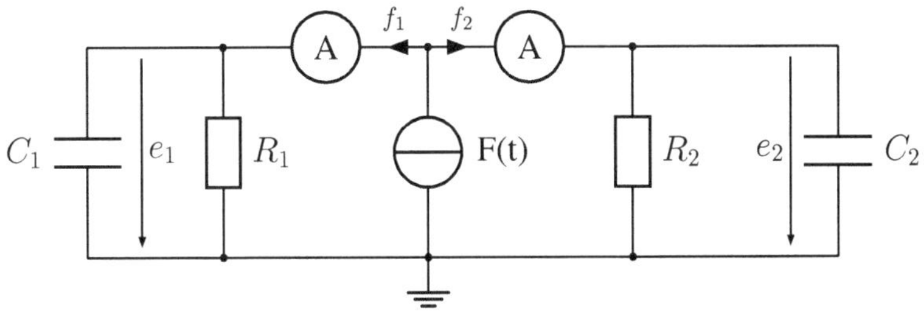

Fig. C.1 Simple circuit

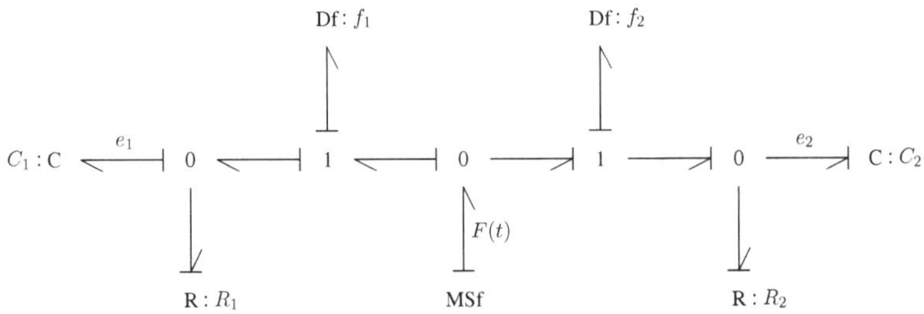

Fig. C.2 Bond graph of the simple circuit in Fig. C.1

Example

Consider the simple circuit in Fig. C.1. From its bond graph in Fig. C.2 the following equations are easily deduced.

$$\dot{e}_1 = \frac{1}{C_1}\left(f_1 - \frac{e_1}{R_1}\right) \tag{C.6a}$$

$$\dot{e}_2 = \frac{1}{C_1}\left(f_2 - \frac{e_2}{R_2}\right) \tag{C.6b}$$

$$e_2 = e_1 \tag{C.6c}$$

$$F = f_1 + f_2 \tag{C.6d}$$

Differentiating the algebraic constraints (C.6c)–(C.6d) with respect to time gives a system in which still an ODE for f_2 is missing.

$$\dot{e}_1 = \frac{1}{C_1}\left(f_1 - \frac{e_1}{R_1}\right) \tag{C.7a}$$

$$\dot{e}_2 = \frac{1}{C_1}\left(f_2 - \frac{e_2}{R_2}\right) \tag{C.7b}$$

$$\dot{e}_2 = \dot{e}_1 \tag{C.7c}$$

$$\dot{f}_1 = -\dot{f}_2 + \dot{F} \tag{C.7d}$$

Differentiating (C.6a)–(C.6b) and equating the second order derivatives yields

$$\frac{1}{C_2}\left(\dot{f}_2 - \frac{\dot{e}_2}{R_2}\right) = \frac{1}{C_1}\left(\dot{f}_1 - \frac{\dot{e}_1}{R_1}\right) \tag{C.8}$$

Substituting $\dot{f}_1$ by means of (C.8) results in an equation that expresses $\dot{f}_2$ as a function of e_1, e_2, f_1, f_2. That is, the initial DAE system (C.6a)–(C.6d) is of index 2 as the DAE system obtained by differentiation of (C.6a)–(C.6d) had also to be differentiated.

Remark C.1 Pantelides has developed an algorithm that finds a consistent set of initial conditions for a DAE system and can also be used for index reduction [6]. It is implemented for instance in the software program Modelica [4]. □

C.2 Convergence of the Gauss-Newton Algorithm

Consider the nonlinear least squares problem (NLSP)

$$\min f(\boldsymbol{\Theta}) = \frac{1}{2}||\boldsymbol{e}(\boldsymbol{\Theta})||_2^2 \tag{C.9}$$

where $\boldsymbol{\Theta} \in \mathbb{R}^m$ is an m-dimensional real vector and $\boldsymbol{e} : \mathbf{R}^m \mapsto \mathbf{R}^{n(q+1)}$ is an n(q+1)-dimensional real vector function of $\boldsymbol{\Theta}$.
The Jacobian of $\boldsymbol{e}$ is denoted by $\mathbf{J}_e(\boldsymbol{\Theta})$. The gradient ∇f and the Hessian $\mathbf{H}$ of $f(\boldsymbol{\Theta})$ is given by

$$\nabla f(\boldsymbol{\Theta}) = \mathbf{J}_e(\boldsymbol{\Theta})^T \boldsymbol{e}(\boldsymbol{\Theta}) \tag{C.10}$$

$$\mathbf{H}(\boldsymbol{\Theta}) = \nabla^2 f(\boldsymbol{\Theta}) = \mathbf{J}_e(\boldsymbol{\Theta})^T \mathbf{J}_e(\boldsymbol{\Theta}) + \mathbf{Q}(\boldsymbol{\Theta}) \tag{C.11}$$

where

$$\mathbf{Q}(\boldsymbol{\Theta}) := \sum_{j=1}^{n(q+1)} \boldsymbol{e}_j(\boldsymbol{\Theta}) \nabla^2 \boldsymbol{e}_j(\boldsymbol{\Theta}) \tag{C.12}$$

Let $\rho(\mathbf{A})$ denote the spectral radius of a quadratic matrix $\mathbf{A}$.

Theorem C.1 (Convergence of the Gauss-Newton iteration). [5] *Assume that*

- *there exists a vector* $\boldsymbol{\Theta}^* \in \mathbb{R}^m$ *such that* $\mathbf{J}^T(\boldsymbol{\Theta}^*)\boldsymbol{e}(\boldsymbol{\Theta}^*) = 0$ *and*
- *the Jacobian matrix* $\mathbf{J}(\boldsymbol{\Theta}^*)$ *at* $\boldsymbol{\Theta}^*$ *has full rank n.*

Let ρ denote the spectral radius of an $n \times n$ matrix $\mathbf{A}$. *If*

$$\rho\left((\mathbf{J}^T(\boldsymbol{\Theta}^*)^T\mathbf{J}^T(\boldsymbol{\Theta}^*))^{-1}\mathbf{Q}(\boldsymbol{\Theta}^*)\right) < 1 \tag{C.13}$$

then the Gauss-Newton iteration converges locally to $\boldsymbol{\Theta}^*$.

C.3 Implicit Function Theorem

In the context of ARRs it is sufficient to consider a simplified version of the implicit function theorem.

Theorem C.2 (Implicit function theorem) [3] *Let $D \subset \mathbb{R}^n$ be a domain, ϕ a real valued function $\mathbb{R}^n \to \mathbb{R}$ that is continuously differentiable on an open set $D_1 \subset D$, $\boldsymbol{x}^0 = (x_1^0, x_2^0, \ldots, x_n^0) \in D_1$ and $\phi(\boldsymbol{x}^0) = 0$. Suppose that $\partial\phi(\boldsymbol{x}^0)/\partial x_1 \neq 0$.*

Then there exists a neighbourhood $U(x_2^0, x_3^0, \ldots, x_n^0) \subset D_1$, an open set $V \subset \mathbb{R}$ containing x_1^0, and a real valued function $\psi : U \to V$ such that

$$x_1^0 = \psi(x_2^0, x_3^0, \ldots, x_n^0) \tag{C.14a}$$

$$0 \equiv \phi(\psi(x_2^0, x_3^0, \ldots, x_n^0), x_2^0, x_3^0, \ldots, x_n^0) \tag{C.14b}$$

References

[1] Gear, C. W., & Petzold, L. R. (1984). ODE methods for the solution of differential/algebraic systems. *SIAM Journal on Numerical Analysis*, *21*, 367–384.

[2] Hairer, E., & Wanner, G. (1996). *Solving ordinary differential equations II, stiff and differential-algebraic problems* (2nd ed.). Berlin: Springer.

[3] Krantz, S. G., & Parks, H. R. (2013). *The implicit function theorem—History, theory, and applications*. Berlin: Springer. ISBN 978-1-4614-5980-4. doi10.1007/978-1-4614-5981-1. (Modern Birkhäuser Classics, Boston, 2003, second printing).

[4] Modelica Association. Modelica and the Modelica Association. URL http://www.modelica.org.

[5] Ortega, J. M., & Rheinboldt, W.C. (1970). *Iterative solution of nonlinear equations in several variables*. New York: Academic Press.

[6] Pantelides, C. C. (1988). The consistent initialization of differential-algebraic systems. *SIAM, Journal of Scientific and Statistical Comutation*, *9*, 213–231.

Glossary

In this glossary, some key notions used in the text are listed in alphabetical order along with short explanations in accordance with a list of definitions compiled by the Safeprocess Technical Committee of IFAC, the International Federation of Automatic Control [1].

Analytical redundancy relations (ARRs) Are mathematical equations that relate known system inputs, known parameters and quantities obtained by measurements from a real system. Their evaluation results in so-called ARR residuals that are identical to zero or close to zero in narrow limits as long as the system is healthy. Residuals that deviate distinguishably from zero serve as fault indicators. If nonlinear constitutive element equations do not permit to eliminate unknown variables in a candidate for an ARR in closed symbolic form then residuals are given implicitly and can be determined by numerically solving a set of equations. As inputs into ARRs may be time derivatives of measured quantities, measurement noise is to be filtered appropriately. The differentiation is carried out in discrete time.

Diagnostic bond graph A system bond graph model with storage elements in derivative causality and sensors in inverted causality from which ARRs can be systematically deduced. In online model-based FDI, initial conditions are difficult to obtain. Therefore, storage elements are in derivative causality. With regard to FDI, measured quantities provided by sensors are known inputs into a diagnostic bond graph model. Its purpose is to provide ARR residuals as fault indicators to a diagnosis module.

Disturbance Is an unknown and uncontrollable system input.

Failure A failure is a permanent interruption of a system's ability to perform a required function. It can only be accommodated by a reconfiguration of the system.

Fault A system fault is a deviation of the system structure or the system parameters from the nominal conditions [2]. Appropriate actions may enable to recover from a component fault without replacing the component. The fault may be accommodated through fault tolerant control.

W. Borutzky, *Bond Graph Model-based Fault Diagnosis of Hybrid Systems*,
DOI 10.1007/978-3-319-11860-4

Fault accommodation Means to assess the severity of a fault and to decide if and what actions can be taken. If a fault cannot be accommodated, a controlled shut-down may become necessary. If it can be accommodated, possible actions may be a change of controller parameters or a controller redesign, or a system reconfiguration, i.e. a malfunctioning component is replaced by one with similar functionality.

Fault detection Means to constantly monitor the behaviour of system and to determine if it has deviated from its normal operation beyond permissible limits.

Fault diagnosis Means to detect and to isolate faults and to analyse their type and magnitude.

Fault isolation If values of fault indicators are beyond acceptable limits and an alarm has been raised then fault isolation means to locate possibly faulty components.

Fault quantification (identification) Means to determine the type of a fault and its severity. If the fault is due to a parametric change that is not too severe then parameter estimation can be used to determine an approximation of the actual parameter value.

Fault tolerant control (FTC) Is the ability to automatically accommodate system component faults so that overall stability and acceptable performance of the faulty system are maintained without replacing hardware.

Hybrid system model A hybrid system model makes use of the abstraction of instantaneous state changes and captures the dynamic behaviour in various system modes as well as discrete events. The latter ones are either controlled by local automata or take place autonomously and cause the system to instantaneously change from one mode into another.

Hybrid system A System is called a hybrid system for short if its dynamic behaviour is appropriately described by a hybrid model.

Malfunction Is an intermittent, i.e. a temporary irregularity in the fulfillment of a system's desired function. A malfunction is due to one or more faults [3].

Modelling uncertainties Denote all kinds of discrepancies between a mathematical model and the actual faultless system due to imperfect modelling [4].

Prognosis Failure prognosis means the ability of an early detection and isolation of incipient faults that may lead to a component failure, to determine the progression of the fault and to predict the remaining useful life (RUL), i.e. the time to failure given the current state of a system.

Supervision Means monitoring a physical system and taking appropriate actions to maintain the system's operation in the case of faults [2].

References

[1] Isermann, R., & Ballé, P. (1997). Trends in the application of model-based fault detection and diagnosis of technical processes. *Control Engineering Practice*, *5*(5), 709–719.

[2] Blanke, M., Kinnaert, M., Lunze, J., & Staroswiecki, M. (2006). *Diagnosis and fault-tolerant control*. Berlin: Springer.

[3] Isermann, R. (2006). *Fault-diagnosis systems an introduction from fault detection to fault tolerance*. Berlin: Springer.
[4] Frank, P. M. (2004). Control systems, robotics and automation. In H. Unbehauen (Ed.), *Fault diagnosis for linear systems* (Vol. XVI). Encyclopedia of Life Support Systems (EOLLS) ed. Oxford, UK: EOLLS Publishers. Retrieved from http://www.eolss.net.

Index

W. Borutzky, *Bond Graph Model-based Fault Diagnosis of Hybrid Systems*,
DOI 10.1007/978-3-319-11860-4

MIX
Papier aus verantwortungsvollen Quellen
Paper from responsible sources
FSC® C105338

If you have any concerns about our products,
you can contact us on
ProductSafety@springernature.com

In case Publisher is established outside the EU,
the EU authorized representative is:
Springer Nature Customer Service Center GmbH
Europaplatz 3, 69115 Heidelberg, Germany

Printed by Libri Plureos GmbH
in Hamburg, Germany